S0-BBO-969

OPTICAL FIBER
TELECOMMUNICATIONS IIIB

OPTICAL FIBER
TELECOMMUNICATIONS IIIB

Edited by

IVAN P. KAMINOW
Lucent Technologies, Bell Laboratories
Holmdel, New Jersey

THOMAS L. KOCH
Lucent Technologies, Bell Laboratories
Holmdel, New Jersey

 **ACADEMIC PRESS**

San Diego London Boston
New York Sydney Tokyo Toronto

This book is printed on acid-free paper. ∞

Copyright © 1997 by Lucent Technologies

All rights reserved.
No part of this publication may be reproduced or transmitted in any form or by any means, electronic or mechanical, including photocopy, recording, or any information storage and retrieval system, without permission in writing from the publisher.

ACADEMIC PRESS
525 B Street, Suite 1900, San Diego, CA 92101-4495, USA
1300 Boylston Street, Chestnut Hill, MA 02167, USA
http://www.apnet.com

Academic Press Limited
24–28 Oval Road, London NW1 7DX, UK
http://www.hbuk.co.uk/ap/

Library of Congress Cataloging-in-Publication Data

Optical fiber telecommunications III / [edited by] Ivan P. Kaminow, Thomas L. Koch.
  p. cm.
Includes bibliographical references and index.
ISBN 0-12-395170-4 (v. A) — ISBN 0-12-395171-2 (v. B)
1. Optical communications. 2. Fiber optics. I. Kaminow, Ivan P. II. Koch, Thomas L.
TK5103.59.H3516 1997   96-43812
621.382'75—dc20   CIP

Printed in the United States of America
97 98 99 00 01 MP 9 8 7 6 5 4 3 2 1

For our dear grandchildren:
Sarah, Joseph, Rafael, Nicolas, Gabriel, Sophia, and Maura — IPK

For Peggy, Brian, and Marianne — TLK

Contents

Contributors

L. A. Coldren (Ch. 6), Department of Electrical and Computer Engineering, University of California, Santa Barbara, California 93106

Daniel A. Fishman (Ch. 3), Lucent Technologies, Bell Laboratories, 101 Crawfords Corner Road, Holmdel, New Jersey 07733

Stephen G. Grubb* (Ch. 7), Lucent Technologies, Bell Laboratories, 600 Mountain Avenue, Murray Hill, New Jersey 07974

Fred Heismann (Ch. 9), Lucent Technologies, Bell Laboratories, 101 Crawfords Corner Road, Holmdel, New Jersey 07733

Charles H. Henry (Ch. 8), Lucent Technologies, Bell Laboratories, 700 Mountain Avenue, Murray Hill, New Jersey 07974

B. Scott Jackson (Ch. 3), AT&T Laboratories, 101 Crawfords Corner Road, Room 3D-418, Holmdel, New Jersey 07733

Charles H. Joyner (Ch. 5), Lucent Technologies, Bell Laboratories, 791 Holmdel-Keyport Road, Room HOH M-229D, Holmdel, New Jersey 07733

Ivan P. Kaminow (Ch. 1), Lucent Technologies, Bell Laboratories, 791 Holmdel-Keyport Road, Holmdel, New Jersey 07733

Howard D. Kidorf (Ch. 2), AT&T Submarine Systems Inc., Roberts Road, Holmdel, New Jersey 07733

Thomas L. Koch (Ch. 4), Lucent Technologies, Bell Laboratories, 101 Crawfords Corner Road, Room 4E-338, Holmdel, New Jersey 07733

Steven K. Korotky (Ch. 9), Lucent Technologies, Bell Laboratories, 101 Crawfords Corner Road, Room HO 4F-313, Holmdel, New Jersey 07733

* Present address: SDL, Inc., 80 Rose Orchard Way, San Jose, California 95134

Yuan P. Li (Ch. 8), Lucent Technologies, Bell Laboratories, 2000 Northeast Expressway, Norcross, Georgia 30071

Edmond J. Murphy (Ch. 10), Lucent Technologies, Bell Laboratories, 9999 Hamilton Boulevard, Breinigsville, Pennsylvania 18031

Jonathan A. Nagel (Ch. 2), AT&T Laboratories–Research, Crawford Hill Laboratory, 791 Holmdel-Keyport Road, Room L-137, Holmdel, New Jersey 07733

B. J. Thibeault (Ch. 6), Department of Electrical and Computer Engineering, University of California, Santa Barbara, California 93106

John J. Veselka (Ch. 9), Lucent Technologies, Bell Laboratories, 101 Crawfords Corner Road, Holmdel, New Jersey 07733

Alice E. White (Ch. 7), Lucent Technologies, Bell Laboratories, 600 Mountain Avenue, Murray Hill, New Jersey 07974

John L. Zyskind (Ch. 2), Lucent Technologies, Bell Laboratories, Crawford Hill Laboratory, 791 Holmdel-Keyport Road, Holmdel, New Jersey 07733

Chapter 1 | Overview

Ivan P. Kaminow

AT&T Bell Laboratories (retired), Holmdel, New Jersey

History

Optical Fiber Telecommunications, edited by Stewart E. Miller and Alan G. Chynoweth, was published in 1979, at the dawn of the revolution in lightwave telecommunications. This book was a stand-alone volume that collected all available information for designing a lightwave system. Miller was Director of the Lightwave Systems Research Laboratory and, together with Rudi Kompfner, the Associate Executive Director, provided much of the leadership at the Crawford Hill Laboratory of Bell Laboratories; Chynoweth was an Executive Director in the Murray Hill Laboratory, leading the optical component development. Many research and development (R&D) groups were active at other laboratories in the United States, Europe, and Japan. The book, however, was written exclusively by Bell Laboratories authors, although it incorporated the global results.

Looking back at that volume, I find it interesting that the topics are quite basic but in some ways dated. The largest group of chapters covers the theory, materials, measurement techniques, and properties of fibers and cables — for the most part, multimode fibers. A single chapter covers optical sources, mainly multimode AlGaAs lasers operating in the 800- to 900-nm band. The remaining chapters cover direct and external modulation techniques, photodetectors and receiver design, and system design and applications. Still, the basic elements for the present-day systems are there: low-loss vapor-phase silica fiber and double-heterostructure lasers.

Although a few system trials took place beginning in 1979, it required several years before a commercially attractive lightwave telecommunications system was installed in the United States. This was the AT&T Northeast Corridor System operating between New York and Washington, DC,

OPTICAL FIBER TELECOMMUNICATIONS,
VOLUME IIIB

Copyright © 1997 by Lucent Technologies.
All rights of reproduction in any form reserved.
ISBN: 0-12-395171-2

that began service in January 1983, operating at a wavelength of 820 nm and a bit rate of 45 Mb/s in multimode fiber. Lightwave systems were upgraded in 1984 to 1310 nm and about 500 Mb/s in single-mode fiber in the United States, as well as in Europe and Japan.

Tremendous progress was made during the next few years, and the choice of lightwave over copper for all long-haul systems was ensured. The drive was to improve performance, such as bit rate and repeater spacing, and to find other applications. A completely new book, *Optical Fiber Telecommunications II (OFT II)*, edited by Stewart E. Miller and me, was published in 1988 to summarize the lightwave design information known at the time. To broaden the coverage, we included some non-Bell Laboratories authors, including several authors from Bellcore, which had been divested from Bell Laboratories in 1984 as a result of the court-imposed "Modified Final Judgment." Corning, Nippon Electric Corporation, and several universities were represented among the contributors. Although research results are described in *OFT II*, the emphasis is much stronger on commercial applications than in the previous volume.

The early chapters of *OFT II* cover fibers, cables, and connectors, dealing with both single- and multimode fiber. Topics include vapor-phase methods for fabricating low-loss fiber operating at 1310 and 1550 nm, understanding chromatic dispersion and various nonlinear effects, and designing polarization-maintaining fiber. Another large group of chapters deals with a wide geographic scope of systems for loop, intercity, interoffice, and undersea applications. A research-oriented chapter deals with coherent systems and another with possible local area network applications, including a comparison of time-division multiplexing (TDM) and wavelength-division multiplexing (WDM) to effectively utilize the fiber bandwidth. Several chapters cover practical subsystem components, such as receivers and transmitters, and their reliability. Other chapters cover the photonic devices, such as lasers, photodiodes, modulators, and integrated electronic and integrated optic circuits, that compose the subsystems. In particular, epitaxial growth methods for InGaAsP materials suitable for 1310- and 1550-nm applications, and the design of high-speed single-mode lasers are discussed.

The New Volume

By 1995, it was clear that the time for a new volume to address the recent research advances and the maturing of lightwave systems had arrived. The contrast with the research and business climates of 1979 was dramatic. System experiments of extreme sophistication were being performed by building on the commercial and research components funded for a proven multibillion-dollar global industry. For example, 10,000 km of high-performance fiber was assembled in several laboratories around the world for NRZ (non-return-to-zero), soliton, and WDM system experiments. The competition in both the service and hardware ends of the telecommunications business was stimulated by worldwide regulatory relief. The success in the long-haul market and the availability of relatively inexpensive components led to a wider quest for other lightwave applications in cable television and local access network markets. The development of the diode-pumped erbium-doped fiber amplifier (EDFA) played a crucial role in enhancing the feasibility and performance of long-distance and WDM applications.

In planning the new volume, Tom Koch and I looked for authors to update the topics of the previous volumes, such as fibers, cables, and laser sources. But a much larger list of topics contained fields not previously included, such as SONET (synchronous optical network) standards, EDFAs, fiber nonlinearities, solitons, and passive optical networks (PONs). Throughout the volume, erbium amplifiers, WDM, and associated components are common themes.

Again, most of the authors come from Bell Laboratories and Bellcore, where much of the research and development was concentrated and where we knew many potential authors. Still, we attempted to find a few authors from elsewhere for balance. Soon after laying out the table of contents and lining up the authors, however, a bombshell and a few hand grenades struck. AT&T decided to split into three independent companies, Bellcore was put up for sale, and several authors changed jobs, including Tom Koch and I. The resulting turmoil and uncertainty made the job of getting the chapters completed tougher than for the earlier volumes, which enjoyed a climate of relative tranquillity.

In the end, we assembled a complete set of chapters for *Optical Fiber Telecommunications III*, and can offer another timely and definitive survey of the field. Because of the large number of pages, the publisher recommended separating the volume into two sections, A and B. This format should prove more manageable and convenient for the reader. The chapters are numbered from Chapter 1 in each section, with this Overview repeated as Chapter 1 in both sections A and B to accommodate users who choose to buy just one book.

Survey of Volumes IIIA and IIIB

The chapters of Volumes IIIA and IIIB are briefly surveyed as follows in an attempt to put the elements of the book in context.

VOLUME IIIA

SONET and ATM (Chapter 2)

The market forces of deregulation and globalization have driven the need for telecommunications standards. Domestically, the breakup of AT&T meant that service providers and equipment suppliers no longer accepted *de facto* standards set by "Ma Bell." They wanted to buy and sell equipment competitively and to be sure that components from many providers would interoperate successfully. The globalization of markets extended these needs worldwide. And the remarkable capability of silicon integrated circuits to perform extremely complex operations at low cost with high volume has made it possible to provide standard interfaces economically.

The digital transmission standard developed by Bellcore and employed in all new domestic circuit-switched networks is SONET, and a similar international standard is SDH (synchronous digital hierarchy). In the same period, a telecommunications standard was devised to satisfy the needs of the data market for statistical multiplexing and switching of bursty computer traffic. It is called *ATM (asynchronous transfer mode)* and is being embraced by the computer industry as well as by digital local access providers. The basics of SONET, SDH, and ATM are given in Chapter 2, by Joseph E. Berthold.

Information Coding and Error Correction in Optical Fiber Communications Systems (Chapter 3)

The ultimate capacity of a communication channel is governed by the rules of information theory. The choice of modulation format and coding scheme determines how closely the actual performance approaches the theoretical limit. The added cost and complexity of coding is often the deciding factor in balancing the enhanced performance provided by this technology. So far, coding has not been required in high-performance lightwave systems. However, as the demands on lightwave systems increase and the performance of high-speed electronics improves, we can expect to see more uses of sophisticated coding schemes. In particular, forward error-correcting codes (FECs) may soon find applications in long-distance, repeaterless undersea systems. A review of coding techniques, as they apply to lightwave systems, is given by Vincent W. S. Chan in Chapter 3.

Advances in Fiber Design and Processing (Chapter 4)

The design and processing of fibers for special applications are presented in Chapter 4, by David J. DiGiovanni, Donald P. Jablonowski, and Man F. Yan. Erbium-doped silica fibers for amplifiers at 1550 nm, which are described in detail in Chapter 2, Volume IIIB, are covered first. Rare-earth-doped fluoride fibers for 1300-nm amplifiers are described later, as are fibers for cladding-pumped high-power fiber amplifiers.

Dispersion management is essential for the long-haul, high-speed systems described in later chapters. The design and fabrication of these fibers for new WDM installations at 1550-nm and for 1550-nm upgrades of 1310-nm systems are also reviewed.

Advances in Cable Design (Chapter 5)

Chapter 5, by Kenneth W. Jackson, T. Don Mathis, P. D. Patel, Manuel R. Santana, and Phillip M. Thomas, expands on related chapters in the two previous volumes, *OFT* and *OFT II*. The emphasis is on practical applications of production cables in a range of situations involving long-distance and local telephony, cable television, broadband computer networks, premises cables, and jumpers. Field splicing of ribbon cable, and the division of applications that lead to a bimodal distribution of low and high fiber count cables are detailed.

Polarization Effects in Lightwave Systems (Chapter 6)

Modern optical fibers possess an extremely circular symmetry, yet they retain a tiny optical birefringence leading to polarization mode dispersion (PMD) that can have severe effects on the performance of very long digital systems as well as high-performance analog video systems. Systems that contain polarization-sensitive components also suffer from polarization-dependent loss (PDL) effects. In Chapter 6, Craig D. Poole and Jonathan Nagel review the origins, measurement, and system implications of remnant birefringence in fibers.

Dispersion Compensation for Optical Fiber Systems (Chapter 7)

Lightwave systems are not monochromatic: chirp in lasers leads to a finite range of wavelengths for the transmitter in single-wavelength systems, whereas WDM systems intrinsically cover a wide spectrum. At the same time, the propagation velocity in fiber is a function of wavelength that can be controlled to some extent by fiber design, as noted in Chapter 4. To avoid pulse broadening, it is necessary to compensate for this fiber chromatic dispersion. Various approaches for dealing with this problem are presented in Chapter 7, by A. H. Gnauck and R. M. Jopson. Additional system approaches to dispersion management by fiber planning are given in Chapter 8.

Fiber Nonlinearities and Their Impact on Transmission Systems (Chapter 8)

Just a few years ago, the study of nonlinear effects in fiber was regarded as "blue sky" research because the effects are quite small. The advance of technology has changed the picture dramatically as unrepeatered undersea spans reach 10,000 km, bit rates approach 10 Gb/s, and the number of WDM channels exceeds 10. In these cases, an appreciation of subtle nonlinear effects is crucial to system design. The various nonlinearities represent perturbations in the real and imaginary parts of the refractive index of silica as a function of optical field. In Chapter 8, Fabrizio Forghieri, Robert W. Tkach, and Andrew R. Chraplyvy review the relevant nonlinearities, then develop design rules for accommodating the limitations of nonlinearities on practical systems at the extremes of performance.

Terrestrial Amplified Lightwave System Design (Chapter 9)

Chungpeng (Ben) Fan and J. P. Kunz have many years of experience in planning lightwave networks and designing transmission equipment, respectively. In Chapter 9, they review the practical problems encountered in designing commercial terrestrial systems taking advantage of the technologies described elsewhere in the book. In particular, they consider such engineering requirements as reliability and restoration in systems with EDFAs, with dense WDM and wavelength routing, and in SONET–SDH rings.

Undersea Amplified Lightwave Systems Design (Chapter 10)

Because of their extreme requirements, transoceanic systems have been the most adventurous in applying new technology. EDFAs have had an especially beneficial economic effect in replacing the more expensive and less reliable submarine electronic regenerators. Wideband cable systems have reduced the cost and improved the quality of overseas connections to be on a par with domestic communications. In Chapter 10, Neal S. Bergano reviews the design criteria for installed and planned systems around the world.

Advances in High Bit-Rate Transmission Systems (Chapter 11)

As the transmission equipment designer seeks greater system capacity, it is necessary to exploit both the WDM and TDM dimensions. The TDM limit is defined in part by the availability of electronic devices and circuits. In Chapter 11, Kinichiro Ogawa, Liang D. Tzeng, Yong K. Park, and Eiichi Sano explore three high-speed topics: the design of high-speed receivers, performance of 10-Gb/s field experiments, and research on devices and integrated circuits at 10 Gb/s and beyond.

Solitons in High Bit-Rate, Long-Distance Transmission (Chapter 12)

Chromatic dispersion broadens pulses and therefore limits bit rate; the Kerr nonlinear effect can compress pulses and compensate for the dispersion. When these two effects are balanced, the normal mode of propagation is a soliton pulse that is invariant with distance. Thus, solitons have seemed to be the natural transmission format, rather than the conventional NRZ format, for the long spans encountered in undersea systems. Still, a number of hurdles have manifested as researchers explored this approach more deeply. Perhaps the most relentless and resourceful workers in meeting and overcoming these challenges have been Linn Mollenauer and his associates. L. F. Mollenauer, J. P. Gordon, and P. V. Mamyshev provide a definitive review of the current R&D status for soliton transmission systems in Chapter 12. Typical of a hurdle recognized, confronted, and leaped is the Gordon–Haus pulse jitter; the sliding filter solution is described at length.

A Survey of Fiber Optics in Local Access Architectures (Chapter 13)

The Telecommunications Act of 1996 has opened the local access market to competition and turmoil. New applications based on switched broadband digital networks, as well as conventional telephone and broadcast analog video networks, are adding to the mix of options. Furthermore, business factors, such as the projected customer *take rate*, far outweigh technology issues.

In Chapter 13, Nicholas J. Frigo discusses the economics, new architectures, and novel components that enter the access debate. The architectural proposals include fiber to the home (FTTH), TDM PON, WDM PON, hybrid fiber coax (HFC), and switched digital video (SDV) networks. The critical optical components, described in Volume IIIB, include WDM lasers and receivers, waveguide grating routers, and low-cost modulators.

Lightwave Analog Video Transmission (Chapter 14)

Cable television brings the analog broadcast video spectrum to conventional television receivers in the home. During the last few years, it was found that the noise and linearity of lightwave components are sufficiently good to transport this rf signal over wide areas by intensity modulation of a laser carrier at 1310, 1060, or 1550 nm. The fiber optic approach has had a dramatic effect on the penetration and performance of cable systems, lowering cost, improving reliability, and extending the number of channels. New multilevel coding schemes make rf cable modems an attractive method for distributing interactive digital signals by means of HFC and related architectures. Thus, cable distribution looks like an economic technology for bringing high-speed data and compressed video applications, such as the Internet, to homes and offices. Now, in the bright new world of deregulation and wide-open competition, cable may also carry telephone service more readily than telephone pairs can carry video. In Chapter 14, Mary R. Phillips and Thomas E. Darcie examine the hardware requirements and network architectures for practical approaches to modern lightwave cable systems.

Advanced Multiaccess Lightwave Networks (Chapter 15)

The final chapter in Volume IIIA looks at novel architectures for routing in high bit-rate, multiple-access networks. For the most part, the emphasis is on wavelength routing, which relies on the novel wavelength-sensitive elements described in Volume IIIB. Such networks offer the prospect of "optical transparency," a concept that enhances flexibility in network design. Commercial undersea and terrestrial networks are already incorporating preliminary aspects of wavelength routing by the provision of WDM add–drop multiplexing. Further, the proposed WDM PON networks in Chapter 13 also employ wavelength routing.

Chapter 15, however, considers a wider range of architectures and applications of this technology. After reviewing optical transparency, it treats WDM rings for local networks, metropolitan distribution, and continental undersea telecommunications (AfricaONE). Then it reviews several multiaccess test beds designed by consortia organized with partial support from DARPA (Defense Advanced Research Projects Agency).

VOLUME IIIB

Erbium-Doped Fiber Amplifiers for Optical Communications (Chapter 2)

A large part of the economic advantage for lightwave systems stems from the development of the diode-pumped EDFA, which replaced the more expensive and limited electronic regenerators. By remarkable coincidence, the EDFA provides near noise-free gain in the minimum-loss window of silica fiber at 1550 nm. It provides format-independent gain over a wide WDM band for a number of novel applications beyond its original use in single-frequency, long-haul terrestrial and undersea systems.

Important considerations in the basics, design, and performance of EDFAs are given in Chapter 2, by John L. Zyskind, Jonathan A. Nagel, and Howard D. Kidorf. Designs are optimized for digital terrestrial and undersea systems, as well as for applications to analog cable television and wavelength-routed WDM networks, which are covered in Chapters 13, 14, and 15 in Volume IIIA. Performance monitoring and the higher order effects that come into play for the extreme distances encountered in undersea systems are also discussed.

Transmitter and Receiver Design for Amplified Lightwave Systems (Chapter 3)

Chapter 3, by Daniel A. Fishman and B. Scott Jackson, defines the engineering requirements for transmitters and receivers in amplified systems, mainly operating at 2.5 Gb/s and satisfying the SONET–SDH standards. Topics that are essential for commercial networks, such as performance monitoring, are included.

Laser Sources for Amplified and WDM Lightwave Systems (Chapter 4)

As lightwave systems have become more sophisticated, the demands on the laser sources have become more stringent than those described in Chapter 13 of *OFT II*. The greater fiber spans and the introduction of EDFA and WDM technologies require both improved performance and totally new functionality. In Chapter 4, Thomas L. Koch reviews lasers and subsystems designed for low-chirp applications, employing direct modulation, external modulation, and integrated laser–modulators. He also covers a variety of laser structures designed to satisfy the special needs of WDM systems for precise fixed wavelengths, tunable wavelengths, and multiple wavelengths. These structures include fixed DFB (distributed feedback) lasers, tunable DBR (distributed Bragg reflector) lasers, multifrequency waveguide grating router lasers (MFL), and array lasers.

Advances in Semiconductor Laser Growth and Fabrication Technology (Chapter 5)

Some of the greatest advances in laser performance in recent years can be traced to advances in materials growth. In Chapter 5, Charles H. Joyner covers such advances as strained quantum wells, selective area growth, selective etching, and beam expanded lasers.

Vertical-Cavity Surface-Emitting Lasers (Chapter 6)

The edge-emitting lasers employed in today's lightwave systems are described in Chapter 4. In Chapter 6, L. A. Coldren and B. J. Thibeault update progress on a different structure. Vertical-cavity surface-emitting lasers (VCSELs) are largely research devices today but may find a role in telecommunications systems by the time of the next volume of this series. Because of their unique structure, VCSELs lend themselves to array and WDM applications.

Optical Fiber Components and Devices (Chapter 7)

Although fiber serves mainly as a transmission line, it is also an extremely convenient form for passive and active components that couple into fiber transmission lines. A key example is the EDFA, which is described in Chapter 4, Volume IIIA, and Chapter 2, Volume IIIB. In Chapter 7, Alice E. White and Stephen G. Grubb describe the fabrication and applications of UV-induced fiber gratings, which have important uses as WDM multiplexers and add–drop filters, narrow band filters, dispersion compensators, EDFA gain equalizers, and selective laser mirrors.

Special fibers also serve as the vehicles for high-power lasers and amplifiers in the 1550- and 1310-nm bands. High-power sources are needed for long repeaterless systems and passively split cable television distribution networks. Among the lasers and amplifiers discussed are 1550-nm Er/Yb cladding-pumped, 1300-nm Raman, and Pr and Tm up-conversion devices.

Silicon Optical Bench Waveguide Technology (Chapter 8)

A useful technology for making passive planar waveguide devices has been developed in several laboratories around the world; at AT&T Bell Laboratories, the technology is called *silicon optical bench* (*SiOB*). Waveguide patterns are formed photolithographically in a silica layer deposited on a silicon substrate. In Chapter 8, Yuan P. Li and Charles H. Henry describe the SiOB fabrication process and design rules suitable for realizing a variety of components. The planar components include bends, splitters, directional couplers, star couplers, Bragg filters, multiplexers, and add–drop filters. Different design options are available for the more complex devices, i.e., a chain of Fourier filters or an arrayed waveguide approach. The latter technique has been pioneered to Corrado Dragone of Bell Laboratories (Dragone, Edwards, and Kistler 1991) to design commercial WDM components known as *waveguide grating routers* (*WGRs*) serving as multiplexers and add–drop filters.

Lithium Niobate Integrated Optics: Selected Contemporary Devices and System Applications (Chapter 9)

More than 20 years have passed since the invention of titanium-diffused waveguides in lithium niobate (Schmidt and Kaminow 1974) and the associated integrated optic waveguide electrooptic modulators (Kaminow, Stulz, and Turner 1975). During that period, external modulators have competed with direct laser modulation, and electrooptic modulators have competed with electroabsorption modulators. Each has found its niche: the external modulator is needed in high-speed, long-distance digital, and high-linearity analog systems, where chirp is a limitation; internal modulation is used for economy, when performance permits. (See Chapter 4 in Volume IIIB.)

In Chapter 9, Fred Heismann, Steven K. Korotky, and John J. Veselka review advances in lithium niobate integrated optic devices. The design and performance, including reliability and stability, of phase and amplitude modulators and switches, polarization controllers and modulators, and electrooptic and acoustooptic tunable wavelength filters are covered.

Photonic Switching (Chapter 10)

Whereas Chapter 9 deals with the modulation or switching of a single input, Chapter 10 deals with switching arrays. These arrays have not yet found commercial application, but they are being engineered for forward-looking system demonstrations such as the DARPA MONET project (Multiwavelength Optical Network), as mentioned in Chapter 15, Volume IIIA. In Chapter 10, Edmond J. Murphy reviews advances in lithium niobate, semiconductor, and acoustooptic switch elements and arrays. Murphy also covers designs for various device demonstrations.

References

Dragone, C., C. A. Edwards, and R. C. Kistler, 1991. Integrated optics N × N multiplexer on silicon. *IEEE Photon. Techn. Lett.* 3:896–899.

Kaminow, I. P., L. W. Stulz, and E. H. Turner. 1975. Efficient strip-waveguide modulator. *Appl. Phys. Lett.* 27:555–557.

Schmidt, R. V., and I. P. Kaminow. 1974. Metal-diffused optical waveguides in LiNbO$_3$. *Appl. Phys. Lett.* 25:458–460.

Chapter 2 | Erbium-Doped Fiber Amplifiers for Optical Communications

John L. Zyskind

Lucent Technologies, Bell Laboratories, Holmdel, New Jersey

Jonathan A. Nagel

AT&T Laboratories–Research, Holmdel, New Jersey

Howard D. Kidorf

AT&T Submarine Systems, Inc., Holmdel, New Jersey

I. Introduction

The erbium-doped fiber amplifier (EDFA) was first reported in 1987,[1,2] and, in the short period since then, its applications have transformed the optical communications industry. Before the advent of optical amplifiers, optical transmission systems typically consisted of a digital transmitter and a receiver separated by spans of transmission optical fiber interspersed with optoelectronic regenerators. The optoelectronic regenerators corrected attenuation, dispersion, and other transmission degradations of the optical signal by detecting the attenuated and distorted data pulses, electronically reconstituting them, and then optically transmitting the regenerated data into the next transmission span.[3]

The EDFA is an optical amplifier that faithfully amplifies lightwave signals purely in the optical domain. EDFAs have several potential functions in optical fiber transmission systems. They can be used as power amplifiers to boost transmitter power, as repeaters or in-line amplifiers to increase system reach, or as preamplifiers to enhance receiver sensitivity. The most far-reaching impact of EDFAs has resulted from their use as repeaters in place of conventional optoelectronic regenerators to compensate for transmission loss and extend the span between digital terminals. Used as a repeater, the optical amplifier offers the possibility of transforming the optical transmission line into a transparent optical pipeline that will support signals independent of their modulation format or their channel data rate. Additionally, optical amplifiers support the use of wavelength-division multiplexing (WDM), whereby signals of different wavelengths are combined and transmitted together on the same transmission fiber.

13

OPTICAL FIBER TELECOMMUNICATIONS,
VOLUME IIIB

Copyright © 1997 by Lucent Technologies.
All rights of reproduction in any form reserved.
ISBN: 0-12-395171-2

The primary applications that have driven EDFAs to commercial development are long-haul, terrestrial transport, and undersea transport systems. AT&T Submarine Systems Inc. decided in early 1990 to develop EDFA-based repeaters (rather than conventional optoelectronic regenerators) for future submarine cables and initiated the first commercial development of high-capacity, optically amplified communications systems. The first undersea systems, Americas-1 and Columbus II, connecting Florida to St. Thomas in the Caribbean Sea were deployed by AT&T Submarine Systems Inc. in 1994. The first optically amplified transoceanic cables crossed the Atlantic in 1995 (built jointly by AT&T Submarine Systems, Inc., and Alcatel), and the Pacific in 1996 (built jointly by AT&T Submarine Systems, Inc., and KDD). Terrestrial optically amplified systems with dense WDM were first deployed in AT&T's long-distance network in 1996. Today the EDFA has replaced the optoelectronic regenerator as the repeater of choice in both terrestrial and submarine systems.

In the remainder of this chapter, we discuss the general properties of EDFAs, then four fields of application. Two areas already revolutionized by EDFAs are terrestrial transport systems and undersea systems. The amplified systems used for terrestrial transport must be adapted to the embedded base of terrestrial transmission fiber. Undersea transmission systems must span transoceanic distances and meet stringent reliability requirements. Two other areas where EDFAs promise to have a significant impact are in the transmission of analog signals and in optical networking. Analog transmission, particularly of video common antenna television (CATV) signals, requires high output power to overcome shot noise limitations. In optical networking applications, the transparency made possible by EDFAs can be exploited to permit wavelength routing and switching.

II. Properties of EDFAs

The simplest EDFA configurations, shown in Fig. 2.1, include an erbium-doped fiber spliced into the signal transmission path of an optical fiber communications system and a source of pump light. The pump light either counterpropagates or copropagates with the signal light. More advanced EDFA architectures are discussed later in this chapter. It is the atomic level scheme of the Er ion (Fig. 2.2) that gives the EDFA its nearly ideal

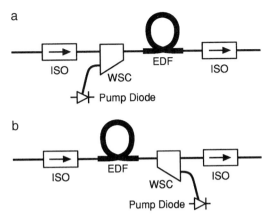

Fig. 2.1 Basic erbium-doped fiber amplifier (EDFA) configurations with pump and signal (a) copropagating and (b) counterpropagating. EDF, erbium-doped fiber; ISO, isolator; WSC, wavelength selective coupler.

properties for optical communications. Light from the pump supplies energy to elevate the erbium ions to the $^4I_{13/2}$ first excited state. The excitation energy of this state corresponds to wavelengths near the minimum optical loss of silica optical fibers ($\sim$1550 nm). Optical signals propagating through the EDFA with wavelengths between about 1525 and 1565 nm induce stimulated emission in excited erbium ions and are thereby amplified.

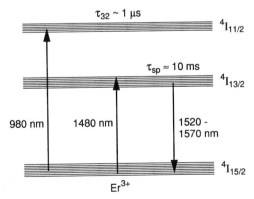

Fig. 2.2 Erbium energy-level scheme.

A. GAIN

Gain is the fundamental characteristic of an amplifier. Optical amplifier gain is defined as the ratio of the output signal power to the input signal power,

$$G(\lambda) \equiv \frac{P_{out}}{P_{in}} = \int_0^L g(\lambda, z) \cdot dz, \qquad (2.1)$$

and it is obtained by integrating the gain coefficient $g(\lambda)$ over the length L of the erbium-doped fiber. The gain coefficient, normally expressed in units of decibels per meter, is the sum of the emission coefficient $g^*(\lambda) = \Gamma_s n_{Er} \sigma_e(\lambda)$ and the absorption coefficient $\alpha(\lambda) = \Gamma_s n_{Er} \sigma_a(\lambda)$ weighted by the fractional populations N_2 and N_1, respectively, of the first excited and ground states of erbium:

$$g(\lambda, z) = \frac{1}{P(\lambda, z)} \cdot \frac{dP(\lambda, z)}{dz} = g^*(\lambda) \cdot N_2(z) - \alpha(\lambda) \cdot N_1(z), \quad (2.2)$$

where Γ_s is the confinement factor of the signal mode in the fiber core, n_{Er} is the concentration of Er ions in the core, and $\sigma_e(\lambda)$ and $\sigma_a(\lambda)$ are, respectively, the signal emission and absorption cross sections as functions of wavelength. The spectra for the fully inverted gain coefficient $g^*(\lambda) = \Gamma_s n_{Er} \sigma_e(\lambda)$ and the small signal absorption coefficient $\alpha(\lambda) = \Gamma_s n_{Er} \sigma_a(\lambda)$ are shown in Fig. 2.3 for an erbium-doped fiber with aluminum and germanium co-doping in the core.

EDFAs can be modeled accurately using rate equations for the populations of the atomic levels and the photon fluxes.[4,5]

B. OUTPUT POWER AND SATURATION

The output power is approximately proportional to the pump power when signal levels are high and the amplifier is saturated, as shown in Fig. 2.4.[6] This is a characteristic of the three-level erbium laser system as can be understood by reference to the erbium energy-level scheme (Fig. 2.1); when the amplifier is saturated, pump absorption from the ground state is balanced by stimulated emission from the first excited state induced by the signal. The higher the pump power is, the higher the signal power at which this balance occurs. This can be verified by using the rate equations describing the populations of the erbium energy levels and the light intensity to calculate the gain coefficient and analyze its saturation characteristics.[7,8] The signal power at which the gain coefficient is reduced to half its small signal value is

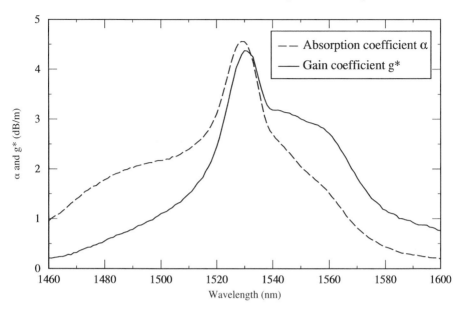

Fig. 2.3 Emission and absorption spectra for an erbium-doped fiber with aluminum and germanium co-doping in the core.

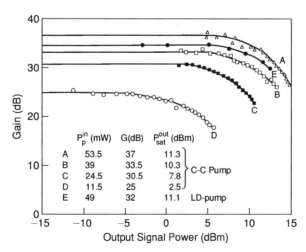

Fig. 2.4 EDFA saturation for different pump powers.[6]

$$P_{sat} = \frac{h\nu_s A_c}{(\sigma_{es} + \sigma_{as})\Gamma_s \tau_{sp}} \left(1 + \frac{\sigma_{as} \cdot P_p}{\sigma_{es} \cdot P_p^{th}}\right), \tag{2.3}$$

where σ_{es} and σ_{as} are the emission and absorption cross sections, respectively, at the signal wavelength; A_c is the core area; τ_{sp} is the first excited state spontaneous lifetime; and P_p is the pump power. The pump threshold for transparency — i.e., the pump power below which the small signal gain coefficient is negative, corresponding to absorption, and above which it is positive, corresponding to gain — is

$$P_p^{th} = \frac{\sigma_{as}}{\sigma_{es}} \cdot \frac{h\nu_p A_c}{\Gamma_p \tau_{sp} \sigma_{ap}}. \tag{2.4}$$

In Eq. (2.4), $h\nu_p$ is the pump photon energy, Γ_p is the pump mode confinement factor, and σ_{ap} is the pump absorption cross section. Equation (2.3) shows that if $P_p \gg P_p^{th}$, then P_{sat} is proportional to P_p/P_p^{th}.

Equations (2.3) and (2.4) are local in character, describing the behavior of the gain coefficient at a particular value of z. The gain characteristics of the complete amplifier are found by solving the rate equation at each point along the length of the erbium-doped fiber length and integrating the gain coefficient as indicated in Eq. (2.1). Because the saturation behavior is typically determined primarily near the output end of the amplifier where the signal power is largest, this local description generally provides a good qualitative understanding of the saturation behavior of a complete amplifier.

C. NOISE FIGURE

The amplification of the EDFA is inescapably accompanied by a background of amplified spontaneous emission (ASE). ASE arises when light emitted by spontaneous decay of excited erbium ions is captured by the optical fiber waveguide and then amplified in the EDFA. This ASE background adds noise that degrades amplified signals. The noise figure, defined as the signal-to-noise ratio (SNR) at the output divided by that corresponding to the shot noise of the signal at the input, is a measure of the degradation of the signal by noise added by the amplifier. The dominant contributions to the noise figure of a well-designed, high-gain amplifier are signal–spontaneous beat noise and signal shot noise, and are given by[9]

$$NF \cong 2n_{sp} \frac{(G-1)}{G} + \frac{1}{G} \approx 2n_{sp}, \tag{2.5}$$

where n_{sp}, the spontaneous emission factor, indicates the relative strengths of the spontaneous and stimulated emission processes. For an EDFA with uniform inversion (defined as $N_2 - N_1$) along its length, $n_{sp} = \sigma_{es}N_2/(\sigma_{es}N_2 - \sigma_{as}N_1)$ where N_1 and N_2 are the fractional populations of the ground and first excited states, respectively. The closer n_{sp} is to 1 (i.e., the better the inversion), the lower the noise figure. Because EDFAs can be efficiently inverted (i.e., $N_2 - N_1 \cong 1$), the noise figure can approach 3 dB, which is the quantum limit for optical amplifiers.

The spontaneous emission factor can be determined from

$$n_{sp} = \frac{P_{ASE}}{h\nu_s \cdot \Delta\nu \cdot (G - 1)}, \tag{2.6}$$

where P_{ASE} is the ASE power in one polarization in bandwidth $\Delta\nu$ (this is one-half the total power in bandwidth $\Delta\nu$ of the ASE, which, in the absence of polarization hole burning, is unpolarized) and $h\nu_s$ is the photon energy. Combining Eqs. (2.5) and (2.6) shows that the signal–spontaneous beat noise contribution to the noise figure is proportional to P_{ASE} and can be viewed as resulting from the addition of ASE by the amplifier. Because the spontaneous emission generated at the EDFA input experiences almost the full gain of the EDFA, when the inversion is not uniform along the length of the EDFA, the inversion near the input has the greatest impact on the noise figure.

D. ERBIUM-DOPED FIBER

The key element in an EDFA is its erbium-doped fiber, a single-mode fiber the core of which is doped with erbium ions. Preforms for silica-based erbium-doped fibers can be made both by the modified chemical vapor deposition (MCVD) and by the vapor axial deposition (VAD) techniques modified to permit addition of erbium as reviewed in Refs. 10 and 11. The use of these vapor phase techniques permits a high degree of control in designing the radial profile of the index of refraction, which can be tailored to obtain optical modes with optimal properties for any given application.

In many applications, pump power is limited by pump laser performance or as a result of system constraints on pump reliability or heat dissipation. In some applications, such as remotely pumped preamplifiers and in-line amplifiers in submarine systems, where pump power is limited by reliability constraints, it is of paramount importance to design the erbium-doped fiber to minimize the transparency threshold and to produce the highest gain with the lowest possible pump power. Unlike a four-level system, in which atoms in the ground state are passive bystanders to the lasing transitions, in a three-level system, such as erbium, atoms remaining in the ground state destroy the gain by absorbing the amplified light. The erbium-doped fiber should be designed to maximize the pump intensity experienced by all erbium ions. The erbium-doped fiber should have a small core and a large difference between the indices of refraction of the core and cladding to minimize the core effective area, A_c, and to maximize the pump optical confinement factor, Γ_p (see Eq. [2.4]). Pump thresholds less than 1 mW have been achieved.[12,13]

High gain efficiency, the ability to produce the most gain with the least pump power, reduces the pump power required and therefore increases the reliability and decreases the cost of the amplifier. One useful figure of merit to compare different erbium-doped fiber designs is the maximum gain efficiency (sometimes incorrectly termed the "gain coefficient") defined as the maximum quotient of the small-signal gain divided by the pump power. At a signal wavelength of 1533 nm, maximum gain efficiencies as large as 11 dB/mW[12] and 6.3 dB/mW[13] have been demonstrated with 980- and 1480-nm pump wavelengths, respectively. Gain exceeding 30 dB can be produced by a few milliwatts of pump power. An amplifier with a gain of 51 dB has been experimentally demonstrated with a 22-m erbium-doped fiber using 180 mW of 980-nm pump power.[14] Rayleigh scattering and ASE will limit the maximum gain achievable in a single-stage amplifier, but multistage designs can be used to increase the maximum gain.

In some applications, high output power is required, as is often the case for terrestrial applications. For three-level laser systems, such as erbium, if $P_p \gg P_p^{th}$, the saturated output power is approximately proportional to the pump power (see Eq. [2.3]). It is desirable to maximize the pump conversion efficiency, defined as $(P_{out} - P_{in})/P_p \approx P_{out}/P_p$. In cases where both the pump and signal powers are strong and much higher than their respective intrinsic saturation powers, $P_{sat}^i(\lambda) = \dfrac{h\nu A_c}{[\sigma_s(\lambda) + \sigma_a(\lambda)]\Gamma_\lambda} \tau_{sp}$, the conversion efficiency is relatively insensitive to the waveguide geometry because the dependences on effective area and confinement factor for the pump and signal tend to cancel (see Eq. [2.3]).

Material considerations such as erbium concentration and core co-dopants are important determinants of an amplifier's saturation characteristics. When the erbium concentration in silica is too high, erbium atoms form clusters, which give rise to cooperative up-conversion and associated nonradiative dissipation of pump power.[15] If more than one atom in a cluster is excited by pump absorption, one of the excited atoms can decay to the ground state, transferring its excitation energy to a nearby ion already excited to the $^4I_{13/2}$ first excited state. This second erbium ion is thereby elevated to a higher excited state and dissipates the extra excitation energy by decaying nonradiatively to the $^4I_{13/2}$ state. The net result is absorption of a pump photon without production of an additional signal photon. It is found that aluminum co-doping of the fiber core permits a higher concentration of erbium atoms (several hundred parts per million) before significant degradation of amplifier performance; for germanium co-doping erbium concentrations must be less than 100 ppm.[16] To permit a lower erbium concentration, erbium-doped fibers with larger cores and smaller core–cladding refractive index differences are commonly used for applications where achieving high output power is more important than producing gain with minimal pump power. However, there is a trade-off. Increasing the core size of the erbium-doped fiber also increases the pump threshold, which exacts a price in pump conversion efficiency (see Eq. [2.3]), so that even for power amplifiers, the erbium-doped fiber is designed with a smaller core and a higher refractive index difference between the core and cladding than for standard transmission fibers.

E. COUPLING LOSS

The mismatch between the smaller optical modes of erbium-doped fibers (typically 2–4 μm in diameter) and the larger modes of transmission fibers (typically 8–10 μm) poses the challenge of achieving acceptable splice losses. Butt-coupling losses for such mismatched modes would be several decibels. These penalties can be avoided by using a fusion splice to couple between the erbium-doped fibers and transmission fibers and optimizing the splicing parameters to diffuse the core dopants in the splice region in such a way as to form a low-loss tapered splice. Losses on the order of a few tenths of a decibel or less can be achieved in this way, even between fibers with severely mismatched optical mode sizes.[17] The total input and output losses in an EDFA are each generally less than 1.5 dB, including the losses of such devices as isolators and pump–signal combiners.

F. POLARIZATION INDEPENDENCE

Because of the circular symmetry of the erbium-doped fiber core and the random orientations of the individual erbium ions in the glass matrix of the fiber core, the gain of EDFAs is polarization independent.[18] This feature is one of the major advantages offered by EDFAs. EDFAs do exhibit polarization hole burning because of the orientations of the individual erbium ions in the glass matrix, which is locally nonisotropic.[19,20] Polarization hole burning (PHB) occurs when a strong polarized signal saturates preferentially those ions aligned with its polarization. As a result, light, including the saturating signal and ASE, with polarizations aligned to that of the saturating signal experiences a slightly lower gain and light with polarizations aligned orthogonal to that of the saturating signal experiences a slightly higher gain. Such polarization hole burning effects are weak but can become significant in systems with many concatenated amplifiers or where the amplifiers are deeply saturated.

G. GAIN DYNAMICS

The gain dynamics of EDFAs are slow because of the extremely long lifetime of the $^4I_{13/2}$ metastable first excited state ($\sim$10 ms). As a result, when the data rate is high enough, the modulation of signals does not cause significant gain modulation of the amplifier, even in deeply saturated amplifiers.[21] The corner frequency for the amplifier can be as low as 100 Hz and increases with pump and signal power, but generally remains less than 10 kHz. Even for intensity-modulated signals with relatively low data rates, the amplifiers do not introduce significant intersymbol interference, cross talk (in the case of multichannel signals), or nonlinear distortions due to intermodulation.

Recent results have shown that for long chains of amplifiers the corner frequency increases with the length of the amplifier chain. Long chains of strongly pumped, deeply saturated amplifiers can be subject to much faster power transients.[22] But, for the high channel data rates used with EDFAs, commonly 622 Mb/s or higher, even the dynamics of such chains are relatively slow.

H. GAIN SPECTRUM

The gain bandwidth of the EDFA extends from about 1525–1565 nm, primarily as a result of the Stark splitting experienced by the high angular momentum ground and first excited states of the erbium ions in the local electric fields in the glass matrix. The gain spectrum, which is determined by the distribution of the Stark split sublevels and the thermal distribution of their populations, is not flat, and its shape changes with the level of inversion. Wysocki has shown that in an amplifier or in an amplified system the wavelength where the gain peaks can be predicted using the average gain per unit length of the erbium-doped fiber to characterize the average inversion.[23,24] In fact, it can be shown from Eqs. (2.1) and (2.2) that the aggregate gain spectrum for an amplifier or system of amplifiers is given simply by the gain coefficient averaged over the length of erbium-doped fiber in the amplifier or system:

$$\overline{g(\lambda)} = g^*(\lambda) \cdot \overline{N_2} - \alpha(\lambda) \cdot \overline{N_1} = [g^*(\lambda) + \alpha(\lambda)] \cdot \overline{N_2} + \alpha(\lambda), \quad (2.7)$$

where the overbars indicate taking the average over the length of all the erbium-doped fiber in the amplifier or system. We have used the fact that $N_1 + N_2 = 1$. The gain spectrum for the system is equal to the spectrum of the average gain coefficient scaled for the total length of erbium-doped fiber in the amplifier or system. The gain spectrum is one case where the gain coefficient applies not just to the local behavior, but the gain coefficient averaged over the length of erbium-doped fiber accurately represents the aggregate behavior of a complete amplifier or even a complete amplified system.

Gain coefficient spectra for different values of inversion are shown in Fig. 2.5 for an erbium-doped fiber with aluminum and germanium co-

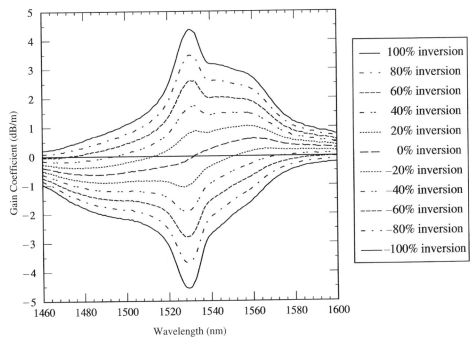

Fig. 2.5 Gain coefficient spectra for different values of inversion for an erbium-doped fiber with Al and Ge co-doping.

doping. For an amplifier, or even for a complete system, the gain divided by the total length of erbium-doped fiber determines the inversion and thus the gain spectrum. Clearly, if the operating wavelengths and operating inversion are not chosen with care, the gain spectrum can be highly nonuniform. For a proper choice of the average inversion the gain is quite flat for wavelengths near 1550 nm.

The calculations shown in Fig. 2.5 are based on the assumption that the transitions are homogeneously broadened. This is not strictly true, but it is a good approximation. Low temperature measurements indicate that homogeneous and inhomogeneous linewidths are comparable.[25,26] Room temperature spectral hole burning, a signature of inhomogeneous saturation, has been observed, but it is even weaker[27] than would be expected from the extrapolated homogeneous and inhomogeneous linewidths, presumably as a result of the rapid thermal redistribution among the sublevels of the Stark split manifolds of the first excited state.

For applications such as WDM systems and multiwavelength networks, amplifiers with flat gain over a substantial spectral range are desired. Depending on the degree of flatness required and the spectral range, flat amplifier gain can be achieved by designing the amplifiers to operate at the appropriate level of inversion or by incorporating gain-flattening filters into the amplifiers.

The gain spectra are strongly dependent on the composition of the erbium-doped core. Erbium-doped silica fibers with aluminum co-doping are capable of flatter and broader gain spectra in the 1545–1560 nm range than are other choices of co-dopants such as germanium or phosphorous (which is necessary in an erbium-doped fiber co-doped with ytterbium fibers to promote efficient energy transfer from the ytterbium to the erbium ions). Erbium-doped fluoride glass fibers produce gain spectra that are flatter in the 1532–1542 nm region.[28]

I. PUMP SCHEMES

The most essential component required for EDFAs, after the erbium-doped fiber, is a pump source to supply light at the correct wavelength (i.e., one of the erbium pump bands) with adequate power to drive the amplifier. The pump sources for the first EDFA demonstrations were an argon ion laser at 514.5 nm[1] and a 670-nm dye laser pumped by an argon ion laser.[2] These lasers are complicated, are expensive, and occupy a large fraction of an optical bench. The pump source for a practical

EDFA should be different: efficient, compact, reliable, and, at least potentially, inexpensive. Fortunately, EDFAs can be pumped with modest optical powers at wavelengths compatible with diode laser technology. The resultant development and commercial availability of suitable diode laser pump sources, particularly those at 1480 and 980 nm, is the key to the rapid acceptance of EDFAs as the first practical optical amplifiers for optical communications.

In addition to 514.5 and 670 nm, erbium has pump bands at 532, 800, 980, and 1480 nm. These wavelengths correspond to the energy differences between the $^4I_{15/2}$ ground state and the first six excited states of the Er^{3+} ion. Absorption of a pump photon at any of these wavelengths raises the Er^{3+} ion to the excited state of the corresponding energy, after which the ion decays nonradiatively (for silica fibers, typically in a time of the order of microseconds) down to the metastable $^4I_{13/2}$ first excited state. Diode lasers have been developed for other purposes at 665 and 800 nm; however, the pumping efficiency at these wavelengths, as well as at 514.5 nm, is degraded by pump excited state absorption (ESA) transitions in which erbium ions in the $^4I_{13/2}$ metastable state can be elevated to a still higher excited state by absorbing pump light.[29]

The most efficient pumping has been demonstrated at 980 and 1480 nm, for which ESA at the metastable level does not occur. High-power diode lasers have been developed at 980 and 1480 nm expressly to meet the need for EDFA pumps, and practical EDFAs are generally pumped at one of these two wavelengths.

Pumping at 1480 nm was first reported by Snitzer et al.,[30] and efficient pumping and high output power were reported by Desurvire et al.[6] Lasers for 1480 nm are made in the InGaAsP/InP material system, the same fundamental technology used for 1.55-μm signal lasers, although modifications must be made to achieve high output powers. Packaged 1480-nm diodes are available commercially with powers in the fiber pigtail exceeding 100 mW.

Efficient pumping at 980 nm was reported by Laming et al.,[31] and progress in developing 980-nm diode lasers, which have InGaAs multiple quantum well active layers grown on GaAs substrates, followed. Packaged 980-nm diodes are also available commercially with powers in the fiber pigtail exceeding 100 mW.

Until recently, commercial EDFAs were generally pumped by 1480 nm diodes because of their high reliability. The dominant failures for 1480-nm lasers are wear-out failures resulting from gradual degradation of the laser.

The dependence of wear-out rates on pump power can be determined so that 1480-nm pump diodes can be run at an appropriate power for any desired degree of reliability. High-power 1480-nm diodes are available with sufficient reliability for terrestrial applications, and at lower power levels, which are still high enough to be useful, they meet the stringent reliability requirements for undersea applications. The InGaAs/GaAs material system used for 980-nm pumps is prone to defects and surface reactivity. In addition, early 980-nm diodes exhibited sudden failures primarily due to catastrophic optical damage at the facet. Advances in coating technology afford sufficient protection that commercial 980-nm pump diodes are now available with demonstrated reliability sufficient for terrestrial applications. With further advances in reliability, 980-nm pumping may also become attractive in undersea applications.

Presuming that pump diodes with comparable reliability and power are available, 980-nm pumping offers several advantages over 1480-nm pumping. The 1480-nm pump band, corresponding to the transition between the $^4I_{15/2}$ ground state and the $^4I_{13/2}$ metastable state, is a special case because it occurs between the same two electronic levels as the lasing transition responsible for amplification. Erbium ions can be pumped at 1480 nm and amplify near 1550 nm because at 1480 nm the absorption cross section, σ_a is larger than the emission cross section, σ_e, by a factor of almost 4, whereas for longer wavelengths, σ_e is comparable to or larger than σ_a (see Fig. 2.3). However, because of stimulated emission at the pump wavelength, complete inversion cannot be achieved, regardless of how great the pump power is. As a result, noise figures less than 4 dB cannot be achieved even with very high pump powers, whereas noise figures approaching the 3-dB quantum limit can be achieved for 980-nm pumping. With 980-nm pumping, noise figures are typically 1 dB lower than those for 1480-nm pumping.[32]

Because it produces a higher level of inversion, pumping at 980 nm also generally produces better spectral characteristics for WDM applications, where flat gain and high output power are both required (as discussed in Section III). Because of their higher quantum efficiencies, 980-nm pumps also require less power to provide injection current and laser cooling; in terrestrial applications, lower power dissipation is an important practical advantage (as discussed in Section III). At high output power, amplifier power conversion efficiency, defined as $(P_{out} - P_{in})/P_p$ is higher for 1480-nm pumping, primarily because of the higher photon energy at 980 nm but also because of the occurrence of ESA at 980 nm from the short-lived $^4I_{11/2}$ pump state at high pump intensities.

Erbium-doped fluoride fibers, which have been proposed for multiwave-length applications because of their superior gain flatness,[33] cannot be pumped at 980 nm. The low phonon energies in fluoride glasses result in inefficient nonradiative decay from the $^4I_{11/2}$ pump state to the $^4I_{13/2}$ metasta-ble state and ESA. In practice, only 1480-nm pumping is used for erbium-doped fluoride fiber amplifiers.

An alternative pump scheme can be used when erbium-doped fibers are co-doped with ytterbium(Yb). The ytterbium ions, which are pumped directly, usually at about 1060 nm, transfer their excitation energy to erbium ions, elevating them to the $^4I_{13/2}$ metastable state. The advantage of this scheme over direct pumping of Er at 980 or 1480 nm is that compact, high-power, 1060-nm pumps can be made by using a Nd:YAG laser pumped by a high-power, multistripe, 800-nm diode laser. Extremely high output powers can be achieved with proper design of the composition of the Er/Yb co-doped fiber.[34]

J. COMPONENTS

In addition to the erbium-doped fiber and the pump source, every EDFA requires a wavelength selective coupler (WSC), or pump–signal combiner, to combine the signal and pump wavelengths. The two most common types of pump–signal combiners are the fused fiber WSC and the interference filter WSC. The fused fiber WSC is a fused fiber device in which the coupling region is designed so that light at the pump and signal wavelengths entering the two respective input ports exits the device on a single output port (the fourth unused port is sometimes terminated internally). The interference filter WSC is most commonly a three-port device in which pump and signal are coupled using graded index (GRIN) lenses onto an interference filter aligned so that the signal light from one input port is reflected to the output port and the pump light entering the other input port is transmitted to the output port. Alternatively, the coupler can be configured so that pump light is reflected and the signal light transmitted to the output port. Interfer-ence filter WSCs typically have lower polarization-dependent loss, lower cross talk, and lower spectral dependence in the signal passband. Fused fiber WSCs can have lower loss and are free of reflections, which can give rise to weak etalon effects.

Optical isolators must also typically be used in EDFAs to prevent excess reflected ASE or even lasing (which would add excess noise), the backward propagation of ASE noise (that saturates the amplifier and degrades the noise performance), and multipath interference (which, in the presence of the high gain of the EDFA, would add noise even for small reflections). More advanced amplifier architectures may include optical filters for shaping the gain spectrum, which are commonly implemented using interference filters or fiber gratings, optical circulators, or other optical elements such as dispersion compensators.

III.   Terrestrial Applications for EDFAs

Digitally regenerated lightwave systems carry essentially all interoffice communications traffic in the United States, with tens of thousands of voice circuits multiplexed onto a single fiber. In the past, increased system capacity was generally achieved by pushing electronic time-division multiplexing (TDM) to higher signaling rates. Optical WDM was not the preferred alternative to increase system capacity because optical losses of multiplexers and demultiplexers significantly reduce the achievable system length compared with single-channel systems. With the advent of EDFAs, which can compensate for these losses, optically amplified WDM transmission has become a practical method of increasing both system capacity and loss budget between terminals.

Within a few years after the demonstration of the EDFA in the laboratory, commercial terrestrial systems began to appear. Optical amplifiers were first commercially applied as power amplifiers in single-channel systems placed following the laser transmitter to boost the optical power launched into the fiber and extend the repeater distance. Repeater distances were increased from 80 to more than 120 km. Next, EDFAs were used to compensate for WDM multiplexer and demultiplexer losses in WDM systems with two to four channels. For this application, an optical preamplifier is usually required at the receive end in addition to the power amplifier at the transmitter. Finally, EDFAs are being used to extend the span length of these WDM systems by adding in-line amplifiers to replace regenerators at sites between the power amplifier at the head end and the preamplifier at the receive end.

A basic optically amplified WDM system with amplifiers serving all these functions is shown in Fig. 2.6. On the transmit end, M lightwave channels are combined in a passive M-to-1 coupler, and the resultant WDM signal is optically amplified. The power amplifier operates in the saturated regime and compensates for the losses in the M-to-1 coupler. At the receive end,

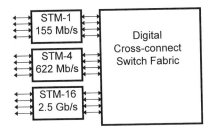

Fig. 2.5   Conceptual diagram of a SONET digital cross-connect switch.

used in North America. The SONET hierarchy starts at exactly one-third the rate of the SDH base signal. In order to make transport systems compatible with the existing signal hierarchies, a number of mappings are defined. At this point, the highest rate with available equipment is STM-64. The signal rates in most common use are shown in Table 2.2.

Higher rate signals in the hierarchy either can be a multiplex of lower rate signals or can themselves constitute a new, high-capacity single channel. The term *concatenated signal* is used to indicate a signal that is not composed of multiple component signals. Concatenated signals are referred to with a *C* after the hierarchy number, such as STM-4C, or STS-12C, to denote a signal with a single payload that must be delivered intact. This feature allows for the transport of future high-capacity signals as the need for them arises.

2.   The SDH Frame Structure

The TDM signal is periodic, with a period of 125 μs. This corresponds to the period of a voice sample from telephony. One octet serves to carry an 8-bit digital sample for one voice channel, and these samples must be taken every 125 μs to faithfully reproduce the frequencies in the 3-kHz analog

Table 2.2   **Selected Signal Rates for the SDH and SONET**

Signal Rate	*SDH Signal*	*SONET Signal*
155.52 Mb/s	STM-1	STS-3
622.08 Mb/s	STM-4	STS-12
2.488 Gb/s	STM-16	STS-48
9.953 Gb/s	STM-64	STS-192

amplifiers is between 20 and 40 dB. At these levels of saturation, the ASE noise output is reduced and the optical SNR at the output of the amplifier is extremely high. Thus, the noise figure is not an important design parameter for power amplifiers. On the other hand, the most important parameter for power amplifiers is maximum saturated output power, for this determines how much power can be launched.

The output power of an amplifier saturates at high input signal levels. The output power for saturated amplifiers is directly proportional to the pump power, as explained in Section II. In power amplifiers, the optical conversion efficiency describing the energy transfer between pump and signal quantifies the ability of the amplifier to produce high output signal power. Power conversion efficiencies of 50% and 80% have been achieved with 980- and 1480-nm pump wavelengths, respectively. The difference in the power conversion efficiencies arises mainly from the quantum defect between the photon energy of the 1550-nm signal light and that of the two pump wavelengths.

The electrical-to-optical power conversion efficiency is also a key figure of merit for power amplifiers. Electrical power is used both to drive and to cool the pump laser, and the efficiency with which electrical power can be successfully converted to amplified signal power is important. Terrestrial systems must operate at ambient temperatures as high as 65°C, so that heat management is an important design consideration and plays a key role in determining the physical size of the amplifier. The heat that a pump laser generates is proportional to the drive current, so that electrical-to-optical conversion efficiency can be improved by reducing the drive current while maintaining the optical output power. Semiconductor diode lasers operating at 980-nm wavelengths typically have significantly better quantum efficiency and require less drive current than 1480-nm diode lasers, thus the electrical-to-optical conversion efficiency of 980-nm pumped amplifiers is usually better than that of 1480-nm pumped amplifiers.

The erbium-doped fiber in a power amplifier is designed for the most efficient conversion of pump energy into signal energy. In a single-stage amplifier, the length of the single erbium-doped fiber is made as long as possible so that the pump light is converted to signal energy as completely as possible. Additional pumps may be used to further increase the net output power. For amplifiers with two or more stages, interstage elements such as filters or isolators may be used to keep ASE from propagating and reducing the net inversion in the following or preceding stage. Pump reflectors, which transmit the signal wavelengths, may be used to return into the gain medium unabsorbed pump light that would otherwise be lost.

Reports of EDFA power amplifiers frequently mention multipumped, multistage architectures. Conventional approaches use 1480-nm bidirectional pumping arrangements that may rely on polarization combiners to increase the total pump power. For example, +22.7-dBm output power has been obtained in a packaged amplifier with four 1480-nm semiconductor diodes.[35]

Amplifiers using a hybrid of 1480- and 980-nm lasers have been designed with high output powers in the range of +20 dBm. The advantages of hybrid pumping over pumping at only 1480 nm include the suppression of pump cross talk and the lowering of power consumption. The major drawback is the difficulty in providing pump redundancy.

Output powers in excess of +20 dBm have also been demonstrated in dual, 980-nm pumped amplifiers. The rather simple design of the reported topologies and their low electrical power consumption make them attractive candidates for systems applications. In addition, the availability of 980-nm high-power (1-W) master oscillator power amplifiers (MOPAs) opens the possibility of very high output power (500-mW) amplifiers.

Output power in excess of +27 dBm has been demonstrated with an Er/Yb co-doped fiber amplifier pumped at 1060 nm with a diode-pumped Nd^{3+} laser.[36] Provided that the high electrical power consumption for this type of amplifier is compatible with system requirements, it offers an excellent solution for power amplifier applications.

B. PREAMPLIFIER

The sensitivity of a direct detection receiver can be improved significantly by using a low noise figure (3- to 5-dB) optical preamplifier. An optical preamplifier is used at the end of a transmission link, just before the photodetector and regenerator. The input power level of the preamplifier is extremely low because the signal has lost power in the transmission link. The output power of the preamplifier needs to be sufficiently high so that at the photodetector the noise is dominated not by its receiver noise but by the signal–spontaneous beat noise of the optical preamplifier. This means operating at least 10 dB higher than the nominal sensitivity of the photodetector–regenerator combination. Typically, this does not place stringent requirements on the preamplifier gain, even after demultiplexing losses have been taken into account.

Thus, an optical preamplifier is designed primarily to achieve a low noise figure with about 20–30 dB of small-signal gain. This requires a low input coupling loss. A low insertion loss, high isolation, and polarization-independent isolator must be placed at the input to minimize degradation of the noise figure and to prevent optical feedback reflections that could result in lasing. It is also important to keep the average inversion level at the input as high as possible. This can be accomplished in a single-stage amplifier by using a short segment of erbium-doped fiber and reverse pumping to minimize input coupling losses. Multistage designs are not required to achieve these goals. However, in WDM systems requiring additional gain to compensate for the demultiplexer losses following the optical preamplifier, multistage preamplifier designs are generally used.

Optical preamplifiers are usually pumped at 980 nm because complete inversion is possible at this pumping wavelength. As long as the amplifier architecture allows sufficient 980-nm pump light near the input so that inversion is high there, quantum-limited internal noise figures approaching 3 dB are possible.

The insertion of interstage components allows the optimum combination of low noise figure and high-gain for a minimum pump power. For example, by using a combination of isolators and band-pass filters, one can easily achieve 30–40 dB of gain with a 3- to 4-dB noise figure for 980-nm pumps.[37] When using optical preamplifiers in combination with narrow band-pass filters and appropriate photodetectors, Livas[38] achieved record sensitivities close to the quantum limit for bit rates up to 10 Gb/s. These hero experiments reveal the need for interstage isolators to suppress backward ASE that degrades the amplifier noise figure and causes light pollution from Rayleigh backscattering in the transmission fiber.

In high-speed (>2.5-Gb/s) transmission systems, where the sensitivity has been limited by the gain-bandwidth product of the photodetector and the thermal noise of the receiver, the use of an optical preamplifier receiver has improved receiver sensitivity by between 5 and 20 dB. Indeed, the clear advantage of optical amplifiers is that there is no gain-bandwidth-product limitation or pulse distortion even after multistage amplification. In addition, system upgrades are readily implemented because the optical preamplifier is wavelength and bit rate independent. Figure 2.7 shows the record sensitivities for PIN and APD direct detection, coherent detection, and optically preamplified receivers. Record sensitivities have been demonstrated at 5, 10, and 20 Gb/s with high-gain tandem amplifier configurations.

All the optically preamplified high-speed receiver sensitivities reported have been demonstrated using multistage amplifier designs in which the first stage is pumped at the 980-nm pumping wavelength to guarantee a near-quantum-limited noise figure. Practical dual-stage preamplifier designs

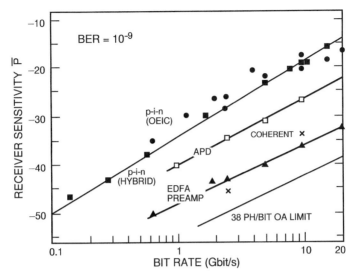

Fig. 2.7 Sensitivity for PIN and avalanche photodiode (APD) direct detection, coherent detection, and optically preamplified receivers. BER, bit error rate; OEIC, optoelectronic integrated circuit.

offer a combination of low noise and high gain that could not be achieved with a single-stage topology.

C. IN-LINE AMPLIFIER

Finally, we consider in-line amplifiers used as repeaters to boost the signal power and extend the transmission distance between digital regenerators. Because in-line amplifiers are used to extend the transmission distance, high output power is required. However, the signals entering in-line amplifiers are weak, so that noise added by each in-line amplifier is important. Therefore, a low noise figure is also required. Thus an in-line amplifier must be both a good preamplifier and a good power amplifier. Finally, because the span losses preceding in-line amplifiers can be different, and because the number of channels present can change from none to the design maximum, the total input power to in-line amplifiers can vary over a wide range. Thus dynamic range is important for in-line amplifiers.

In-line amplifiers must be designed with high gain, high output power, and a low noise figure, all realized for a wide dynamic range of input signals. It is difficult to optimize simultaneously all three parameters using a single-stage design. For example, high gain is reached by using a long segment of erbium-doped fiber. In contrast, the best noise figure is obtained by maintaining a high inversion at the input and is usually best achieved with a short segment of erbium-doped fiber. The solution is to use multistage designs in which the first stage is designed to function as an efficient preamplifier and the succeeding stages as a power amplifier.

To maintain a high enough optical SNR, the input powers to in-line amplifiers cannot be too small. High input powers saturate the amplifiers. Also, the more channels that are present, the higher the total input power to each amplifier, leading to further saturation. Typically, the small-signal gain required for an in-line amplifier is 5–15 dB more than the span loss for which it is compensating. Thus in-line amplifiers in practical systems operate 5–15 dB into compression.

The desire to maximize the repeater spacing and transmission distance has led to the design of multistage optical repeaters with high output power (up to 17 dBm), high gain (up to 45 dB for small signals), and a low external noise figure (as low as 3.5 dB). Such a combination of performance corresponds to a potential 33-dB optical budget margin or the equivalent of 150 km of low-loss (0.2-dB) optical fiber between repeaters for high-capacity (eight 2.5-Gb/s channels) WDM transmission. Several two-stage amplifier topologies meeting these performance criteria have been reported and investigated in transmission experiments.[39]

Gain flatness is also an important characteristic for optical amplifiers, especially for in-line amplifiers, used in WDM applications. The gain of a single, unsaturated optical amplifier is significant over a wide spectral range of between 40 and 50 nm. However, if many amplifiers are connected in series and operated significantly into gain saturation, the effective gain of the entire amplifier chain narrows as a result of concatenation of gain spectra, which are each individually narrowed (see Section II). For example, Fig. 2.8 shows the calculated effective gain region as a function of span number for a chain of amplifiers separated by twelve 80-km spans. The effective gain is calculated by sweeping the spectrum with a small-signal probe. After 12 spans, the 3-dB bandwidth is only about 5 nm. Because these amplifiers are deeply saturated and therefore poorly inverted, a signal at 1530 nm, the small-signal gain peak, will lose power in a chain of amplifiers if there are other signals present near the 1558-nm gain peak.

Because of the challenges of demultiplexing closely spaced channels, as well as impairments introduced by four-wave mixing for closely spaced channels, channels in WDM systems should not be spaced too closely. To

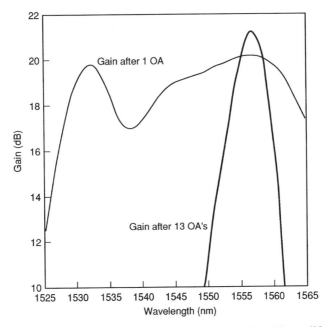

Fig. 2.8 Gain region after the first and last amplifiers in a 13-amplifier chain.

maximize WDM capacity, it is desirable that the optical bandwidth of the system be as wide as possible. The variations in the gain spectrum result in channel-to-channel variations in the optical SNR and absolute signal power. Because system performance is limited by the SNR of the worst performing wavelength, a large gain variation can severely limit system length. Clearly, the peaking of the concatenated gain shown in Fig. 2.8 restricts the range of usable WDM wavelengths, and thus the WDM capacity.

Gain-flattened amplifiers for a limited dynamic range can be designed by selection of the host material and the operating inversion and possibly by use of gain equalization filters. To extend the dynamic range, several approaches can be used. One is to use feedback to lock the gain, so that the net inversion and gain curvature remain fixed over a wide dynamic range. Another method is to use separate gain modules customized for the actual span losses, so that the average inversion is constant.

Most optically amplified, long-distance transmission experiments have used dispersion-shifted fiber (DSF) to avoid limitations at high data rates imposed by chromatic dispersion. However, most installed terrestrial fiber is conventional single-mode fiber, which has a relatively high dispersion near 1550 nm ($\approx$17 ps/nm-km). The use of dispersion-compensating fiber (DCF) offers the possibility of upgrading the embedded fiber network with multigigabit 1.5-μm amplified repeatered transmission systems. However, the additional transmission loss incurred by the addition of DCF needs to be overcome by additional in-line amplifiers. As an example, a recent field trial demonstrating 10-Gb/s transmission through 360 km of non-dispersion-shifted installed fiber uses three in-line EDFA repeater modules, each comprising a DCF sandwiched between two tandem optical amplifier configurations.[40] The spacing between repeaters was 120 km, corresponding to a 33-dB optical span margin.

D. SYSTEM DESIGN ISSUES

A terrestrial interoffice lightwave route consists of terminal sites at each end and repeater huts every 40–120 km along the transmission line. At the terminal sites, the incoming signals are optically demultiplexed and regenerated, and outgoing signals are optically multiplexed and transmitted. At the repeater site, the signal is either optically amplified (in the case of optical amplifiers), or amplified, retimed, and regenerated (in the case of conventional optoelectronic digital regenerators). The distance between terminal sites is typically as long as 600 km. The distance between repeater sites in commercial systems has increased from 2 km in the early days of multimode fiber systems to 120 km with single-mode fiber and 1.5-μm optics.

The most important system design parameter for optically amplified systems is the optical SNR measured for each signal channel at the output of the last optical amplifier. The optical signal coming out of the final amplifier in the chain is degraded by the accumulated ASE. The SNR is the ratio of signal power to ASE power in a fixed bandwidth. ASE-induced noise is converted to an electrical noise signal in the photodetector, giving rise to bit errors. The bit error rate must be kept less than a limit determined by system performance requirements, typically 10^{-15} for terrestrial networks. End-to-end system performance can be estimated from the optical SNR and a model of the total electrical noise in the receiver. System performance requirements then determine a minimum optical SNR for each signal. The entire system and the individual amplifiers must be designed so that the minimum SNR is maintained over all possible operating conditions.

The noise figure of the amplifiers in a chain determines the rate of ASE noise buildup. The system requirement of a minimum SNR then determines how many amplifiers can be chained together with fixed span losses before regeneration is required. The lower the noise figure, the longer the chain. The noise figure also affects the maximum span loss that can be supported by the chain.

System design for an optically amplified system is a two-step approach that begins at the receiver and allocates receiver margins in a way similar to regenerated system design. The two steps can be summarized as follows:

(1) Determine the minimum optical SNR required out of the last amplifier to maintain a low bit error rate over the range of receiver degradations.

(2) Determine how much additional SNR is required so that the minimum SNR into the receiver is maintained for all expected degradations in the amplifier chain.

An approximate relationship for the SNR in dB after N optically amplified spans is given by

$$SNR \approx 58 + P_{out} - L - NF - 10 \log_{10} N, \qquad (2.8)$$

where SNR is the optical signal-to-noise ratio in decibels after N spans, measured in a 0.1-nm bandwidth; P_{out} is the output power per channel in dBm (i.e., decibels referenced to 1 mW); NF is the noise figure in dB, including input coupling loss; and L is the span loss in dB. This result shows that the SNR can be increased decibel for decibel by increasing the output power per channel, by decreasing the noise figure, or by decreasing the span loss. The SNR falls proportionally to $\log_{10}(N)$, where N is the number of amplified spans.

Raising the per-channel signal power is a good way to increase the SNR, but the useful signal power is limited by optical nonlinearities in transmission fiber, discussed in detail in Chapter 8 in Volume IIIA. For most terrestrial systems, the dominant nonlinear transmission effect is self-phase modulation, which limits per-channel signal power to less than 8–12 dBm, depending on system length. Decreasing the noise figure to increase the SNR is possible subject to a limit set by the 3-dB quantum limit for minimum noise figure.

The most effective way to increase the SNR is to decrease the loss between spans. When this is done, the number of spans, N, increases because $N \cdot L$ is a constant set by the distance between terminals. Because the SNR decreases linearly with L and only logarithmically with N, amplified systems should logically be designed with many short spans N with small losses L.

However, this strategy conflicts with route design imperatives, driven by economics and operational considerations, to minimize the number of repeaters. In practice, for optically amplified terrestrial systems, the repeater lengths are made as long as possible consistent with maintaining an adequate SNR.

As an example, Fig. 2.9 shows the SNR versus system gain $N \cdot L$ for several values of span loss. A fixed per-channel output power of 10 dBm and a noise figure of 6 dB have been assumed. All nonlinear transmission effects have been neglected. The region of applicability for terrestrial systems is for system gain $N \cdot L < 200$ dB. One can see from the figure that a SNR of 20 dB can be achieved for span losses of 30–35 dB. This translates into repeater span lengths of 100–160 km, a dramatic increase compared with the 40- to 80-km spacings used in typical regenerated systems. The figure also shows the region of applicability for undersea systems, with system gains greater than 1000 dB. Clearly, to achieve these high system gains, shorter repeater spans with less than 15 dB of loss are required. For longer systems transmission impairments due to nonlinear effects reduce the acceptable output power and further limit the repeater span.

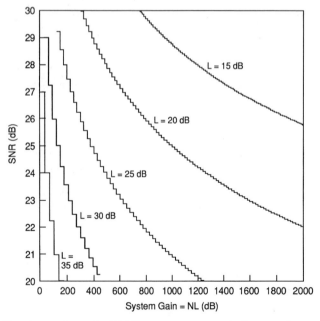

Fig. 2.9 Signal-to-noise ratio (SNR) versus system gain for span losses (L) of 15, 20, 25, 30, and 35 dB.

So far, we considered only how power and noise figure affect system performance. In practical WDM systems, gain shape is important. For example, Fig. 2.10 shows the optical spectrum at the end of an eight-channel system with $N = 12$ and $L = 22$. Note that the concatenation of 13 amplifiers has resulted in a net gain shape with considerable curvature over the bandwidth of interest. One way of dealing with this is to use preemphasis.[41] By adjusting the transmitter powers of the individual channels so that the input channel power spectrum is complementary to the gain or SNR spectrum of the amplifier chain, one can equalize either the output power or the SNR at the output

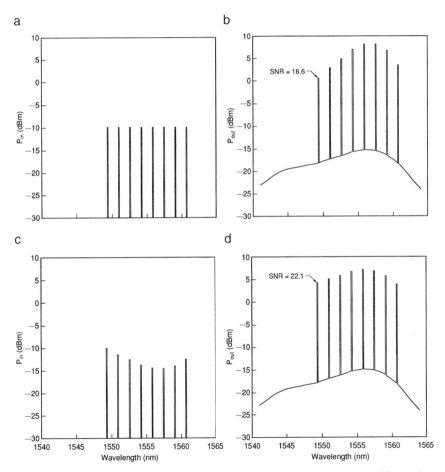

Fig. 2.10 Optical spectra for an eight-channel, 12- × 22-dB system. (a) At the input, without preemphasis; (b) at the output without preemphasis; (c) at the input, with preemphasis; and (d) at the output, with preemphasis.

of an amplifier chain. Another method is to use optical filters within the chain to correct for gain curvature. A third approach, as discussed in Section II, is to choose amplifier designs with flat gain spectra — for example, by controlling the inversion or by using erbium-doped fluoride fiber.

E. TELEMETRY CHANNEL

A lightwave system requires a telemetry channel to transmit information needed for system management, system maintenance, and fault location, as well as to provide voice communications to remote repeater sites. In conventional digitally regenerated systems, the telemetry channel uses overhead bits on one signal channel. These bits are accessible through digital demultiplexers. In optically amplified systems, there is no digital demultiplexing; therefore, other means must be found to transport a telemetry channel.

One solution is to create a new channel, just for telemetry, using WDM methods. Unlike the signal channels, the telemetry channel must be added and dropped at each repeater site. A transmitter, a receiver, and a regenerator for the telemetry channel are needed at each site. The required telemetry channel capacity is usually less than 1.5 Mb/s, a slow rate for lightwave communications.

Figure 2.11 shows several possible telemetry channel configurations. Perhaps the simplest approach would be to use existing local telephone service to provide telemetry access (Fig. 2.11a). However, this method has the disadvantage that local service is not always available at remote repeater sites and can be expensive for 1.5-Mb/s links. Another approach that has been suggested is to use gain modulation for telemetry purposes (Fig. 2.11b). This method has the advantage that no additional optical multiplexing is required, but it has the disadvantage that the efficiency of gain modulation at 1.5 Mb/s is extremely low due to the dynamic response of erbium-doped fiber.

All the other configurations shown in Fig. 2.11 use a separate WDM wavelength for the telemetry channel. For example, telemetry channels at wavelengths of 1300, 1480, 1510–1520, 1530, 1580, and 1600 nm have been proposed. For amplifiers using 1480-nm pumps, it is possible to modulate the pump laser and use the pump feedthrough to transmit telemetry information to the next amplifier site (Fig. 2.11c). As long as the modulation rate is more than a few megahertz, the gain of the amplifiers is not affected because of the slow response time of erbium-doped fiber. This method has the additional advantage that no extra laser source or WDM coupler is

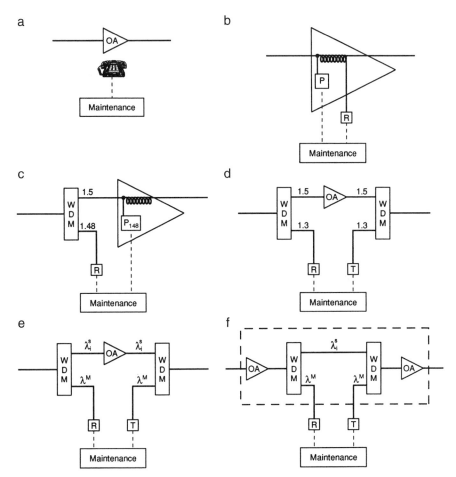

Fig. 2.11 Telemetry channel configurations: (a) Dial-up or cellular phone, (b) gain modulation, (c) pump modulation, (d) 1.3-μm wavelength-division multiplexing (WDM) (outside OA), (e) 1.5-μm WDM (outside OA), (f) 1.5-μm WDM (inside OA).

required. However, it has the disadvantage that the pump feedthrough depends on the compression of the amplifier, and this can change with the number of channels present and the span losses between amplifiers.

Just as 1300- and 1500-nm channels have been used for traffic, they can also be used for telemetry (Figs. 2.11d and 2.11e). A 1300-nm telemetry channel has the advantage that the WDM couplers for adding and dropping the telemetry channel have been used for many years and are a mature technology. On the other hand, fiber losses at 1300 nm are higher and spans of 120 km or more may be difficult at this wavelength. If the 1500- to 1600-nm region is used for telemetry transport, fiber losses are low. However, care must be taken in this region to avoid interference with the signal channels in the erbium window of 1530–1565 nm.

Some configurations split off the telemetry channel before the input to the amplifier, and add the outgoing signal after the output of the amplifier. These configurations have the advantage of decoupling the telemetry channel from the amplifier design, but they have the disadvantage that the system gain is reduced by the coupling losses at both the input and the output. The configuration shown in Fig. 2.11f uses a midamplifier drop and add of the telemetry channel, which has a minimal impact on the system gain. On the other hand, amplifier complexity is increased by this type of telemetry channel, and it is more difficult to isolate the telemetry channel from the signal channels.

Factors such as cost, reliability, and system performance affect the choice of the telemetry channel. For example, the telemetry channel used in digitally regenerated systems uses overhead bits to transmit telemetry data. This type of telemetry channel reduces system capacity by the ratio of telemetry bits to total bits, usually less than 0.05%. Therefore, the telemetry channel implementation in an optically amplified system should be expected to cause only a small reduction (0.05%) in system capacity.

IV. Submarine Systems

The reliability required of undersea communications systems differentiates them from the systems developed for terrestrial applications. Repairs to undersea systems are extremely expensive and require many days to complete. Therefore, for hardware to be placed under the sea, the primary design focus is on reliability. The demand for highly reliable systems leads to a desire for simplicity, limits the complexity of designs and hence the features available, and restricts the choices available for components.

The appeal of an EDFA in undersea applications is tremendous. The EDFA epitomizes simplicity, which promises reliability; only three components are required to form an optical amplifier, although practical amplifiers usually use more. An added appeal of EDFAs is that they are bit-rate and transmission-format insensitive. This fact has tremendous appeal in the submarine transmission marketplace, where (1) the cost of new undersea designs is high; (2) the economic appeal of high bit rates is alluring but the difficulty and expense of developing and qualifying higher and higher speed electronics for undersea deployment is huge; and (3) new marketplaces demand varied and ever-increasing bit rates.

This section discusses the design of EDFAs used in undersea applications. To gain a fuller understanding of the design and operation of undersea repeaters, the reader may refer to Chapter 10, Undersea Amplified Lightwave Systems Design, in Volume IIIA, which contains a detailed discussion of the design considerations for long-haul undersea systems.

A. DESIGN REQUIREMENTS

EDFAs used in undersea repeaters are critical parts of optical transmission systems that span from less than 2000 to 9000 km and, as such, must meet critical design requirements for the following:

- Architectural simplicity and reliability
- Gain (output power, gain compression, gain shape)
- Noise generation
- Polarization-dependent loss and gain (PDL and PDG)
- Polarization mode dispersion (PMD)
- Ability to participate in a performance-monitoring scheme

Furthermore, the design of the repeater must meet these requirements over the range of temperatures that can be expected undersea and during manufacture (0–35°C), over the range of expected transmission wavelengths, and for the full range of parametric variability that is found in component and subsystem manufacture.

B. ARCHITECTURAL SIMPLICITY AND RELIABILITY

Undersea systems are designed to survive for 25 years. During this period, no more than three system repairs are accepted as a result of component failure or any failure not due to external aggression (e.g., fishing trawlers, anchors, sea quakes). Because of the cost and time necessary to carefully design and test components to ensure this high level of reliability, an amplifier design is required that uses the minimum number of components.

Failures of components in submarine applications are categorized into *hard failures* and *soft failures*. A hard failure is caused by a complete signal path failure, such as a fiber break. Every occurrence of a hard failure requires a repair of the undersea system by a cable ship. A soft failure is caused by a change in component characteristics and can result in degradation to the performance of the transmission line. As is discussed later in this chapter, mechanisms are designed into the amplifier to control the impact of soft failures. Only in cases where the degradation is so severe that it exceeds the available performance margin is a ship repair required.[42]

1. Amplifier Architectures

The architectures of amplifiers used in undersea applications vary. All designs share at least one attribute: simplicity. Figure 2.12 shows one architecture used in submarine repeaters: an amplifier pair.[43] The amplifier pair contains two EDFAs to provide amplification on two transmission paths in opposite directions. A repeater consists of up to four of these amplifier pairs.

The basic amplifier uses a counterpropagating pump to pump an erbium-doped fiber. A single isolator is used to prevent the accumulation of counter-propagating ASE, to reduce multipath interference, and to limit the possibility of lasing.

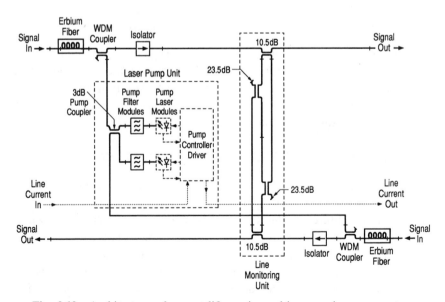

Fig. 2.12 Architecture of an amplifier pair used in an undersea repeater.

To maximize the reuse of fused-fiber coupler technology, a fused-fiber WSC is used. Because fused-fiber WSCs typically have an isolation of only 13 dB, a filter is placed between the pump laser and the WDM coupler to limit the coupling of the transmission signal into the pump laser and to eliminate the reflection of the signal off the end face of the pump pigtail. Without this pump filter, the backward-reflected signal would be reflected again in the direction of transmission by the Rayleigh backscattering in the transmission fiber (after three passes through the EDFA). This time-delayed copy of the signal can cause a significant transmission penalty.[44]

2. Reliability and Sparing

High reliability is demanded of the pump laser to maximize repeater reliability. If the pump laser were placed in a configuration where a pump failure caused a hard failure, extremely high pump reliability would be required — i.e., approximately 1 FIT (failure in time) (1 FIT is defined as the expected number of failures in 10^9 h). Because this level of reliability is not currently demonstrable in pump lasers, redundancy is required. A novel sparing approach[45] is employed in the architecture shown in Fig. 2.12, where two shared lasers pump both EDFAs in an amplifier pair through a 3-dB coupler. Similarly, additional reliability may be obtained by bidirectionally pumping a single EDF (Fig. 2.13).

With these sparing approaches, if one laser fails, the remaining laser still provides pump power to its amplifiers. Although the gain of these amplifiers will decrease, the deep gain compression of the subsequent amplifiers in the system will cause the signal level along the transmission path to "recover" to

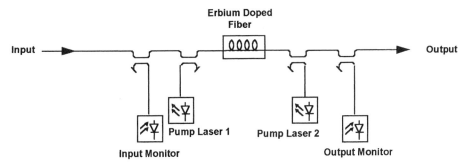

Fig. 2.13 Bidirectionally pumped amplifier architecture equipped for the command–response supervisory system (after Ref. 50).

its original level (see the next section). The penalty caused by the resultant increase in noise is small and is accounted for in the system design.

C. Gain and Noise Generation

Signal-level control is the mechanism by which the optical power level is established along the optical transmission path. Two means have been used:[46]

(1) In the active power control scheme, the signal power at each repeater is sampled by a photodetector and used to control the pump power of the amplifier.

(2) The amplifier is operated in compression. As shown in Fig. 2.14, the small-signal gain is chosen to be several decibels larger than required to make up for the span loss. The input power is therefore no longer in the small-signal regime (i.e., the gain is not independent of input power), and gain compression results. This establishes the nominal operating point for the amplifier. If the input power is then decreased (e.g., by the increase of loss in a component or a pump failure in the previous repeater), the gain of the subsequent amplifier will increase. Similarly, if the input power of an amplifier is increased, the output power of the subsequent amplifier will decrease.

In submarine applications, EDFAs have been used to provide repeaters with gain from 7 to 22 dB and output power from 3 to 6 dBm as dictated

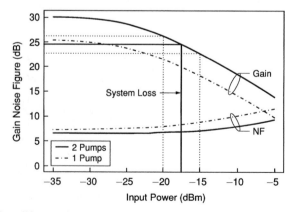

Fig. 2.14 Amplifier performance in compression with one or two pumps operating.[47] NF, noise figure.

by the application's demands. The gain and output power of the amplifiers are determined during the manufacturing process by selection of the proper length of the EDF and the adjustment of the pump power.

In long systems (6000–9000 km), low gain amplifiers are used because noise generation must be minimized. (Note that the noise generated by an amplifier is linearly dependent on its gain, and therefore exponentially dependent on the distance between amplifiers. Hence, a system composed of high-gain amplifiers generates more noise than one composed of a greater number of low-noise amplifiers.) Additionally, the transmitted power level must be kept low to avoid optical nonlinearity (that is dependent on the interaction length). Low-output power amplifiers are used (~3 dBm). The selection of output power is a trade-off between the desire to minimize the output power (to avoid optical nonlinearity) and the desire to maximize the output power (to maintain an adequate input power to the amplifier).

In short systems (<6000 km), the maximum output power that can be transmitted is less constrained. Therefore, higher output power amplifiers (often limited by the amount of pump power available) and higher gain amplifiers (18–22 dB) are used.

Figure 2.14 shows the gain and noise characteristics of the amplifiers designed for short systems (with a span length of approximately 70–100 km). The natural gain compression of the EDFA is used to provide power-level control for the amplifier chain. Although the small-signal gain is greater than 30 dB, the amplifier will be operated with more than 5.0 dB of compression. This provides a gain "recovery" of 0.6-dB gain per decibel change of P_{in}. The "recovery" is used to compensate for power fluctuations along the transmission line resulting from such causes as pump failures, fiber loss increase due to hydrogen absorption, component degradation, and manufacturing variation.

Figure 2.14 also shows the amplifier's gain characteristics if a single pump laser fails. Note that the amplifier still has substantial gain. This results in a system that is tolerant to a limited number of pump failures.

The noise figure of a saturated high-gain amplifier is less than 7.0 dB. The average low-gain amplifier (suitable for use in long systems) has a noise figure of less than 6.0 dB.

Because the system is designed to be tolerant of pump failures, the increase in noise figure due to reduced pumping must also be considered. As shown in Fig. 2.14, the noise figure increase of an amplifier with a failed pump is less than 2.0 dB.

In long chains of optical amplifiers, the accumulation of wideband (~30–40 nm) ASE causes a transmission impairment because it steals part of the population inversion (i.e., gain) of the EDFA. Note that the architecture shown does not contain a filter to remove this ASE. The system design relies on the natural shape of the gain spectrum of the saturated EDFA to provide the filtering necessary to limit ASE saturation (see Fig. 2.8). This characteristic is often referred to as *self-filtering* and is desirable in long chains of amplifiers because it obviates the need for an optical filter. In long systems, the gain peak is extremely narrow. It is therefore critical that the system design match the gain peak of the amplifier chain to the wavelength of the shore transmitter and the zero-dispersion wavelength of the transmission fiber. Although this amplifier gain peaking is desirable for single-channel systems, it presents a severe limitation on the design of amplifiers for WDM systems unless separate gain equalization filters are used.

D. POLARIZATION-DEPENDENT LOSS AND GAIN

PDL is also an important consideration in the design of the amplifier. PDL contributes to optical SNR fluctuations and SNR degradation by inducing a loss variation in the signal (which has a high degree of polarization) while leaving the unpolarized ASE noise unaffected. Although the PDL in each amplifier is small (<0.1 dB), PDL of just 0.1–0.2 dB has been shown to contribute an unacceptable system impairment in 9000-km systems.[49]

The largest potential PDL contributions are from the WSC and isolators. The fused-fiber WSC has one wavelength at which it has zero PDL. By careful design and controlled manufacture, this PDL minimum coincides with the wavelength selected for transmission (the use of a thin-film WSC, eliminates this concern). Similarly, the careful design of the isolator ensures a low polarization dependence. The PDL of the amplifier shown in Fig. 2.13 is less than 0.08 dB.[43]

PHB (see Section II.F) is another polarization-dependent effect in undersea transmission systems that is contributed by the amplifier.[19] As a result of the inhomogeneous saturation behavior of the amplifier, ASE noise orthogonal to the signal experiences a higher gain. This causes the ASE noise to accumulate faster and contributes to a degraded SNR upon detection. The deleterious effect of PHB can be avoided by proper control of the transmitted signal state of polarization. Use of an unpolarized transmitter or use of one where the state of polarization is rapidly changing has been shown to reduce the PHB impairment.[20]

E. POLARIZATION MODE DISPERSION

PMD can contribute to pulse dispersion and therefore causes a transmission impairment. Although PMD is usually thought of as a fiber transmission effect, the PMD contributed by the repeater must also be considered because the total PMD in a transmission path is made up of contributions from both the transmission fiber and the repeater. The goal for the repeater design is to make the PMD contribution of the repeater small in comparison with that of the fiber (which is less than $0.15 \text{ ps}/\sqrt{\text{km}}$). The components with the largest potential for PMD are the isolator and the erbium-doped fiber. Isolators have been designed to contain a PMD compensation element that greatly reduces the PMD from that found in standard single-stage devices (usually >1 ps). Dual-stage isolators also have greatly reduced PMD. Care must also be taken during the manufacture of the erbium-doped fiber to minimize its PMD. On average, a repeater's PMD can be expected to be less than 0.30 ps.[43] The couplers do not contribute significantly to the repeater's PMD.

F. PERFORMANCE MONITORING

In long-haul systems, the capability to remotely monitor the performance of any repeater and to locate the cause of system degradation and faults is essential. Past submarine systems were based on optoelectronic regenerative repeaters; hence, each repeater had access to the data signal. The advantage that optically amplified systems have over regenerative systems whereby their operation is independent of the data stream is a disadvantage in designing a fault-detection scheme. A new supervisory paradigm must be found without access to the base-band signal.

Several options have been proposed to remotely monitor the undersea system. These options can be placed in two categories: command–response systems and passive monitoring.

1. Command–Response Performance Monitoring

In command–response systems, shore terminals provide signaling that is interpreted by the repeater. A response is provided by the repeater to ensure that the command is received and to provide return telemetry data.

Various methods have been proposed to implement command–response channels.[50,51] Proposals include the following:

(1) An independent optical system using the fiber's 1300-nm transmission window (this requires supervisory signal regeneration in each repeater)

(2) An electrical supervisory system using the power feed path (this provides a very low bandwidth over a limited distance)

(3) Low-frequency AM of the optical signal, in which an optical modulator provides the return path from the repeater (this requires more components in the repeater)

(4) Low-frequency AM of the optical signal, in which a separate 1330- or 1550-nm light source provides the return path from the repeater (this requires more components in the repeater)

(5) Low-frequency amplitude modulation (AM) of the optical signal, in which gain modulation provides the return path from the repeater

Of the proposed approaches, commercial systems have been implemented using only the last option: a low-frequency AM command channel and a response channel implemented with gain modulation. With this technique, an outbound signal is provided by a low-level intensity modulation placed on the high-speed data by the shore terminal equipment. The tone is detected in the repeater with a photodetector followed by a low-frequency receiver. Low-speed electronic circuits in the repeater interpret the received command and generate a response. The return signal is provided by modulating the pump power with the low-speed supervisory signal.

For this supervisory system to work, two important aspects of the amplifier's performance must be understood. These characteristics are related to the extremely long lifetime of the erbium first-excited state. First, the response of the amplifier must be satisfactory such that the pump modulation required for a response is transformed to output power modulation. The slow recovery time of the EDFA causes high-frequency fluctuations of the pump to be attenuated. For example, at a pump modulation frequency of 10 kHz, the response of the amplifier would cause the gain modulation by the pump to be substantially attenuated. That is, the pump would fail to effectively modulate the gain of the amplifier. Second, the response of the amplifier chain to low-frequency modulation must be adequate so that the low-frequency signal is amplified by subsequent amplifiers so as to be sufficient to be detected at its terrestrial destination. This also requires attention because the chain of amplifiers is largely insensitive to low-frequency changes in signal level. The high-pass response of an EDFA begins to roll off at about 10 kHz. This high-pass nature of the transmission path must be balanced with the low-pass nature of the pump modulation scheme and the modulation frequency selected appropriately. Figure 2.15 shows the low-pass response to pump modulation and the high-pass response of the following amplifiers.

The advantage of the gain modulation scheme is that queries can be made of the repeater for detailed performance information (e.g., pump

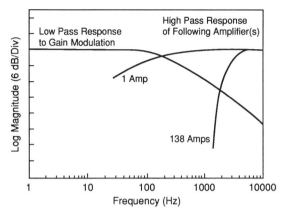

Fig. 2.15 Low-pass response to gain modulation and high-pass response of the following amplifiers. (Reproduced from Ref. 50, with permission of Nortel Technology.)

bias current, pump back-face current, received optical power, transmitted optical power). An important disadvantage of this system is that it requires additional components that add to the complexity, add to the cost, and decrease the reliability of the repeater.

2.   Passive Performance Monitoring

Systems that rely on passive monitoring lack repeaters that can interpret commands. Instead, they incorporate a mechanism that provides an indication of system health to the shore terminals by providing a path in every repeater to loop back a portion of the signal. The coupler arrangement would be implemented as shown in Fig. 2.16.

The primary function of the loop-back couplers is to inject a small portion of each transmitted signal into the opposite direction's transmission fiber. This small signal is then returned from each repeater like an echo. Equipment in the terminal detects the magnitude of the echoes from each of the repeaters using a sensitive correlation technique to enhance the SNR.[52] In systems being installed currently, the relative signal level of the injected signal has been selected to be −45 dBc. This level is large enough to be detected at the shore terminal in a reasonable amount of time and small enough to minimize the transmission impairment caused by its interference with the primary signal on the other fiber.

An additional function of the loop-back coupler module is to provide a path through the repeater so that Rayleigh backscattered light from the

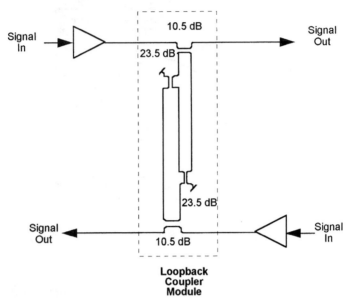

Fig. 2.16 Repeater architecture showing a loop-back coupler module for passive performance monitoring.

transmission fiber can pass through and be coupled into the opposite transmission path. This feature allows the use of optical time domain reflectometry (OTDR)-like examination of the transmission line to locate faults between repeaters.

The advantage of passive monitoring is the relative simplicity of the components required in the repeater; only four couplers are required.

G. UNDERSEA APPLICATIONS

The advances in optical amplifier technology have opened new possibilities for undersea transmission. The global undersea network is called upon to span a wide range of distances. Repeaterless transmission systems span rivers and lakes, connect islands, and form festoons along coastlines. Repeaterless systems that span more than 500 km have been demonstrated in laboratory experiments. Regional networks require transmission that spans 500 to 5000 km. Transoceanic systems are required to span up to 9000 km.

1. Repeaterless Systems

For systems less than 500 km in length, repeaterless systems are an appealing choice for the connection of two terminals separated by water. By eliminating repeaters, the expense of reliable power equipment, repeater deployment, complicated undersea performance monitoring equipment, and spare parts is eliminated.

Recent results at a bit rate of 2.488 Gb/s span transmission distances of more than 529 km, overcoming a fiber loss of 93.8 dB.[53] Many technologies have enabled this astounding result:

- High-power pump sources: High-power lasers provide more than 1 Watt to create a Raman amplifier in the transmission fiber. In addition, these lasers are used to remotely pump midspan EDFAs through a dedicated, low-loss, pure silica-core fiber.
- Low-loss silica-core fiber: By using a dedicated pump fiber with very low loss, the remotely pumped EDFAs can be placed further from the terminal stations.
- Dispersion compensation fiber (DCF): Fiber with a large negative chromatic dispersion (> -8000 ps/nm) is used in the receiver to offset the large positive dispersion accumulated by transmission through silica-core fiber.
- Forward error correction (FEC): In undersea transmission systems it is common practice to use data encoding to improve transmission performance. FEC encoding in current undersea systems uses a (255,239) Reed-Solomon forward error correcting code that can correct randomly distributed bit errors from a BER of 10^{-4} to a BER of 10^{-12}.

Figure 2.17 shows the architecture used to obtain unregenerated transmission over more than 529 km.

2. Moderate Distances

Performance of systems that span less than approximately 2000 km can be estimated by analyzing noise accumulation. For amplifiers with modest output power (limited to about 6 dBm by the available pump power), the accumulation of nonlinear transmission impairments due to the intensity-dependent refractive index can be ignored because the transmission distance is short.

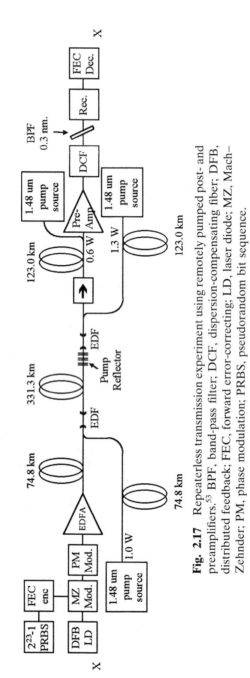

Fig. 2.17 Repeaterless transmission experiment using remotely pumped post- and preamplifiers.[53] BPF, band-pass filter; DCF, dispersion-compensating fiber; DFB, distributed feedback; FEC, forward error-correcting; LD, laser diode; MZ, Mach–Zehnder; PM, phase modulation; PRBS, pseudorandom bit sequence.

The optical SNR of a transmission path made up of identical amplifiers in compression (i.e., gain equals loss) is given by

$$SNR = \frac{P_{in}}{(NF)h\nu_s \cdot \Delta\nu \cdot N}, \tag{2.9}$$

where P_{in} is the amplifiers' input power, NF is the amplifiers' noise figure (see Eq. 2.5), $\Delta\nu$ is the optical bandwidth over which the SNR is measured, and N is the number of amplifiers in the transmission path. In the absence of nonlinearity-induced degradation, the optimum system design (i.e., to maximize the SNR and reduce the number of amplifiers) is achieved by maximizing the ratio P_{in}/NF for the amplifier. The primary limitation on increasing the ratio P_{in}/NF of EDFAs is the amount of pump power available. Although commercial pump sources are available that provide more than 100 mW into an optical fiber pigtail, the severe reliability requirements imposed on components that are deployed undersea limits the pump power available for use. To achieve a target of less than 50 FITs,[54] the pump power available to the erbium-doped fiber is typically less than 30 mW.

3. Transoceanic Systems

Transoceanic systems are called upon to span distances of up to 9000 km,[55,56] for example, from North America to Japan. At distances this long, the accumulation of impairments due to fiber nonlinearity becomes a serious problem and must be managed in the overall system design. (Various impairments are caused by the nonlinear index of refraction of the fiber. Four-wave mixing causes mixing between the signal and the ASE noise. Self-phase modulation combined with chromatic dispersion in the transmission fiber causes uncorrectable waveform distortion.)

The effect of fiber nonlinearity increases rapidly as the amplifier output power and transmission distance increase. Therefore, nonlinear transmission impairments can be decreased by reducing the optical power in the transmission fiber. In long transmission systems there is a trade-off between the desire to increase the output power (either to decrease the number of amplifiers in the system or to increase performance) and the impairment caused by fiber nonlinearity. This trade-off results in an amplifier design where a specific output power is targeted. Deviation from the target output power causes transmission impairment due either to the drop in amplifier input power (for lower output power) or to the increase in impairments caused by fiber nonlinearity (for higher output power).

The amplifier design for long transmission systems differs from most other amplifier designs (that are mostly concerned with minimizing noise figure and maximizing output power). The appropriate selection of input power, erbium-doped fiber length, and pump power must be found to provide the targeted output power (typically 2–4 dBm), in addition to the desired gain, noise figure, and degree of compression.

2. WDM Systems

Wavelength-division multiplexing (WDM) technology is extremely important for fiber optic submarine networks. WDM technology can be used to significantly increase system transmission capacity and to provide a mechanism for creating multipoint networks. Many experiments have been performed in laboratory environments and on commercially installed undersea networks that demonstrate the feasibility of the technology.[54a] These results will drive the designs of the next generation of undersea fiber optic systems. Results of transmission experiments in the laboratory at capacities up to 100 Gb/s per fiber have been reported by AT&T over distances sufficient for transoceanic networks.[54b]

The most important concern in the design of amplifiers for WDM transmission is transmission band flatness. Since the passband provided by concatenated erbium-doped fiber amplifiers varies by tens of dBs as a function of wavelength, the amplifiers used in WDM systems require careful design to ensure adequate performance for all channels. Techniques such as transmitter pre-emphasis can provide some channel equalization, but often additional measures must be taken. A common method for equalizing the channels is to include passband equalization filters (filters that approximate the inverse characteristic of the combination of the EDFA and the fiber span) in the amplifier. Various technologies exist for creating these filters: ultraviolet induced fiber gratings, thin film interference filters or even samarium-doped fibers (to correct only the slope of the passband).

WDM transmission imposes an additional demand on the amplifiers used in undersea systems. In single-channel, long-distance transmission systems it is often necessary to limit the output power of the amplifiers due to the introduction of optical nonlinearities (e.g., to about +3 dBm for 9000-km transmission). When providing the optical carriers at many wavelengths, sometimes it would be desirable that the power in each carrier be as large as it was in a single-channel system. For an eight-channel WDM system, this would require amplifiers capable of providing greater than 10-dBm total output power.

The advantage of using WDM to increase the capacity of undersea networks is not unique to systems that traverse long distances. Systems that are short enough to forgo the need of repeaters are also natural applications for WDM technology.

In a laboratory experiment, AT&T has demonstrated the transmission of 8 WDM carriers at 10 Gb/s over a distance of 352 km.[54c] This result was achieved by using the same technologies that allowed transmission of a single carrier over 529 km (see the section on "Repeaterless Systems"). Since repeaterless systems do not contain a chain of concatenated amplifiers, extraordinary measures to protect the band shape are not required. No gain equalization device was required and a transmitter pre-emphasis of only 0.9 dB was required.

The large potential bandwidth of optical fiber transmission systems has been understood for several decades; however, only recently has this potential been translated into tangible results. Wavelength-division multiplexing techniques along with erbium-doped fiber-amplifier technology are making the utilization of these enormous bandwidths possible. Exciting results are now being reported for both short-haul and long-haul systems. The next generation of undersea fiber optic networks will use WDM techniques to greatly increase their capacity and network flexibility. The WDM transmission techniques being developed today promise to satisfy the demand for international telecommunication capacity well into the next decade and further enhance global connectivity.

V. Optical Amplifiers for Analog Video Transmission

EDFAs are also being used for analog transmission, which requires very high signal powers. The primary commercial application for analog transmission is in trunk distribution systems of CATV video signals. The CATV signals are commonly transmitted in the AM–vestigial sideband (AM-VSB) format, with 6-MHz channel spacing (in the United States) starting at about 50 MHz.

The trunk system transmits video signals from the CATV head end to feeder systems that distribute the signal to customers over coaxial lines. The lengths of conventional CATV trunk systems based on coaxial cable transmission lines are limited by the high transmission loss of the coaxial cables and the splitting losses inherent in the trunk and branch architecture. This results in the need for long chains of closely spaced electronic radiofrequency amplifiers. The noise of these RF amplifiers limits the length and architecture of coaxial trunk systems; their high failure rates also impair the service quality and reliability of the trunk.

Optical fiber offers an alternative, low-loss transmission medium that can replace coaxial trunk lines while improving noise performance and reliability. However, optical transmission of analog signals such as AM-VSB video signals requires a very high carrier-to-noise ratio (CNR) (typically more than 49 or 50 dB for fiber optic backbone lines) and extremely high linearity. As a result, high optical powers must be achieved at the receiver, typically of the order of 1 mW, to avoid unacceptable degradations from shot noise and receiver noise. This severely limits loss budgets, given the limited powers available from transmitters with the high linearity required for AM-VSB transmission.

EDFAs can be used as power amplifiers to boost transmitter power and permit larger loss budgets. When used for this purpose, EDFAs must deliver high power to increase the loss budget, but must also meet stringent requirements on added noise and distortion. The CNR in an amplified analog system is given by[57]

$$CNR = \frac{m^2 I_{RS}^2}{2B_e \left(2eI_{RS} + i_c^2 + RIN \cdot I_{RS}^2 + \dfrac{4I_{RS}I_{RA}}{\Delta\nu} + \dfrac{4I_{RA}^2}{\Delta\nu} \right)}, \quad (2.10)$$

where the terms in the denominator represent the contributions to the CNR of shot noise, receiver noise, relative intensity noise (RIN), signal–spontaneous beat noise and spontaneous–spontaneous beat noise, respectively. In Eq. (2.10), m is the modulation index of the AM-VSB signal (typically not more than 5% for a 36-video-channel system and 3.5% for a 77-video-channel system because of the fundamental clipping limit), I_{RS} is the signal photocurrent at the receiver, B_e is the electrical bandwidth (4 MHz for AM-VSB), e is the charge of an electron, i_c is the thermal circuit noise of the receiver in A/Hz$^{1/2}$, RIN is the relative intensity noise, I_{RA} is the ASE photocurrent at the receiver, and $\Delta\nu$ is the optical bandwidth (~25 nm if no filter is used). In this case, $I_{RS} = P_s GLe\eta/h\nu_s$ and $I_{RA} = P_{ASE}Le\eta/h\nu_s \cong n_{sp}GLe\eta\Delta\nu$, where P_s is the transmitter power, P_{ASE} is the ASE power in one polarization, G is the amplifier gain, L is the system loss, η is the receiver quantum efficiency, $h\nu_s$ is the signal photon energy, and n_{sp} is the EDFA's spontaneous emission factor. When the signal power is sufficiently high and the RIN is sufficiently low, the CNR is dominated by the signal–spontaneous beat noise term, which is determined by the noise figure (or equivalently n_{sp}) of the amplifier. Reflections in the transmission path may give rise to additional RIN due to multipath interference.[58] The sensitivity to reflections is even greater in the amplifier because of its gain and it is thus of great importance to achieve very low reflectivities.

As an example of the role of an EDFA in enhancing analog system performance, Fig. 2.18 shows the CNR as a function of amplifier output power (or equivalently in this case amplifier gain) for a system with a 15-dB loss budget. This loss budget might correspond, for example, to an eight-way split and 20 km of transmission fiber with some allowance for outside plant margin. A transmitter with power of 3 dBm and RIN of -155 dB/Hz, a receiver with equivalent noise of 6 pA/$\sqrt{\text{Hz}}$ and an amplifier with a noise figure of 4.5 dB were assumed. At low amplifier output power, the CNR, dominated by receiver thermal noise, is unacceptably low. As the amplifier output power increases, the CNR increases until for $P_{out} \geq 15$ dBm the CNR is dominated by RIN and signal–spontaneous beat noise. For systems with appreciable link loss, high amplifier output power and a low noise figure are required to achieve a 50-dB CNR. This clearly is demanding, requiring close to 15 dBm of output power. It is possible to concatenate amplifiers to increase the splitting ratios and extend

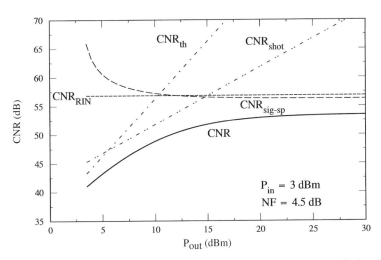

Fig. 2.18 Carrier-to-noise ratio (CNR) of an amplified analog transmission link with a 15-dB loss. RIN, relative intensity noise.

the transmission range, but only a few stages can be added before the CNR decreases to unacceptable levels.

In addition to the output power and the noise figure, the gain slope is important when the transmitter is a directly modulated distributed feedback (DFB) diode laser. Modulation of the DFB laser introduces chirp, which gives rise to nonlinear distortion when the amplifier gain is not flat.[59] Controlling the composite second order (CSO) intermodulation products to the required level (typically −65 dBc) requires that the amplifier's gain slope at the operating point be held to within strict limits, typically on the order of one-tenth of a decibel per nanometer.

Because of the slow gain dynamics of EDFAs, gain saturation does not introduce significant distortion in analog transmission.

Amplifiers pumped at 980 nm offer the best combination of high output power, low noise figure, and small gain slope. An alternative is to use an Er:Yb-doped fiber pumped at 1060 nm by Nd-doped lasers pumped in turn by high-power 800-nm diode arrays. The gain spectrum of Er:Yb-doped fiber is not suitable, but the gain slope can be altered within acceptable limits by use of a gain equalization filter.[60,61]

VI. Optical Amplifiers for Optical Networking

The transparency of optically amplified transmission lines permits transmission capacity to be dramatically increased through WDM. The success of WDM transmission in point-to-point systems has already resulted from the development of EDFAs with high output power, low noise figure, and wide optical bandwidth, as discussed in Section III. The advent of WDM systems suggests the further possibility of achieving optical networking by using their wavelengths to route and switch the different channels. If switching and routing functionality now provided by electrical cross-connects are replaced with fixed wavelength add–drops and subsequently with fully reconfigurable wavelength cross-connects (as envisioned by the MONET [multiwavelength optical network][63] project, for example), multiwavelength optical networks promise cost savings and increases in capability and routing flexibility.[62]

Multiwavelength networks that are under study range from fixed wavelength add–drop networking capability on WDM transport systems to a national-scale reconfigurable network encompassing local exchange subnetworks linked by a national-scale long-distance subnetwork (the MONET project). Optical amplifiers will be needed to compensate for the losses of transmission spans (with requirements that may be more diverse than those in point-to-point systems), as well as to compensate for the loss of network elements such as wavelength add–drop sites or wavelength selective cross-connects.

Optical amplifiers for optical networking will be similar to those for WDM transmission, only more so. The bandwidth and gain flatness requirements on amplifiers for optical networking, already demanding for point-to-point systems, will be even more demanding for multiwavelength optical networks for several reasons:

(1) *No preemphasis:* Preemphasis, which provides some tolerance to spectral gain nonuniformity in point-to-point systems, cannot be used in networks for optical SNR equalization because of the diverse and changing paths through the network followed by different wavelength channels.

(2) *More amplifiers:* For a large network, such as the national-scale network envisioned by MONET with widely separated local exchange subnetworks linked by long-distance networks, a single channel will in general traverse more amplifiers without electrical regeneration than in terrestrial point-to-point systems. (The MONET network is designed to support paths that traverse as many as 100 amplifiers.) A channel's path through the network will change as the network is reconfigured.

(3) *Granularity:* Whereas in a point-to-point system the number of channels required is determined by the required capacity and the available channel data rate (typically 2.5 Gb/s or perhaps in the future 10 Gb/s), the capabilities of an optical network may be enhanced by subdividing the traffic into more channels (which may operate at lower channel data rates) to permit more flexible routing. This will require greater optical bandwidth.

The requirements on noise figure and output power will also be demanding to control noise accumulation for amplified paths with as many as 100 amplifiers. It will be necessary to maintain the per-channel input signal power to every amplifier above a certain level, which will increase logarithmically with the number of amplifiers in the chain (see Eq. [2.8]).

In-line amplifiers for optical networks would have architectures with strong similarities to those for high-performance amplifiers for WDM transmission systems discussed in Section II. Generally, two-stage amplifiers with a low noise figure, a highly inverted first stage, and a power-converting second stage will be required. Because of the more demanding requirements for gain flatness, amplifiers will be optimized to operate at the inversion with flattest gain for all signal wavelengths; gain equalization filters, such as long period fiber gratings[60] may also be required. Another possibility is to use erbium-doped fluoride fibers, which offer flatter gain near 1535 nm but are more difficult to work with and may still require gain filtering to hold gain excursions within limits acceptable for networks.

EDFAs will also be needed for network elements, such as cross-connects, which may have large optical losses. The input amplifier of a high-loss network element will be designed primarily for maximum output power to ensure adequate input power at the network element's output amplifier. One possible architecture is to use a high-power MOPA pump now commercially available with 500 mW of fiber-coupled output power at 980 nm to pump the amplifier's power amplifier stage.[64]

Optical amplifiers in multiwavelength optical networks will also experience variable channel loading. This can result either from network reconfigurations or from faults, such as cable cuts or other failures interrupting signals feeding into a cross-connect. This can cause a loss of signal for the interrupted channels on the input line of the cross-connect, or, in the case of reconfigurations, appearance of additional channels. The "surviving channels," those that traverse the same amplifiers but do not participate directly in the fault or reconfiguration, will suffer changes in their power levels as a result of EDFA cross-saturation. In large networks with long chains of amplifiers, the power transients can be extremely fast; their speed is proportional to the number of amplifiers comprising the affected amplifier chain.[22] Protection of service on the surviving channels will require dynamic gain control.

One proposed technique to accomplish this is automatic gain control achieved by incorporating the amplifier in a resonant cavity to induce lasing at a wavelength within the erbium gain band but outside the band of signal channels.[65] The EDFA simultaneously amplifies the signal channels and serves as the gain of the laser, which clamps the EDFA's gain. Other possible techniques include pump control,[66,66a] in which the input power or the gain is monitored and the pump power is adjusted to maintain the gain constant, or use of a saturated control channel for each amplifier[67] or collectively for the amplifiers of a link between two wavelength routing network elements.[68]

VII. Conclusions

EDFAs offer a unique combination of features that are revolutionizing lightwave communications systems. Among these features are high gain, high optical power, low noise, diode pumps, polarization independence, fiber compatibility, linearity, wavelength transparency, and gain dynamics sufficiently slow that intersymbol interference, interchannel cross talk, or intermodulation distortion is not induced. However, amplified systems, unlike conventional optoelectronically regenerated systems, are fundamentally analog systems carrying digital data. Certain problems become more serious. Each amplifier introduces noise and, for WDM systems, spectral gain variations that accumulate over the full length of the system. The management of these effects is central to the design of EDFAs for each application and to the design of optically amplified systems. The success in meeting the challenges described in this chapter has enabled the optical communications industry to take full advantage of the EDFA's features, providing the basis for new system architectures providing increased capacity at significantly reduced costs.

EDFAs are now the basis for the design of essentially all terrestrial and undersea transport systems. EDFA-based systems are serious contenders for analog video distribution. And the capabilities of EDFAs are perhaps the key enabler for the intense work now going on in multiwavelength optical networking that may mark the path for the next major advance in optical communications.

References

1. Desurvire, E., J. R. Simpson, and P. C. Becker. 1987. High-gain erbium-doped fibre amplifier. *Opt. Lett.* 12(11):888.
2. Mears, R. J., L. Reekie, I. M. Jauncey, and D. N. Payne. 1987. Low-noise erbium-doped fibre amplifier operating at 1.54 μm. *Electron Lett.* 23(19):1026–1028.
3. Henry, P. S., R. A. Linke, and A. H. Gnauck. 1988. Introduction to lightwave systems. In *Optical communications systems,* ed. S. E. Miller and I. P. Kaminow, 781. New York: Academic Press.
4. Giles, C. R., and E. Desurvire. 1996. Modeling erbium-doped fiber amplifiers. *IEEE J. Lightwave Tech.* 9(2):271.
5. Saleh, A. A. M., R. M. Jopson, J. D. Evankow, and J. Aspell. 1990. Modeling of gain in erbium-doped fiber amplifiers. *IEEE Photon. Tech. Lett.* 2(10):714.

6. Desurvire, E., C. R. Giles, J. R. Simpson, and J. L. Zyskind. 1989. Efficient erbium-doped fiber amplifiers at a 1.53 μm wavelength with a high output saturation power. *Opt. Lett.* 14(22):1266.

7. Zyskind, J. L. 1991. Advances in erbium-doped fiber amplifiers for optical communications. In *Fiber laser sources and amplifiers II*, ed. M. J. Digonnet, 80–92. SPIE Proceedings No. 1373. Bellingham, WA: SPIE.

8. Kogelnik, H. 1995. Optical communications. In *Encyclopedia of applied physics*, vol. 12, 119.

9. Desurvire, E. 1994. *Erbium-doped fiber amplifiers: Principles and applications.* New York: Wiley.

10. Simpson, J. R. 1990. Fabrication of rare-earth doped glass fibers. In *Fiber laser sources and amplifiers*, ed. M. J. Digonnet, 2. SPIE Proceeding No. 1171. Bellingham, WA: SPIE.

11. DiGiovanni, D. J. 1991. Fabrication of rare earth doped optical fiber. In *Fiber laser sources and amplifiers II*, ed. M. J. Digonnet, 2. SPIE Proceedings No. 1373. Bellingham, WA: SPIE.

12. Shimizu, M., M. Yamada, M. Horiguchi, T. Takeshita, and M. Oyasu. 1990. Erbium-doped fibre amplifiers with an extremely high gain coefficient of 11.0 dB/mW. *Electron Lett.* 26(20):1641; corresponding to 30 dB for 3 mW.

13. Kashiwada, T., M. Shigematsu, T. Kougo, H. Kanamori, and M. Nishimura. 1991. Erbium-doped fiber amplifier pumped at 1.48 μm with extremely high efficiency. *IEEE Photon. Tech. Lett.* 3(8):721.

14. Zervas, M. N., K. Dybdal, and L. C. Larsen. 1992. Gain limit in erbium-doped fiber amplifiers due to internal Rayleigh backscattering. *IEEE Photon. Tech. Lett.* 4(6):559.

15. Ainslie, B. J., S. P. Craig-Ryan, S. T. Davey, J. R. Armitage, C. G. Atkins, and R. Wyatt. 1989. Optical analysis of erbium-doped fibres for efficient lasers and amplifiers. In *Proceedings of the Seventh International Conference on Integrated Optics and Optical Fiber Communications, IOOC'89*, Kobe, Japan, 22. Paper 20A3-2.

16. Shimizu, M., M. Yamada, M. Horiguchi, and E. Sugita. 1990. Concentration effects on optical amplification characteristics of erbium-doped silica single-mode fibers. *IEEE Photon. Tech. Lett.* 2(1):43.

17. Tam, H. Y. 1991. Simple fusion splicing technique for reducing splicing loss between standard single-mode fibres and erbium-doped fibre. *Electron Lett.* 27(17):1597.

18. Giles, C. R., E. Desurvire, J. R. Talman, J. R. Simpson, and P. C. Becker. 1989. 2-Gbit/s signal amplification at $\lambda = 1.53$ μm in an erbium-doped single-mode fiber amplifier. *IEEE J. Lightwave Tech.* 7(4):651.

19. Taylor, M. G. 1993. Observation of new polarization dependence effect in long haul optically amplified systems. *IEEE Photon. Tech. Lett.* 5:1244–1246.

20. Mazurczyk, V. J., and J. L. Zyskind. 1994. Polarization dependent gain in erbium-doped fiber amplifiers. *IEEE Photon. Tech. Lett.* 6(5):616–618.

21. Desurvire, E., C. R. Giles, and J. R. Simpson. 1989. Gain saturation effects in high-speed, multichannel erbium-doped fiber amplifiers at $\lambda = 1.53$ μm. *IEEE J. Lightwave Tech.* 7(12):2095.

22. Zyskind, J. L., Y. Sun, A. K. Srivastava, J. W. Sulhoff, A. J. Lucero, C. Wolf, and R. W. Tkach. 1996. Fast power transients in optically amplified multiwavelength optical networks. In *Optical Fiber Communication Conference,* vol. 2, PD31-1. Postdeadline paper PD31. 1996 OSA Technical Digest Series. Washington, DC: Optical Society of America.

23. Wysocki, P. F., J. R. Simpson, and D. Lee. 1994. Prediction of gain peak wavelength for Er-doped fiber amplifiers and amplifier chains. *IEEE Photon. Tech. Lett.* 6(9):1098.

24. Wysocki, P. F., D. Lee, and J. R. Simpson. 1994. Simple theory of gain peaking in erbium-doped amplifier chains for long-haul communications. In *Fiber laser sources and amplifiers V,* ed. M. J. Digonnet, 146. SPIE Proceedings No. 2073. Bellingham, WA: SPIE.

25. Desurvire, E., J. L. Zyskind, and J. R. Simpson. 1990. Spectral gain hole-burning at 1.53 μm in erbium-doped fiber amplifiers. *IEEE Photon. Tech. Lett.* 2(4):246.

26. Zyskind, J. L., E. Desurvire, J. W. Sulhoff, and D. DiGiovanni. 1990. Determination of homogeneous linewidth by spectral gain hole-burning in an erbium-doped fiber amplifier with GeO_2–SiO_2 core. *IEEE Photon. Tech. Lett.* 2(12):869.

27. Srivastava, A. K., J. L. Zyskind, J. W. Sulhoff, J. D. Evankow, Jr., and M. A. Mills. 1996. Room temperature spectral hole-burning in erbium-doped fiber amplifiers. In *Optical Fiber Communication Conference,* vol. 2, 33. OSA Technical Digest Series. Washington, DC: Optical Society of America.

28. Ronarc'h, D., M. Guibert, H. Ibrahim, M. Monerie, H. Poignant, and A. Tromeur. 1991. 30 dB Optical net gain at 1.543 μm in Er^{3+} doped fluoride fibre pumped around 1.48 μm. *Electron. Lett.* 27(11):908.

29. Miniscalco, W. J. 1991. Erbium-doped glasses for fiber amplifiers at 1500 nm. *IEEE J. Lightwave Tech.* 9(2):234.

30. Snitzer, E., H. Po, F. Hakimi, R. Tumminelli, and B. C. McCollum. 1988. Erbium fiber laser amplifier at 1.55 μm with pump at 1.49 μm and Yb sensitized Er oscillator. In *Proceedings of the 1988 Optical Fiber Communications Conference.* Postdeadline paper PD2. Washington, DC: Optical Society of America.

31. Laming, R. I., M. C. Farries, P. R. Morkel, L. Reekie, and D. N. Payne. 1989. Efficient pump wavelengths of erbium-doped fiber optical amplifiers. *Electron. Lett.* 25:12–14.

32. Yamada, M., M. Shimizu, M. Okayasu, T. Takeshita, M. Horiguchi, Y. Tachikawa, and E. Sugita. 1990. Noise characteristics of Er^{3+}-doped fiber amplifiers pumped at 0.98 and 1.48 μm laser diodes. *IEEE Photon. Tech. Lett.* 2:205.

33. Sugawa, T., T. Kokumai, and Y. Miyajima. 1990. Optical amplification in Er^{3+}-doped single-mode fluoride fibers. *IEEE Photon. Tech. Lett.* 2(7):475.

34. Grubb, S. G., W. F. Humer, R. S. Cannon, S. W. Vendetta, K. L. Sweeney, P. A. Leilabady, M. R. Keur, J. G. Kwasegroch, T. C. Munks, and D. W. Anthon. 1992. +24.6 dBm Output power Er/Yb codoped optical amplifier pumped by diode-pumped Ng : YLF laser. *Electron. Lett.* 28(13):1275.

35. Takenaka, H., H. Okuno, M. Fujita, Y. Odagiri, T. Sunohara, and I. Mito. 1991. Compact size and high output power Er-doped fiber amplifier modules

pumped with 1.48 μm MQW LD. In *Proceedings, Optical Amplifiers and Their Applications, Snowmass, CO*, 251. Washington DC: Optical Society of America.

36. Grubb, S. G., and A. Leilabady. 1993. High power erbium optical amplifier. In *Proceedings, Optical Amplifiers and Their Applications, Yokohama, Japan*, 84. Washington, DC: Optical Society of America.

37. Delavaux, J-M., J. A. Nagel, and D. J. DiGiovanni. 1995. Optimized two-stage in-line balanced optical amplifier design. *Opt. Fiber Tech.* 1(3):239.

38. Livas, J. C. 1996. High sensitivity optically preamplified 10 Gb/s receivers. In *Proceedings of the Optical Fiber Conference, San Jose*. Paper PD4. Washington, DC: Optical Society of America.

39. Delavaux, J-M., and J. A. Nagel. 1995. Multi-stage erbium-doped fiber amplifier designs. *IEEE J. Lightwave Tech.* 13(5):703.

40. Chen, C. D., J-M. Delavaux, B. W. Hakki, O. Mizuhara, T. V. Nguyen, R. J. Nuyts, K. Ogawa, Y. K. Park, C. Skolnik, R. E. Tench, J. J. Thomas, L. D. Tzeng, and P. D. Yeates. 1994. A field demonstration of 10 Gb/s, 360 km transmission through embedded standard (non DSF) fiber cables. In *OFC'94, San Jose, CA*. Washington, DC: Optical Society of America.

41. Chraplyvy, A. R., R. W. Tkach, K. C. Reichmann, P. D. Magill, and J. A. Nagel. 1993. End-to-end equalization experiments in amplified WDM lightwave systems. *IEEE Photon. Tech. Lett.* 4(4):428.

42. Bubel, G. M., and R. L. Easton. 1993. Reliability for the SL2000 optical amplifier systems. In *2nd International Conference on Optical Fiber Submarine Telecommunication Systems, March 29–April 2*, 191–195. Paris: Societé des Electriciens et des Electronicians.

43. Kidorf, H. D. 1996. Fiber-amplifier repeater design for undersea system application. In *Optical Amplifiers and Their Applications, Yokohama, Japan, July 4–6*, 114–117. Paper MB2. Washington, DC: Optical Society of America.

44. Duff, D. G., D. A. Fishman, and J. A. Nagel. 1990. Measurements and simulation of multipath interference for 1.7 Gb/s lightwave systems utilizing single and multi frequency lasers. *J. Lightwave Tech.* 8(6):894–905.

45. Bergano, N. S., R. F. Druckenmiller, F. W. Kerfoot, and P. R. Trischitta. 1992. Pump redundancy for optical amplifiers. U.S. patent no. 5,173,957, awarded December 22, 1992.

46. Kerfoot, F. W., and S. M. Abbott. 1993. Optical amplifier system technology. In *2nd International Conference on Optical Fiber Submarine Telecommunication Systems, March 29–April 2*, 181–185. Paris: Societé des Electriciens et des Electronicians.

47. Costelloe, J. R. Unpublished test results.

48. J. P. Gordon, and L. F. Mollenauer. 1991. Effects of fiber nonlinearities and amplifier spacing on ultra-long distance transmission. *J. Lightwave Tech.* 9(2):170–173.

49. Malyon, D. J., and Lord T. Widdowson. 1993. Assessment of the polarisation loss dependence of transoceanic systems using a recirculating loop. *Electron. Lett.* 29:207–208.

50. Davis, F., N. E. Jolley, M. G. Taylor, J. Brannan, N. Baker, and M. C. Wright. 1993. Optical amplifiers for submerged systems. In *2nd International Conference on Optical Fiber Submarine Telecommunication Systems, March 29–April 2,* 375–379. Paris: Societé des Electriciens et des Electronicians.

51. Brannan, J., A. Hadjifotiou, and I. J. Hirst. 1992. Supervisory system for fibre amplifier systems. In *IEEE Colloquium on Optical Amplifiers for Communications, May 20,* 9(1–9):7.

52. Jensen, R. A., H. L. Lang, and M. D. Tremblay. 1993. New technology for operating and maintaining SL2000 systems. *SUBOPTIC 1993,* 523–526. Paris: Societé des Electriciens et des Electronicians.

53. Hansen, P. B., L. Eskildsen, S. G. Grubb, A. M. Vengsarkar, S. K. Korotky, T. A. Strasser, J. E. J. Alphonsus, J. J. Veselka, D. J. DiGiovanni, D. W. Peckham, E. C. Beck, D. Truxal, W. Y. Cheung, S. G. Kosinski, D. Gasper, P. F. Wysocki, V. L. da Silva, and J. R. Simpson. 1995. 2.488-Gb/s Unrepeatered transmission over 529 km using remotely pumped post- and pre- amplifiers, forward error correction, and dispersion compensation. In *Conference on Optical Fiber Communication (OFC '95), San Diego, CA, February.* Postdeadline paper PD25. Washington, DC: Optical Society of America.

54. Schesser, J., S. M. Abbott, R. L. Easton, and M. S. Stix. 1995. Design requirements for the current generation of undersea cable systems. *AT&T Tech. J.* 74(1):16–32.

54a. Feggeler, J. C., D. G. Duff, N. S. Bergano, C. C. Chen, Y. C. Chen, C. R. Davidson, D. G. Ehrenberg, S. J. Evangelides, G. A. Ferguson, F. L. Heismann, G. M. Homsey, H. D. Kidorf, T. M. Kissell, A. E. Meixner, R. Menges, J. L. Miller, Jr., O. Mizuhara, T. V. Nguyen, B. M. Nyman, Y.-K. Park, W. W. Patterson, and G. F. Valvo. 1995. 10 Gb/s WDM transmission measurements on an installed optical amplifier undersea cable system. *Elect. Lett.* 31(19):1676–1678.

54b. Bergano, N. S., C. R. Davidson, D. L. Wilson, F. W. Kerfoot, M. D. Tremblay, M. D. Levonas, J. P. Morreale, J. D. Evankow, P. C. Corbett, M. A. Mills, G. A. Ferguson, A. M. Vensarkar, J. R. Pedrazzani, J. A. Nagel, J. L. Zyskind, and J. W. Sulhoff. 1996. 100 Gb/s error free transmission over 9100 km using twenty 5 Gb/s WDM data channels. In *Conference on Optical Fiber Communication, San Jose, CA.* Postdeadline paper PD23. Washington, DC: Optical Society of America.

54c. Hansen, P. B., L. Eskildsen, S. G. Grubb, A. M. Vengsarkar, S. K. Korotky, T. A. Strasser, J. E. J. Alphonsus, J. J. Veselka, D. J. DiGiovanni, D. W. Peckham, D. Truxal, W. Y. Cheung, S. G. Kosinski, and P. F. Wysocki. 1995. 8 × 10 Gb/s WDM repeaterless transmission over 352 km. In *Proceedings of IOOC '95, Hong Kong,* vol. 5, PD 2-4, 27–28. Hong Kong: The Chinese University Press.

55. Bergano, N. S., J. Aspell, C. R. Davidson, P. R. Trischitta, B. M. Nyman, and F. W. Kerfoot. 1991. A 9000 km 5 Gb/s and 21,000 km 2.4 Gb/s feasibility demonstration of transoceanic EDFA systems using a circulating loop. In *OFC '91.* Postdeadline paper PD13. Washington, DC: Optical Society of America.

56. Bergano, N. S., C. R. Davidson, G. M. Homsey, D. J. Kalmus, P. R. Trischitta, J. Aspell, D. A. Gray, and R. L. Maybach. 1992. 9000 km, 5 Gb/s NRZ transmission experiment using 274 erbium-doped fiber amplifiers. In *Topical meeting on optical amplifiers, Santa Fe, NM, June 24.* Postdeadline paper. Washington, DC: Optical Society of America.

57. Habbab, I. M. I., and L. J. Cimini, Jr. 1991. Optimized performance of erbium-doped fiber amplifiers in subcarrier multiplexed lightwave AM-VSB CATV systems. *J. Lightwave Tech.* 9(10):1321.

58. Yoshinaga, H., K. Kikushima, and E. Yoneda. 1992. Influence of reflected light on erbium-doped fiber amplifiers for optical AM video signal transmission systems. *J. Lightwave Tech.* 10(8):1132.

59. Kuo, C. Y. and E. E. Bergmann. 1991. Erbium-doped fiber amplifier second-order distortion in analog links and electronic compensation. *IEEE Photon. Tech. Lett.* 3(9):829.

60. Vengsarkar, A. M., P. J. Lemaire, J. B. Judkins, V. Bhatia, T. Erdogan, and J. E. Sipe. 1996. Long-period fiber gratings as band-rejection filters. *J. Lightwave Tech.* 14(1):58.

61. Park, N., T. Nielsen, J. Simpson, P. Wysocki, R. Pedrazzani, A. Vengsarkar, D. DiGiovanni, S. Grubb, D. Peckham, M. Haner, and K. Walker. 1996. Dispersion precompensated, high-power Er-Yb linear amplifier with gain tilt optimization over 11 nm. In *OFC '96,* 280. Paper ThR3. Technical Digest. Washington, DC: Optical Society of America.

62. Saleh, A. A. M. 1996. Overview of the MONET, multiwavelength optical networking program. In *Optical Fiber Communication Conference,* vol. 2, 240. 1996 OSA Technical Digest Series. Washington, DC: Optical Society of America.

63. Wagner, R. E., R. C. Alferness, A. A. M. Saleh, and M. S. Goodman. 1996. MONET: Multiwavelength optical networking, *J. Lightwave Tech.* 14(6):1349.

64. Srivastava, A. K., *et al.* 1996. High power, low noise EDFA for switching elements in optical networks.

65. Zirngibl, M. 1991. Gain control in erbium-doped fiber amplifiers by an all-optical feedback loop. *Electron. Lett.* 27(7):560.

66. Giles, C. R., E. Desurvire, and J. R. Simpson. 1989. Transient gain and crosstalk in erbium-doped fiber amplifier. *Opt. Lett.* 14(16):880.

66a. Srivastava, A. K., Y. Sun, J. L. Zyskind, J. W. Sulhoff, C. Wolf, and R. W. Tkach. 1996. Fast gain control in an erbium-doped fiber amplifier. In *Proceedings of Optical Amplifiers and Their Applications, Monterey, CA.* Paper PDP 4. Washington, DC: Optical Society of America.

67. Desurvire, E., M. Zirngibl, H. M. Presby, and D. DiGiovanni. 1991. Passive gain compensation in saturated erbium-doped fiber amplifiers. *IEEE Photon. Tech. Lett.* 3(5):453.

68. Zyskind, J. L., A. K. Srivastava, Y. Sun, J. C. Ellson, G. W. Newsome, R. W. Tkach, A. R. Chrvaplyvy, J. W. Sulhoff, T. A. Strasser, J. R. Pedrazzani, and C. Wolf. Fast link control protection for surviving channels in multiwavelength optical networks. In *Proceedings of the European Conference on Optical Communications,* Kjeller, Norway: Telenor R&T.

Chapter 3 | Transmitter and Receiver Design for Amplified Lightwave Systems

Daniel A. Fishman

Lucent Technologies, Bell Laboratories, Holmdel, New Jersey

B. Scott Jackson

AT&T Laboratories, Holmdel, New Jersey

I. Introduction

In Chapter 2 of Volume IIIA, the synthesis of synchronous optical network (SONET), synchronous digital hierarchy (SDH), and asynchronous transfer mode (ATM) digital formats was discussed. These electrical signals are composed of binary bits with time slot widths extending from 100 ps to 10 ns. Even 10-ns bits can travel up to only several kilometers in coaxial cable before bit distortions necessitate regeneration. However, when these bits are optically encoded, hundreds to thousands of kilometers of transmission would be possible. This demonstrates the enormous transmission bandwidth available in state-of-the-art fiber optic networks. Building upon this bandwidth, the advent of optical amplification offers dramatic cost and performance improvements in the design of these optical transmission systems by extending the transmission limits imposed by fiber loss.

The key to the success of these systems is high-performance transmitters and receivers. Amplified multigigabit transmission systems present unique challenges to the design of laser transmitters and electrooptic receivers. In addition, transmitters and receivers interact with optically amplified systems in subtly different ways than they do with nonamplified or regenerative systems. This chapter extends the treatment that transmitter and receiver design was given in Chapters 18, 19, and 21 of *Optical Fiber Telecommunications II* (Miller and Kaminow 1988); it focuses on trans-

OPTICAL FIBER TELECOMMUNICATIONS,
VOLUME IIIB

Copyright © 1997 by Lucent Technologies.
All rights of reproduction in any form reserved.
ISBN: 0-12-395171-2

mitter and receiver design concerns and their influence on amplified optical system performance.

1.1 TRANSMITTER AND RECEIVER OPTICAL TECHNOLOGIES EMPLOYED IN MULTIGIGABIT SYSTEMS

The transmission fiber and amplifiers can ideally be thought of as a light pipe that allows the transmission of optical signals indefinitely with no net loss. Actually, there are limitations on how far the optical signals can go before digital regeneration is necessary. The degradations take two forms: (1) a degraded *optical signal-to-noise ratio* (SNR_O) primarily caused by fiber loss and amplifier noise, and (2) *intersymbol interference* (*ISI*) — i.e., pulse distortions due to the fiber and receiver frequency response. In the following sections we show how careful transmitter design can reduce the limits imposed by SNR_O as well as the effects of the fiber finite frequency response — e.g., chromatic dispersion. In Sections 3 and 4, we show how the receiver performance is critically dependent on how well it accommodates optical amplifier noise. Section 5 discusses transmitter and receiver design considerations for multiple channel systems. The chapter concludes with a discussion of optical system performance monitoring schemes.

1.2 SINGLE-MODE OPTICAL FIBER CHARACTERISTICS: THE TRANSMISSION MEDIA AND SYSTEMS APPLICATIONS

Transmission in single-mode fiber is generally limited by loss and dispersion. The primary loss mechanism results from Rayleigh scattering, which falls off as λ^4 and results in a fiber loss that diminishes with increasing wavelength up 1600 nm. At wavelengths greater than 1600 nm the fiber loss increases as a result of intrinsic lattice absorption. In addition, most fiber will exhibit a large loss near 1400 nm, termed the *water peak*, which results from hydrogen absorption. In undoped silica single-mode fiber, the dispersion is dominated by the material dispersion of fused silica, which passes through a zero around 1270 nm. With a germanium dopant the dispersion zero λ_0 is shifted toward the longer wavelengths, typically between 1300 and 1320 nm, which is referred to as the *zero-dispersion window* in standard non-dispersion-shifted fiber (non-DSF).

Clearly, it is advantageous to design lasers for operation near λ_0 because the effects of laser chirp and mode partitioning would be minimal (Agrawal, Anthony, and Shen 1988). On the other hand, transmission losses can be reduced by as much as 50% by operating in the 1550-nm low-loss but high-dispersion regime. However, because the fiber chromatic dispersion results from the material and waveguide dispersion, it is possible to tailor the waveguide dispersion so as to shift λ_0. So-called dispersion-shifted fiber (DSF) has a dispersion zero around 1550 nm. These fibers generally have a smaller modal cross section, which makes DSF more vulnerable to optical fiber nonlinearities (Chapter 8, Volume IIIA). DSF has been designed that reduces the often severe degradations that result from optical nonlinearities, by increasing the modal cross section and shifting λ_0 away from the signal wavelengths, which has the effect of reducing four-photon mixing (Tkach *et al.* 1995; Yanming, Antos, and Newhouse 1996; Chapter 8, Volume IIIA) in DSF (e.g., TrueWave® fiber).* In the next few sections we discuss various systems applications that utilize these fiber characteristics.

1.2.1 Loss-Limited Systems

Applications that operate near the dispersion zero are considered *loss limited* by virtue of the fact that the loss is relatively high while the penalties due to the residual dispersion are generally small. Typically, the fiber loss is 0.4 dB/ km near 1310 nm, with less than ± 4 ps/nm-km dispersion. This is not to say that the effects of fiber dispersion can be neglected. At high bits rates, the existence of laser mode partitioning and jitter in multimode (Agrawal, Anthony, and Shen 1988; Hakki, Bosch, and Lumish 1989) and distributed feedback (DFB) lasers (Fishman 1990; Langley and Shore 1992) can severely compromise system performance. For 2.5-Gb/s systems, single longitudinal mode (SLM) DFB lasers are required for transmission distances exceeding 20 km with as much as ± 200-ps/nm dispersion [Consultative Committee of International Telegraph and Telephone (CCITT) G.957 recommendation for S-16.1, S-16.2, and L-16.1-3 systems]. Systems operating at 1.7 Gb/s using multilongitudinal mode (MLM) 1310-nm lasers have transmission distances that are limited to 40 km, and they must operate within several nanometers of the fiber dispersion zero to ensure less than ± 10-ps/nm dispersion (Fishman *et al.* 1986). These lasers must be designed and tested to ensure minimum mode partitioning, reflection sensitivity, and chirp.

1.2.2 Dispersion-Limited Systems

In the 1550-nm regime the loss is typically less than 0.25 dB/km, with a dispersion of 17 ps/nm-km. Currently, most 2.5-Gb/s 1550-nm laser transmitters utilize low-chirp, directly modulated, multiple quantum well (MQW)-DFB

* TrueWave® is a registered trademark of Lucent Technologies Network Cable Systems Inc.

lasers, because they are generally more available, cost less, and have higher optical power than externally modulated lasers (Uomi *et al.* 1992). In some cases the directly modulated 1550-nm DFB lasers have been capable of transmission through more than 160 km of standard single-mode fiber (Kuo *et al.* 1990). On the other hand, transmission distances of more than 670 km have been achieved using an integrated laser and electroabsorption modulator transmitter (Reichmann *et al.* 1993), and 710 km with a laser modulated with a LiNbO$_3$ Mach–Zehnder modulator that is designed to have virtually no pattern-dependent chirp (Edagawa *et al.* 1990). These transmitters are said to have *transform-limited pulses*, where the effective laser linewidth, $\Delta\nu$, is limited by the pulse width T, where $\delta\nu T \leq 1$. This will ultimately limit the transmission distance to approximately 1200 km at 2.5 Gb/s, and 75 km at 10 Gb/s, in standard single-mode fiber (Fishman 1993).

1.2.3   Dispersion-Managed Systems

Ultralong transmission systems, such as undersea systems, use dispersion management to minimize the dispersion penalty and to reduce the penalties from nonlinear effects (Chapter 10, Volume IIIA). These systems operate at about 1558 nm, near the fiber loss minimum and within the narrow (~1-nm) gain band that results from concatenating large numbers of optical amplifiers (>250 amplifiers are concatenated in the first transoceanic systems) (Giles and Desurvire 1991). These systems also achieve a net zero dispersion at the end of the system to minimize the dispersion penalty. Zero dispersion is not maintained at all points along the system, however, to avoid penalties caused by nonlinear effects, particularly four-wave mixing (Marcuse, Chraplyvy, and Tkach 1991). Instead, dispersion is allowed to accumulate over the system length and is then periodically corrected to zero dispersion according to a dispersion map, as shown in Fig. 3.1. This is accomplished as follows: DSF with λ_0 slightly above the operating wavelength is used for a majority of the system, and negative dispersion accumulates. Shorter lengths of standard fiber are interspersed to periodically bring the cumulative dispersion to zero. Peak system performance occurs when the system gain peak, the signal wavelength, and the dispersion zero coincide (Taga *et al.* 1994); optimum dispersion maps are often determined empirically.

1.2.4   Soliton, Dispersion-Supported, and Noncoherent Frequency Shift Keyed Transmission

Three other transmission techniques that are not mainstream approaches to dealing with fiber dispersion warrant brief mention because they have a profound impact on transmitter and receiver design. Soliton transmission

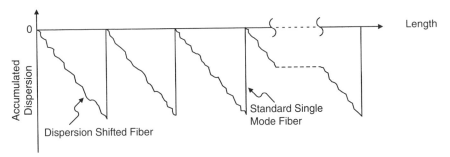

Fig. 3.1 Dispersion map for a dispersion-managed system.

is discussed in Chapter 12 in Volume IIIA. In dispersion-supported transmission (DST), the laser frequency modulation (FM) is converted to amplitude modulation (AM) by fiber chromatic dispersion and receiver filtering (Wedding 1992). The primary advantage of DST is that the laser modulation is minimal, just enough for the fiber dispersion to convert the residual optical FM to an AM signal that can be optimally filtered at the receiver, and the receiver bandwidth is generally a fraction of the baud rate (Wedding 1992; Bungarzeanu 1994). Noncoherent frequency shift keyed (NC-FSK) transmission is similar to DST in the transmitter design, but an optical frequency discriminator, generally in the form of a fiber Fabry–Perot filter, is used to provide the optical FM to optical AM conversion (Chraplyvy *et al.* 1989). The receiver for the NC-FSK system is identical to that of an intensity modulated/direct detection (IM/DD) system. As with DST, NC-FSK has been shown to benefit from the FM-to-AM conversion introduced by the fiber dispersion (Fishman 1991). In this chapter, we focus on IM/DD issues because for the foreseeable future IM/DD transmission will dominate fiber optic systems.

1.3 OPTICAL AMPLIFICATION: OVERCOMING THE LIMITATIONS OF FIBER LOSS

Two types of optical amplifiers are prevalent today: The semiconductor optical amplifier (SOA) and the fiber amplifier (FA). The SOA is similar to a laser, but instead of reflective cleaved facets, the SOA facets are coated with antireflection coatings. A quantum well SOA with fiber-to-fiber unsaturated gain of 27 dB and a 3-dB gain bandwidth of greater than 60 nm has been demonstrated (Tiemeijer *et al.* 1994). SOAs can be fabricated to provide gain in the 1300- or 1550-nm transmission window.

Although SOAs predate FAs, they are not widely used as optical repeaters in amplified transmission systems. There are two major reasons for this: First, the SOA is highly nonlinear in saturation (Agrawal and Habbab 1990), which results in significant optical cross products when two or more channels are simultaneously amplified. Second, the fiber-to-chip coupling loss is generally higher than 5 dB for each coupling, which greatly reduces the available SOA gain. In addition, problems like TE–TM gain variability and gain flatness in saturation reduce its attractiveness as a line amplifier. However, SOAs are useful in unrepeatered single-channel systems as a booster and/or optical preamplifier. SOAs at the transmitter and receiver can improve the sensitivity by as much as 13 dB resulting in a 38.6 dB loss budget at 10 Gb/s (Tiemeijer *et al.* 1995). Another SOA application uses its nonlinear characteristics to provide all-optical wavelength conversion (Durhuus *et al.* 1994; Mikkelsen *et al.* 1996).

Fiber amplifiers consisting of a length of fused silica fiber doped with rare-earth ions, typically Er^{3+}, and pumped with intense (>40-mW) 980- or 1480-nm light provide an attractive alternative (Chapter 4 and Chapter 2 in Volume IIIB). These so-called erbium-doped fiber amplifiers (EDFAs) provide more than 35 dB of fiber-to-fiber gain in the 1540- to 1560-nm wavelength range. Fiber amplifiers that consist of fiber doped with the praseodymium ion Pr^{3+} (PDFAs) can provide gain at 1310 nm. Optimization of PDFA designs has resulted in internal amplifier gains exceeding 26 dB with 135 mW of pump power at 1017 nm (Yanagita *et al.* 1995).

In the 1550-nm wavelength window, EDFAs can be used to extend transmission distance by using them as power boosters, line amplifiers, and preamplifiers. As system designs become more ambitious, consideration of the limitations of fiber dispersion, optical nonlinearities, and amplifier noise must be considered in transmitter and receiver designs.

2.   Design Intricacies of Laser Transmitters

In long amplified systems the presence of amplified spontaneous emission (ASE) noise (Olsson 1989; Marcuse 1990), multiple optical reflections (Fishman, Duff, and Nagel 1990), and large amounts of dispersion (Yamamoto *et al.* 1987) have made transmitter design more challenging because designs must now address these additional degradations. In addition, the design of a high-performance laser transmitter must include careful consideration of laser reliability, thermal and mechanical design, circuit design, and manufacturability.

The SONET optical standards (*Synchronous Optical Network Transport Systems Common Generic Criteria* TA-TSY-000253) provide performance guidelines for components and systems to ensure that equipment from different manufacturers work together. Although these guidelines are useful in providing product uniformity and in reducing transmitter–receiver mismatch in low bit-rate systems, these criteria cannot adequately characterize multigigabit systems such as the 2.5-Gb/s OC-48/STM-16 systems. In the design of these components, one must go beyond the SONET/SDH standards to ensure a successful design. In the following sections we discuss some of the subtle, and not so subtle, transmitter design issues that are particularly relevant to optically amplified systems.

2.1 IMPORTANT TRANSMITTER CHARACTERISTICS

There are three major transmitter characteristics that are important in the design of multigigabit laser transmitters used in IM/DD systems:

(1) What is the minimum acceptable extinction ratio for the system application?
(2) What are the requirements for the modulated pulse shape?
(3) How much pulse distortion can be tolerated?

In the next section each of these characteristics is discussed in greater detail in the context of state-of-the-art device technologies.

2.1.1 Extinction Ratio

An important transmitter parameter is the laser extinction ratio, which is the ratio between the unmodulated optical power and the modulated optical power. In directly modulated lasers the extinction ratio is largely determined by the modulation amplitude and the laser zero-state bias point with respect to the laser threshold. The extinction ratio is given by

$$r = \frac{P_{thr} + \eta_e(I_{bias} - I_{thr})}{\eta_e(I_{mod} + I_{bias} - I_{thr})},\tag{3.1}$$

where I_{bias} is the laser current that corresponds to the OFF state, I_{mod} is the laser current corresponding to the ON state, I_{thr} is the lasing threshold, P_{thr} is the laser spontaneous emission when the laser is biased at threshold, and η_e is the laser slope efficiency. For systems with a high SNR (e.g., >25 dB), the extinction penalty approaches that of unamplified systems (Duff 1984):

$$Penalty = 10 \log_{10} \left(\frac{1 + r}{1 - r} \right), \tag{3.2}$$

where r is the extinction ratio. Ideally, a laser biased at I_{thr} will have an extinction ratio determined entirely by the laser spontaneous emission; this extinction ratio is typically greater than 15 dB. However, in practice there are electrical reflections arising from driver to laser impedance mismatch that result in some light in the zero state. In addition, lasers that are modulated at several gigabits per second need to be biased above I_{thr} to avoid mode partitioning and excess jitter that results from random turn-on delay (Andersson and Akermark 1992). Typically, extinction ratios for directly modulated lasers extend from 9 to 14 dB depending on the transmitter application. In externally modulated lasers, the extinction ratio is determined by the modulator OFF-to-ON contrast. Externally modulated lasers with extinction ratios exceeding 15 dB are commonplace with LiNbO$_3$ Mach–Zehnder modulators. Devices that employ integrated laser and electroabsorption modulators (see, for example Aoki *et al.* 1993) typically have 11- to 13-dB extinction ratios.

2.1.1.1 Impact of Extinction Ratio on Amplified System Performance

In unamplified systems a degraded extinction ratio will degrade the sensitivity of the receiver because some of the received light is unmodulated, but it will have little impact on eye margin (Duff 1984). In other words, it will not affect the received eye at the decision circuit. In an amplified system, a poor extinction ratio can degrade system margin because the signal in the OFF state will contribute to base-band noise as a result of signal–spontaneous beat noise (Olsson 1989; McDonald, Fyath, and O'Reilly 1989). The extinction ratio penalty in amplified systems depends on the ASE noise level — i.e., the SNR$_O$. In systems with significant ASE noise resulting in a low SNR$_O$ (e.g., <15 dB), the penalty approaches

$$Penalty = 10 \log_{10} \left(\frac{1 + r}{1 - r} \right) \left(\frac{1 + \sqrt{r}}{1 - \sqrt{r}} \right), \tag{3.3}$$

which is equal to the penalty of a conventional avalanche photodiode (APD) receiver where the shot noise dominates. For example, if the laser extinction ratio is 10 dB ($r = 0.1$), the penalty will be between 0.9 and 3.7 dB, depending on the SNR_O.

2.1.2 Rise–Fall Time and Pulse Shape (Meeting the SONET Standard)

Typical 2.5-Gb/s (OC-48/STM-16) transmitters with directly modulated lasers have 20–80% rise–fall times that are less than 150 ps. In addition, the SONET mask specification requires that the electrically filtered transmitter optical signal meet an "open eye" template, which simulates the effect of receiver filtering. Realistically, only several thousand waveforms can be measured within a reasonable amount of time using a sampling scope, hence degradations that affect low bit error (BERs) — i.e., $<10^{-4}$ BER — cannot be detected. The mask requirement has limited use because most transmitter–receiver degradations affect BERs less than 10^{-4}. Alternatively, the performance of the transmitter can be baselined by measuring the BER characteristics using a reference receiver that consists of an optical-to-electrical (O/E) converter, electrical filtering, amplification, and a decision circuit. This has proven to be the best way of certifying design compliance for the ideal dispersion-free system.

Causes of pulse degradation include electrical reflections between the laser and the driver, pulse ringing in directly modulated lasers due to laser electron–photon dynamics (Marcuse 1984), and for both externally and directly modulated lasers, pulse rise–fall time degradation due to laser or package bandwidth limitations. The effects of pulse ringing and turn-on jitter can be mitigated in directly modulated lasers by specifying a large relaxation oscillation frequency and biasing the laser above threshold. As a rule-of-thumb: (1) the relaxation oscillation frequency at $I_{mod}/2$ of lasers modulated at 2.5 Gb/s should exceed 6 GHz to minimize the effect of ringing and turn-on jitter, and (2) the packaged laser small-signal bandwidth should exceed the bit rate by at least 25% to ensure "clean and open" non-return-to-zero (NRZ) optical pulses.

2.1.2.1 *Effect of Pulse Shape in Amplified Systems*

In long amplified systems the transmitted optical pulse must adhere to more stringent requirements than what would be acceptable in an unamplified system. More stringent laser transmitter specifications provide more decision circuit eye margin (Duff 1984) for SNR_O and dispersion degradations, thus allowing longer amplified systems. For example, at 2.5 Gb/s the 20–80% optical rise–fall time may be 150 ps for an unamplified system, but it should be less than 100 ps for an amplified system. In addition, the pulse overshoot should be significantly lower in long amplified systems to avoid degradations due to the often coincident wavelength chirping.

An additional feature of amplified systems is the potential of pulse compression that results from self-phase modulation (SPM), an important optical fiber nonlinearity discussed in Chapter 8 in Volume IIIA. In systems that employ a booster fiber amplifier, the fiber launch optical power may exceed +20 dBm to ensure a good SNR_O at the optical receiver, and the effects from accompanying nonlinearities must be considered. Generally, pulse compression has the potential of reducing the fiber dispersion penalty by offsetting the effects of dispersion-induced pulse broadening. However, significant system degradations may result from excessive pulse narrowing in lasers with negative chirp (Yamada et al. 1995), if the pulses have a low duty cycle, or if the launch power is too large (Hamide, Emplit, and Gabriagues 1990; Ogata et al. 1992; Suzuki and Ozeki 1993; Velschow et al. 1995).

2.1.3   Chirp, Mode Partitioning, and Polarization Mode Dispersion

One of the more challenging aspects in designing laser transmitters for long multigigabit systems is the difficulty in specifying a laser parameter that correlates well with the dispersion penalty. It is now known that the time-averaged laser spectrum has little to do with the ultimate fiber performance in systems with chromatic dispersion (see, for example, CCITT Recommendation G.957 ITU-T Laser Chirp Correspondence Group: Contribution from Telecom Australia 21/7/93). Alternatively, time-resolved chirp measurements (TRCMs) have been used to characterize directly modulated (Linke 1985) and externally modulated (Fishman 1993) lasers. TRCM provides a bit-by-bit measurement of the pattern-dependent instantaneous laser chirp. The use of TRCM to characterize lasers for use in highly dispersive systems is gaining popularity (Kataki and Soda 1995; Fells et al. 1995a).

A key TRCM parameter that correlates well with the dispersion penalty is the peak-to-peak time-resolved chirp (Runge and Bergano 1988). A typical time-resolved chirp plot is shown in Fig. 3.2 for an integrated electroabsorption modulator laser (EML) grown by selective area metalorganic vapor-phase epitaxy (MOVPE) (Johnson et al. 1994).

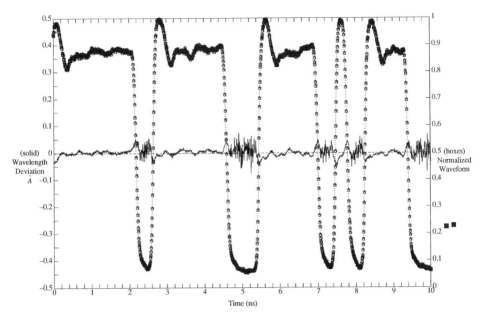

Fig. 3.2   Time-resolved chirp measurement (TRCM) showing maximum wavelength excursions during data transitions.

As shown, the wavelength excursions typically occur at the bit transitions, and the polarity of the excursion for the rising edge is reversed for the falling edge. In this case, the peak-to-peak wavelength excursion is about 0.11 Å. The "noisy" wavelength excursion coincident with the OFF state is due to the inaccuracies in the wavelength measurement when the optical signal is small. Because very little of the pulse power is contained in the OFF state, the presence of this is not detrimental to laser performance. The TRCM data can be used to extract key laser performance parameters, such as pulse rise and fall times, the magnitude and polarity of effective rising and falling edge alpha parameters (Koch and Bowers 1984), and the location of the dominant chirp on the pulse. These parameters can be used to determine the suitability of a device for its intended application.

A sample of the results of an analysis of 128 time-resolved bits is shown in Table 3.1. The analysis indicates that this laser is capable of over 600-km (11,000-ps/nm dispersion) transmission with a less than 2-dB dispersion

Table 3.1   **TRCM Characteristics for a 2.5-Gb/s EML**

⟨Rise time⟩ = 74.2, rms = 3.1 ps	Extinction ratio = 11.4 dB
⟨Fall time⟩ = 82.8, rms = 3.5 ps	Peak-to-peak chirp = 0.11 Å
Rising edge ⟨α_{eff}⟩ = 0.32	α_{eff} rms = 0.17
Falling edge ⟨α_{eff}⟩ = 0.52	α_{eff} rms = 0.07

penalty. The peak-to-peak chirp represents the wavelength fluctuations for all pulses that are 4% or more of the average ones level. The effective alpha, α_{eff}, is computed using the formula

$$\alpha_{eff} = -\frac{\delta\lambda(t)4\pi c}{\lambda_c^2\{d \ln [I(t)]/dt\}}, \tag{3.4}$$

where $I(t)$ is the laser intensity waveform, $\delta\lambda(t)$ is the corresponding pattern-dependent wavelength fluctuations, λ_c is the laser wavelength, and c is the speed of light. The derivative $d \ln [I(t)]/dt$ is computed at its maximum value, which corresponds to the peak $\delta\lambda(t)$ value. Note that the average falling edge α_{eff} is larger than the rising edge value, which is consistent with Fig. 3.2.

The time-resolved chirp data have been correlated to measured dispersion penalties; thus, TRCM is a good method of selecting lasers that will operate in highly dispersive fiber. Figure 3.3 shows a plot of the peak-to-peak TRCM chirp versus the measured dispersion penalty for 2.5-Gb/s EML devices through more than 330 km (6000-ps/nm dispersion) of standard fiber. Some of the scatter results from lasers that have a negative α_{eff}, which results in a smaller transmission penalty.

Alternative laser testing techniques entail direct measurement of the effective modulator alpha parameter (Dorgeuille and Devaux 1994; Suzuki and Hirayama 1995), and indirect measurement of alpha by measurement of the pulse width versus chromatic dispersion (Kataoka *et al.* 1992). Direct alpha measurements are not suitable for integrated laser and electroabsorption modulator devices because the effects of optical and electrical coupling to the laser cannot be ignored (Marcuse and Wood 1994; Fells *et al.* 1995a). The effects of optical fiber nonlinearities (Agrawal 1989; Chapter 8, Volume IIIA) must be taken into account when one is measuring the laser pulse width using the indirect α_{eff} technique.

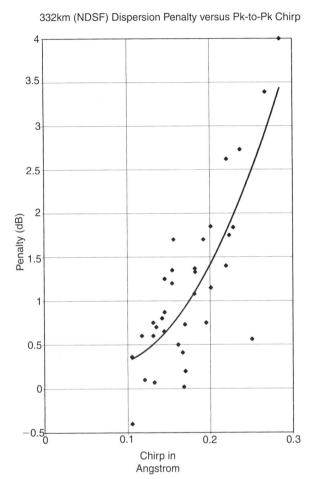

Fig. 3.3 Dispersion penalty versus peak-to-peak chirp in a 332-km non-DSF system.

2.1.3.1   Effect of Laser Chirp on System Performance

Laser chirping will result in pulse distortions at the receiver that are proportional to the total fiber dispersion, the peak-to-peak wavelength chirp, and the duration and location of the chirp (Yamamoto *et al.* 1987). Generally, these pulse distortions limit the transmission distance (Fishman 1993). However, in amplified systems the effect of fiber nonlinearities, particularly SPM, can alter the pulses in such a way as to cancel the chirp, depending on the polarity of α_{eff} and the dispersion coefficient (Ogata *et al.* 1992; Suzuki and Ozeki 1993). If carefully controlled, this results in improved system performance.

Polarization mode dispersion (PMD) may result in significant system degradations in long-haul systems (Chapter 6 in Volume IIIA; Poole *et al.* 1991). Because of the statistical nature of PMD, it is difficult to predict system penalties. However, it is clear that the PMD penalty is dependent on laser chirp, with lower chirp producing a smaller penalty (Poole and Giles 1988). Hence, laser transmitters with low peak-to-peak chirp, or negative α_{eff}, should produce smaller PMD penalties.

2.1.4   Transmitter Design Issues at 10 Gb/s (OC-192/STM-64)

Although we have concentrated on 2.5 Gb/s, similar design considerations must be applied at 10 Gb/s, but with even more stringent chirp and pulse shape requirements. It appears that Mach–Zehnder modulators would be most suitable at 10 Gb/s because of their adjustable chirp (Gnauck *et al.* 1991); however, EMLs have been shown to be suitable as well (Yamada *et al.* 1995). In addition, transmitter performance can be optimized in several ways, such as by introducing FM to effectively reduce the chirp by half (Binder and Kohn 1994). A redeeming aspect of 10-Gb/s systems is that it is generally assumed that some form of dispersion compensation will be available for transmission distances beyond 50 km; this assumption eases the chirp requirements.

2.2   CONSIDERATION OF OPTICAL FIBER NONLINEARITIES IN TRANSMITTER DESIGN

In this section we focus on the transmitter design implications of optical fiber nonlinearities, the general topic covered in Chapter 8 in Volume IIIA. With judicious transmitter and receiver design, penalties arising from some fiber nonlinearities can often be reduced. For example, the effect of linear and nonlinear pulse distortions can be reduced by use of an equalizing decision circuit (Winters and Kasturia 1992). Also, the laser wavelength can be set far from the fiber dispersion zero to avoid four-photo mixing, and the channel spacing of multichannel systems can be made large enough to avoid cross-phase modulation. In cases where high optical launch power is necessary, the effect of SPM in fiber with anomalous dispersion can be mitigated by choosing a laser with a positive α_{eff} and avoiding transmitters that have a low duty cycle (e.g., <20%). One fiber nonlinearity that critically

depends on the laser linewidth and has important transmitter design implications is discussed next.

2.2.1 Linewidth, Stimulated Brillouin Scattering (SBS), and SBS Suppression

Laser linewidth is generally an issue for only very long transmission systems
in the 1550-nm window or for systems that launch high power into the
fiber. Lasers that are used in highly dispersive systems — e.g., >6000-ps/nm
dispersion — must have narrow linewidths to avoid degradations that result
from phase-to-amplitude conversion noise (Chraplyvy *et al.* 1986). Typically, the laser linewidth should be less than 50 MHz to ensure less than
1 dB of penalty in a system with 10,000-ps/nm dispersion (Fishman 1993).
On the other hand, if high-power optical signals are launched into low-loss
fiber — e.g., >+13 dBm into non-DSF — lasers with narrow linewidths are
likely to stimulate Brillouin scattering (SBS), which has the effect of reducing the launch power and creating potentially severe system degradations
(Fishman and Nagel 1993). In directly modulated lasers, SBS is not an issue
because of the chirp of the carrier. Fortunately, in the case of externally
modulated lasers, SBS can be easily suppressed by FM dithering the laser
with a low-frequency tone, typically greater than 5 kHz. The tone broadens
the laser linewidth and prevents SBS buildup in the fiber. The FM dither
can be accomplished by amplitude modulating the laser bias. Optical phase
modulators, typically implemented using $LiNbO_3$, can also be used to
spread the optical spectrum and reduce SBS (Korotky *et al.* 1995).

2.3 TRANSMITTER POLARIZATION MANAGEMENT

Several polarization-related transmission impairments arise from the transmission of highly polarized signals in long optically amplified systems (see
Chapter 6 in Volume IIIA). Polarization hole burning (PHB), which provides preferential gain for noise over the signal (Mazurczyk and Zyskind,
1993), and polarization-dependent loss (PDL), which produces signal fades
as the transmission signal polarization varies slowly with time (Lichtman
1995) are two of the most important. Adjusting the signal's launched polarization state to maximize the SNR_O can reduce the effects of PDL-induced
fading, but it requires a reverse communications channel (from the system
receive end to the transmit end). A more practical option is to scramble the
polarization of the transmitted signal (Bergano, Mazurczyk, and Davidson
1994). Scrambling the signal at a rate faster than the optical amplifier

response time eliminates PHB effects but produces unwanted AM as the changing optical polarization interacts with system PDL. However, if the scrambling rate is fast with respect to the receiver bandwidth, say twice the bit rate, the AM does not pass through the receiver channel filter and does not adversely affect the BER.

Timing effects are also observed as the polarization state interacts with the PMD of the transmission path, because group delay of the transmitted signal varies with the signal polarization state. This timing variation causes jitter at the receiver, which will result in alignment jitter or will be passed through a receiver, depending on the receiver timing recovery jitter bandwidth. The magnitude of scrambling-induced jitter depends on the method used to scramble polarization and the PMD of the transmission system, the latter of which wanders slowly with time as environmental changes influence the composite system PMD (Bergano and Davidson 1995).

Polarization scrambling is most commonly implemented using $LiNbO_3$ polarization modulators (Heismann *et al.* 1994). These modulators are operated such that a signal with fixed linear polarization, as comes from a laser transmitter, is rotated through a great circle of the Poincaré sphere, producing a minimum degree of polarization. Polarization scrambling with $LiNbO_3$ imparts a residual phase modulation to the optical carrier that produces AM upon interaction with chromatic dispersion in the transmission fiber. If this AM is phase synchronous and properly aligned with the transmitted data, improvements in the eye opening at the receiver — and thus improvements in system margin — can be achieved (Bergano *et al.* 1996). For that reason, many polarization scrambler implementations synchronize the polarization scrambler with the transmitted data.

Pure phase modulation, which modulates the carrier phase and does not affect signal polarization, also interacts with system dispersion and can be used to improve eye margin (Bergano *et al.* 1996).

3. Receivers for Optically Amplified Systems

The receiver in digital systems must detect and convert optical digital information into electrical information with minimum influence on the content of the transmitted data. For newer generation SONET or SDH systems, which incorporate chains of optical amplifiers, this means a receiver capable of producing the minimum BER while tolerating the noise and distortion typical in an optically amplified NRZ digital signal. A receiver

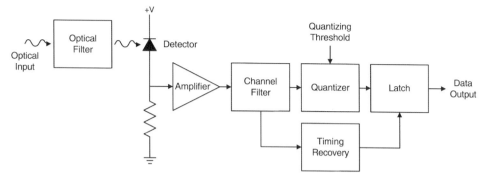

Fig. 3.4 Receiver for optically amplified systems.

must achieve a low BER while maximizing performance in other metrics demanded by a system design: sensitivity, Q factor, dynamic range, and so forth.

As the advent of optical amplification has revolutionized the design of optical transmission systems, it has also radically affected receiver design. This section touches on the fundamentals of receiver design as a basis for discussion of receiver designs for optically amplified systems. Also introduced are receiver topologies incorporating optical amplifiers that provide for specialized receiver performance enhancements.

3.1 BASIC RECEIVER

Figure 3.4 shows the topology of a basic digital receiver for optically amplified systems.* This receiver comprises seven functions, the salient features of which are described next.

1. *Optical filter.* The optical filter filters the incoming signal, reducing ASE power generated by the amplifier chain arriving at the detector. The optical filter's function lies in a gray area between system and receiver design, but discrete optical filters are often packaged with and considered part of the receiver. For this reason, and given their bearing on overall receiver performance, they are included in this chapter as part of the receiver.

---

* For a complete discussion of classic receiver design, the technology choices available for receiver front ends, and the fundamentals of noise tolerance (maximum sensitivity designs), see Kasper (1988).

2. *Detector.*   The detector converts the optical signal to an electrical signal. This is typically a photodiode (PIN or APD) with a square-law detection characteristic. This square-law characteristic creates as base-band signals information modulated onto the carrier by the transmitter, as well as noise and other distortions impressed on the optical carrier by the transmission system.

3. *Amplifier.*   The amplifier provides gain and band shaping to signals from the detector. Flat gain and linear phase response are the desired design features for the receive amplifier because this simplifies the channel filter design. The electrical noise contribution from the detector and amplifier is insignificant for amplified systems, as is shown later.

4. *Channel filter.*   The channel filter equalizes the channel to the desired characteristic, typically compensating for the band shapes of the detector and amplifier to produce a channel shape that minimizes ISI and maximizes the electrical SNR at the quantizer. To reduce noise and ISI, classic receiver design dictates a Nyquist channel narrower than the baud rate (see Bell Telephone Laboratories, 1982). These are good starting points for the channel design of a receiver for optically amplified systems. Unfortunately, deviations from ideal channel band shape and phase linearity, and the tolerance of the decision circuit to the subtleties of the distortions in the received optical signal, usually make empirical optimization of the channel filter a necessity for peak performance. As a practical matter, the channel filter is often incorporated in the amplifier.

5. *Quantizer.*   The quantizer combines high gain and fast slew rates to quantize into two distinct amplitudes data from the channel filter. Input voltages above a threshold are quantized into one state (a mark, or one), whereas those below the threshold are quantized into the other state (a space, or zero). In a typical amplified system, the BER is established in the quantizer because errors due to amplitude variation (from system noise and ISI) dominate those due to timing effects (jitter).

6. *Timing recovery.*   Timing recovery extracts a clock synchronous with and at the rate of the received signal. Controlling the static phase of this clock determines the timing of the decision point in the received data bit stream. Control of the dynamic behavior of the phase (e.g., data to clock phase versus frequency transfer function) and phase noise on the recovered clock determine the jitter transfer function and the jitter generated, respectively, by the receiver. Because SDH and SONET use an NRZ format, clock extraction typically comprises a nonlinear element

to generate a frequency component at the clock rate, followed by a phase-locked loop or appropriate filters and amplifiers. The timing recovery filter design dominates the jitter transfer function and the timing jitter generation (Trischitta and Varma 1989).

7. **Latch.**   The latch retimes the quantized data signal with the recovered clock to produce a data signal with minimum amplitude and phase distortion. The quantizer and latch functions are often combined into a single unit called a *decision circuit*.

3.2   ORIGIN OF ERRORS IN A RECEIVER

Consider Fig. 3.5, which shows the amplitude probability density functions (PDFs) for an arbitrary NRZ bit stream, sampled at the temporal midpoint of the eye, as it appears at the input to the quantizer. The width of the distributions $P(1)$ and $P(0)$ is caused by signal distortion from transmission

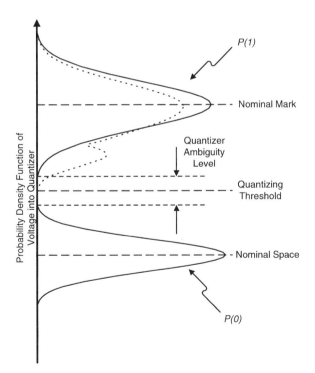

Fig. 3.5 Voltage probability density functions (PDFs) of non-return-to-zero (NRZ) data, quantizer ambiguity, and quantization level.

effects (dispersion, ASE power mixing with signal at the detector, distortion from nonlinear effects) and by noise and ISI contributions from the receiver detector, amplifier, and channel filter. For optimum BER performance, the quantizing threshold V_{quant} must be set such that $\int_{-\infty}^{V_{quant}} P(1) = \int_{V_{quant}}^{\infty} P(0)$, resulting in a minimum total error rate. The ability of a real quantizer to resolve an exact threshold is limited by noise and circuit biases within the quantizer itself. This threshold error is called the *ambiguity level* of the quantizer and is a measure of quantizer performance (it is shown in Fig. 3.5 as a range about V_{quant}). A typical high-speed (2.5-Gb/s) quantizer has an ambiguity level of 5–10 mV.

Although the quantizer and latch are often combined into a decision circuit to reduce parts count and costs, there can be advantages in separating the two. Keeping the quantizer in a separate package with well-isolated supply voltages and carefully controlled impedances can reduce the ambiguity level (by reducing cross talk from the digital circuitry of the latch) and ISI (from RF reflections).

Deviation from the ideal amplitude and phase transfer function in the amplifier–channel filter combination is often the root cause of the transmission penalty contributed by a receiver. These deviations produce a distribution of amplitudes for ones and zeros. Distortion typical of these effects is often visible on an oscilloscope as multiple traces in an eye diagram. A good example is isolated ones that do not make it to the upper rail because of limited channel bandwidth — broadening $P(1)$ (this example is shown as the dotted distribution in Fig. 3.5). There are an arbitrary number of these subdistributions making up $P(1)$ and $P(0)$, depending on channel distortions. The total error rate is the sum of the contributions from the individual bits, with those nearest V_{quant} making the largest contribution (Mazurczyk and Duff 1995).

3.3   INFLUENCE OF OPTICAL AMPLIFIERS ON RECEIVER DESIGN

The design of receivers used in optically amplified systems departs from traditional receiver designs in two ways: First, receivers must accommodate the noise and distortions that arise when optical amplifiers are used as transmission line amplifiers. Second, optical amplifiers can be incorporated into receiver designs to enhance some receiver characteristics.

3.3.1 Optical Noise in Amplified Systems

Accepting low optical powers and minimizing ISI impairment from dispersion or transmitter effects are characteristics needed by receiver designs for nonamplified systems. The introduction of a chain of optical amplifiers as gain elements increases the maximum bit rates that can be practically implemented in a long-haul transmission system. A challenge for the receiver designer is achieving the low-distortion, wide bandwidth electrooptic path between the receiver input and the quantizer. However, optical amplifiers also strongly influence the treatment of noise in receiver design.

The unpolarized noise power P_n from an optical amplifier is

$$P_n = 2n_{sp}h\nu(G - 1)B_O, \tag{3.5}$$

where n_{sp} is the amplifier spontaneous emission factor, $h\nu$ is the energy in a photon, G is gain, and B_O is optical bandwidth. Amplifiers are typically designed to operate in gain compression; the resulting automatic gain controlled (AGC) action ensures that the gain of the amplifier equals the span loss. If we consider the noise contribution from each amplifier in a system of N amplifiers with output power P_{out}, the SNR_O can be expressed as

$$SNR_O = \frac{P_{out}}{2n_{sp}h\nu(G - 1)B_O N}. \tag{3.6}$$

Note that SNR_O is inversely proportional to both the number of amplifiers and the amplifier gain. For a given system length, the system designer can trade off the number of amplifiers with the gain of each to produce a given SNR_O. To minimize costs, span lengths are increased (gains are increased) and the number of amplifiers is decreased until the minimum acceptable SNR_O is achieved at the end of the system. The noise performance of a system is thus dominated by the optical noise produced in the optical amplifier chain, and not by the noise in the receiver. This results in a fundamental shift in receiver design goals for amplified versus nonamplified systems. Receiver noise — once of paramount importance in receiver design — is a small contributor to overall system noise (and thus BER) performance. Noise generated within the receiver can be traded off for other desirable receiver characteristics (lower ISI, lower cost, or improved dispersion tolerance).

The square-law behavior of the photodetector in a receiver mixes the signal and the ASE produced by the optical amplifiers. Two mixing products from this process are of interest: power arising from signal mixing with nearby ASE, called *signal–spontaneous beat noise*, or N_{s-sp}, and power arising from ASE mixing with itself, called *spontaneous–spontaneous beat noise*, or N_{sp-sp}. The total noise power associated with any mark at the receiver, N_{mark}, is

$$N_{mark} = N_{short} + N_{th} + N_{s-sp} + N_{sp-sp}, \qquad (3.7)$$

where N_{shot} is signal shot noise and N_{th} is thermal noise from the receiver. In receiver designs for regenerative systems, N_{shot} and N_{th} dominate, but in typical amplified systems, N_{s-sp} is by far the dominant term (Olsson 1989).

The width of the mark distribution in a system with no ISI is largely determined by N_{s-sp}, whereas the width of the space distribution is set by the extinction ratio and N_{s-sp}, with the extinction ratio the dominant influence. In a typical optically amplified system, with good extinction ratio at the transmitter (>10 dB), the postdetector noise is asymmetrical, with the mark distribution broader than that for the spaces (see Fig. 3.6). This asymmetry forces the optimum quantization threshold below the midpoint of the eye.

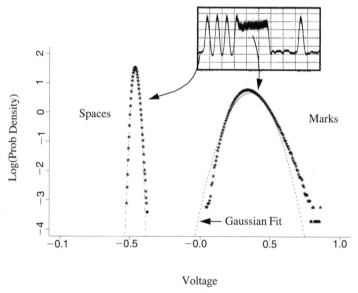

Fig. 3.6 Asymmetrical noise on the detected electrical signal as a result of signal–spontaneous beat noise, resulting in an optimum quantization threshold below the midpoint of the signal.

Moving the quantizer threshold below the midpoint of the eye reduces the error rate from noise on the upper rail, but, owing to finite rise and fall times, changes the duty cycle of the quantizer output. A change in duty cycle affects the DC content of the data stream and can adversely affect the performance of the data latch or clock extraction circuitry (if the clock is extracted from quantized data) unless action is taken to provide DC compensation. The optimum quantization threshold is therefore dependent on both the transmission system design and on the response of receiver circuitry to duty cycle asymmetry. As a practical matter, selection of the quantization threshold can be a challenge. Whereas it is possible to adjust the threshold for small-volume or high-performance applications, high-volume "set and forget" quantization thresholds require thorough characterization of receive amplifier and quantizer performance when the receiver is operating with the expected system noise. As a rule of thumb, setting the threshold for a minimum BER with maximum acceptable noise on the system (often achieved by reducing transmitter power into the system) results in optimum performance for high noise and robust transmission performance when noise levels are lower.

3.3.2   Using Optical Amplifiers in a Receiver

One can capitalize on the features that optical amplifiers offer by incorporating them directly into a receiver design. The most prevalent approaches are to use an optical amplifier as a preamplifier stage directly in front of a detector and to "remote" an amplifier some distance ahead of a detector.

3.3.2.1   *Optical Amplifiers as Receiver Preamplifiers*

There are three primary reasons that optical amplifiers are incorporated into receiver designs: (1) high sensitivities can be achieved, (2) dynamic range can be improved, and (3) lower cost, lower complexity high-performance receivers are possible.

Rühl and Ayre (1993) developed explicit expressions for receiver sensitivity of optically preamplified receivers. Making the conservative assumption that Gaussian statistics apply to all amplifier and system noise (Marcuse 1990), they derived the minimum average power P_{AV} required to meet a target BER performance. With this expression in hand and assuming a high-gain optical amplifier and a perfect extinction ratio, they asserted (with some rearrangement) the following:

$$P_{AV} = \left(\frac{Q^2}{2}f_b h\nu\right) \cdot \left(\frac{2n_{SP}}{L_I}\right) \cdot \left(\frac{2B_{EL}}{f_b} + \frac{\sqrt{2pB_{EL}B_{OP}}}{Qf_b}\right), \qquad (3.8)$$

where f_b is the bit rate, $h\nu$ is the energy of a signal photon, n_{sp} is the EDFA spontaneous emission factor, L_I is the optical amplifier input coupling loss, B_{EL} is the electrical bandwidth, p is the number of polarization states detected, and B_{OP} is the optical bandwidth. Q is related to the desired BER, where

$$Q = \frac{\sqrt{S(1)} - \sqrt{S(0)}}{\sqrt{N(1)} + \sqrt{N(0)}} \qquad (3.9)$$

with $S(\cdot)$ and $N(\cdot)$ being the signal and noise powers detected during a one or a zero.

Note in Eq. (3.8) that the first term is the shot noise limit, the second term is penalties arising from optical amplifier noise and input losses, and the last term is penalties from excess electrical and optical bandwidths. At nominal gains, the optical preamplifier noise dominates. A high-performance receiver is therefore achieved through careful design of the optical preamplifier.

Many optical amplifier architectures offer high gains with noise figures approaching 3 dB. Optically preamplified receivers achieving −37 dBm sensitivities at 10 Gb/s have been reported, and outperform the best nonpre-amplified receivers by 10 dB. A comprehensive treatment of optically pre-amplified receiver design can be found in Park and Granlund (1994).

Optical amplifiers can also be incorporated into receiver designs to improve dynamic range performance. Optical amplifiers can provide distortion-free variable gain with very wide equivalent electrical bandwidths. So, instead of designing high-performance ultrawideband amplifiers to follow the detector in a receiver, a variable gain optical amplifier (an "optical AGC" amplifier) can be placed before the detector, and a fixed-gain wideband amplifier can follow the detector (Fig. 3.7). The designer is thus given the freedom to choose receiver topologies incorporating less expensive or more readily available components. For example, high-speed receiver designs incorporating low-impedance voltage amplifiers following the detector can be designed using readily available microwave amplifiers without resorting to design of custom high-speed transimpedance amplifiers and without sacrificing dynamic range.

Optical preamplifier gain can be controlled to provide constant optical power to a detector, which allows the use of existing receiver modules in

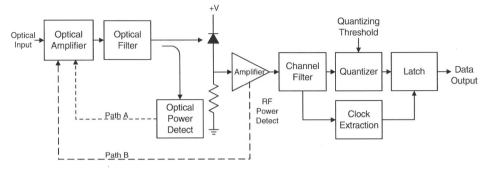

Fig. 3.7 Receiver incorporating an optical preamplifier.

new applications (path A in Fig. 3.7). The detected RF signal level can also be used to control gain (path B in Fig. 3.7), which provides for more constant signal amplitudes to the quantizer as input SNR and total power vary. This latter topology provides for even greater dynamic range, presenting the quantizer with more consistent signal levels and reducing impairments that can arise from using fixed quantization thresholds. Optical AGC can also improve performance by decreasing linear channel changes arising from power-dependent receiver bandwidths.

3.3.2.2 *Remotely Pumped Amplifiers and the Extended Receiver*

Another interesting optical amplifier–receiver combination uses a remotely pumped optical amplifier some distance from the detector (daSilva *et al.* 1995). Figure 3.8 shows such a receiver. In this case, pump power propagates counter to the signal direction through the transmission fiber to excite the

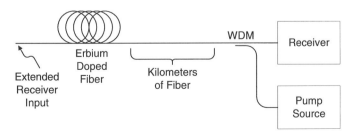

Fig. 3.8 Extended receiver incorporating a remotely pumped optical amplifier.

EDFA, which provides gain for an incoming signal. With prudent selection of pump wavelength, power, and fiber type, distributed stimulated Raman gain can also be maximized to further boost the incoming signal within the transmission fiber carrying the pump power (daSilva and Simpson 1994). As a result, this interesting hybridization of system and receiver produces an effective receiver interface at the input of the remote amplifier some distance from the detector.

Often called an *extended receiver*, this receiver can be designed for the typical receiver parameters (sensitivity, Q, dynamic range, etc.) at the remote amplifier input. The extended receiver then contributes to the overall length of the transmission system. Several variants of this topology, some using separate fiber to deliver the pump power, have been demonstrated. Receivers operating at 2.5 Gb/s, extending as long as 118 km, and delivering -40 dBm sensitivities have been demonstrated (Hansen *et al.* 1995).

4. Systems Performance Metrics

The primary performance metric on a digital communications link is the BER. A low BER requires an electrical signal at the decision circuit that offers sufficiently low noise and signal distortion to allow near error-free decision circuit performance. Yet, following a long chain of optical amplifiers, an optical signal is degraded by optical noise, dispersion, nonlinear effects, and polarization effects that have accumulated along the length of the system. The signal also suffers from distortion and noise contributed by the transmitter and the receiver. These impairments can make it difficult for the decision circuit to quantize the transmitted data without error. The challenge for a system designer is to find an appropriate trade-off in variables affecting the design of the optical amplifiers, transmission fiber, transmitter, and receiver that delivers the best performance at the lowest cost.

4.1 SYSTEMS LOSS BUDGET FOR UNAMPLIFIED SYSTEMS

An example of a typical loss budget for a long-range 1550-nm SONET OC-48 (2.5-Gb/s) transmission system is shown in Table 3.2. Note the penalties allocated for transmitter–receiver mismatch, dispersion, and end-of-life aging margins. The loss budget is the basic blueprint from which a particular fiber optic system route can be designed. For example, if a particular fiber optic route has 40 dB of loss at 1300 nm, but only 25 dB

Table 3.2 **Loss Budget for a 1550-nm Unamplified System**

Parameter	Value	Notes
Transmitter power	−1 dBm	Nominal, room temperature
Receiver sensitivity (10^{-10} BER)	−35 dBm	Excluding connector
System gain	34 dB	
Transmitter allocation	2.3 dB	Temperature, aging, reflections
Receiver allocation	4 dB	Temperature, aging, mismatch
Dispersion penalty	2.0 dB	2500 ps total
Allowed fiber loss	25.7 dB	At 1550 nm; includes cables (~100 km)

of loss with less than 2500-ps/nm dispersion at 1550 nm, a standard 1550-nm system can be installed on this route.

A loss budget presupposes that the system degradation can be offset by increasing the optical power into the receiver. This is the case only if the receiver eye margin (Duff 1984) increases with increased optical power, which occurs when the system margin is limited by the receiver noise. In optically amplified systems this is generally not the case because the system margin is limited by optical amplifier noise (Olsson 1989), and the eye closure due to ISI (Duff 1984).

4.2 SYSTEMS MARGIN BUDGET FOR AMPLIFIED SYSTEMS

In amplified systems, impairments are allocated as degradations in the system's SNR_O margin. For example, a system that is supposed to operate at its end of life at a 10^{-15} error rate (corresponding to $Q = 18$ dB) might require an SNR_O of more than 15 dB if the received eye is undistorted, whereas a signal with significant ISI may require an SNR_O of 21 dB for a 10^{-15} BER. Each system impairment — e.g., dispersion, multipath interference, nonoptimum decision circuit threshold — can be allocated in terms of an SNR_O degradation. Table 3.3 shows a typical SNR_O allocation for a 2.5 Gb/s amplified system consisting of several concatenated EDFAs.

4.3 Q FACTOR AND NOISE MARGIN

To study system design trade-offs, the designer needs tools that can measure system performance of various designs or in the presence of degradations to system components. Unfortunately, direct measurement of the key metric,

Table 3.3 **SNR$_O$ Budget for a 1550-nm Amplified System**

System Degradation	Allocated Margin (dB)
Receiver ISI margin	3.5
Receiver bandwidth margin	0.5
Transmitter ISI margin	1.5
Extinction ratio margin	1.0
Optical filter margin	0.5
V_d voltage margin	1.5
Receiver noise penalty	0.5
Ideal SNR (without ISI)	12.1
SNR Required	19.6

BER, is impractical; the low BERs demanded by current standards (10^{-15} is common) and the margin that must be designed in to accommodate aging and manufacturing variation result in systems that are essentially error free at installation. What is needed, then, are margin measurement techniques that are practical, measurable, and sensitive to the fundamental processes that affect transmission (noise, ISI, dispersion, etc.). Alternatives to measuring the BER to assess system performance have been developed. The most notable among these methods measure the Q factor and the noise-loaded BER.

4.3.1 Q Factor and Its Measurement

Q-factor measurement (Bergano *et al.* 1993) was developed to address the need for a practical and accurate way of measuring the system margin. The method builds upon work by Smith and Personick (1980) that defines the SNR Q factor at a decision circuit as

$$Q = \frac{|\mu_1 - \mu_0|}{\sigma_1 + \sigma_0}, \tag{3.10}$$

where $\mu_{0,1}$ are the means and $\sigma_{0,1}$ are the standard deviations on the zeros and the ones.

If one assumes that the noise on the rails of a signal at the decision circuit is Gaussian distributed (an approximation that leads to only slightly pessimistic Qs [Humblet 1991]) and that ISI effects are negligible compared with the noise from ASE, then Q can be derived by measuring the μ and σ associated with each rail and substituting into Eq. (3.10).

Bergano's method capitalizes on the ability to move a decision circuit threshold to measure the shape of the voltage distribution at each rail. The implementation is straightforward: The decision threshold voltage, V_d, is incremented from its nominal (lowest error rate) value toward the upper rail and then the lower rail of the incoming signal. The BER for each V_d is recorded. Recalling the Gaussian-distributed noise assumption, one can compute $\mu_{0,1}$ and $\sigma_{0,1}$ using

$$BER(V_d) = \frac{1}{2}\left[erfc\left(\frac{|\mu_1 - V_d|}{\sigma_1}\right) + erfc\left(\frac{|V_d - \mu_0|}{\sigma_0}\right)\right], \quad (3.11)$$

where

$$erfc(x) = \frac{1}{\sqrt{2\pi}} \int_x^\infty e^{-\alpha^2/2} d\alpha \cong \frac{1}{x\sqrt{2\pi}} e^{-x^2/2} \quad (3.12)$$

Note that the left-hand $erfc(\cdot)$ term in Eq. (3.11) applies only to bit errors recorded as V_d traverses the upper rail of the eye, whereas the right-hand $erfc(\cdot)$ term applies only to the lower rail measurement. This independence allows one to divide the measured $BER(V_d)$ characteristic into two distinct upper and lower rail measurements and to curve fit each separately with Eq. (3.12). The resulting $\mu_{0,1}$ and $\sigma_{0,1}$ values are used in Eq. (3.10) to compute the Q factor (Chapter 10 of Volume IIIA).

The method has advantages of speed (every bit is measured, so data are gathered quickly), repeatability (0.1 dB is typical), and wide dynamic range (Q factors from 10 to 30 dB can be measured in minutes). Because the Q factor is directly related to the BER, it is sensitive to all factors that determine BER performance in a system, the quality it needs to be a successful system measurement tool.

One must exercise care, however, when measuring the Q factor and interpreting the results. When this method is used to measure the SNR_O from a chain of amplifiers (as opposed to overall system performance including the receiver performance), one must be careful to maintain a linear channel from the receive detector through all components up to the decision point. Compression in this channel, as may result from using limiting amplifiers or amplifiers in compression prior to the quantization point, leads to optimistic (high) measured Q factors. Conversely, ISI from gain ripple in the transmitter and receiver causes pessimistic (smaller) Q factors.

This effect is reduced for low-Q systems (Mazurczyk and Duff 1995). When used with a carefully designed receiver and at high noise levels, though, the Q factor is a useful measure of system performance.

4.3.2 Noise Margin: Using a Noise-Loaded BER as a Metric

An alternative technique for determining the system margin involves adding noise to the transmitted signal, effectively varying the SNR_O, and measuring the resulting BER. The setup for such a measurement is shown in Fig. 3.9.

The noise source in Fig. 3.9 is typically an optical amplifier with no input, or a reflector on its input, that produces wideband ASE. The optical filter shapes the ASE noise to approximate that expected from a chain of optical amplifiers. This optical noise is summed, through an attenuator, with an optical signal and presented to the receiver through another attenuator. By varying the attenuators, one can achieve an arbitrary SNR_O at an arbitrary power level at the receiver input. Using this technique, one can curve fit the $BER(SNR_O)$ using Eq. (3.12), and the SNR_O at which a target BER is achieved can be determined.

This technique, however, provides only an estimate of system performance because it ignores distributed system effects (dispersion, polarization effects, etc.). One approach to overcome this limitation is to characterize a population of transmitters and receivers using both noise-loaded BER and BER versus SNR_O from the intended system design. By comparing the performance between the two, one can extract a correction factor that can be used to convert the simpler noise-loaded BER result to an equivalent result as if an actual system measurement were done. Although this method is not highly accurate, it is suitable for use where adequate SNR_O design

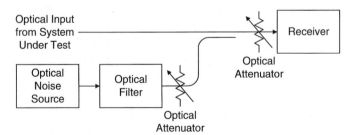

Fig. 3.9 Setup for noise-loaded bit error rate (BER) measurement.

margins exist. For high-volume production, where simple transmitter and receiver performance measurements are a necessity, system designs can allocate the margin needed for measurement error.

Care must be exercised in the selection of the optical filter used to shape the noise. Because the noise at the decision circuit is dominated by optical signal–spontaneous beat noise, subtleties in the filter shape adjacent to the carrier strongly influence the SNR_O; merely matching the optical filter bandwidths on different test setups will not produce identical results — the detailed filter shape near the carrier must be considered.

5. Multiple-Wavelength Systems

A natural extension of the use of optical amplifiers for the transmission of a single channel is their use for simultaneous amplification of several channels. Important constraints include channel spacing, channel wavelength tolerances, and demultiplexing technology (a discussion of the optical components employed for multiplexing and demultiplexing can be found in Chapter 8 in Volume IIIB). Channel spacing is constrained by the optical amplifier gain bandwidth, demultiplex capabilities, and fiber nonlinearities such as four-photon mixing, and cross-phase modulation (Chapter 8 in Volume IIIA). The fiber amplifier gain bandwidth can be increased by alternative fiber designs and by gain equalization using optical filters (Chapter 2 in Volume IIIB).

5.1 WDM TRANSMITTER CONSIDERATIONS

To take full advantage of the EDFA's 12–14 nm of optical bandwidth, it is essential that the channel spacing be as small as possible. Systems that have up to eight 2.5-Gb/s channels are being deployed on commercial networks (*AT&T News* 1995). When considering the number of transmitters per fiber pair, and the necessity of having spares, we find that the key to the success of multichannel systems is the availability of low-cost laser transmitters. This precludes lasers that rely on optical references for stabilization (Chung *et al.* 1991; Verdiell *et al.* 1993) or external-cavity lasers. Alternatively, DFB lasers have exhibited wavelength aging characteristics of less than 0.25 nm for 25 years of aging (Sessa and Wagner 1992; Chung, Jeong, and Cheng 1994; Vodhanel *et al.* 1994). Hence, laser transmitters with DFB lasers, tuned to the desired channel frequency and then temperature controlled, are the best candidates for WDM systems.

Long WDM amplified systems require some form of gain equalization to avoid penalties resulting from inhomogeneous gain saturation in EDFAs. Adjusting the transmitter optical power before transmission is a particularly convenient method of gain equalization (Chraplyvy *et al.* 1993). This may be accomplished by adjusting the laser power output directly or with variable in-line optical attenuators.

5.2 RECEIVER WDM CONSIDERATIONS

WDM complicates the system design, and the role of the optical filter changes dramatically, but the remainder of the receiver is largely unaffected. Among the challenges facing the optical filter on WDM systems are the following:

- The optical filter must perform the same ASE noise reduction that it does in single-carrier systems.
- It must reject unwanted energy from adjacent carriers. This requires large out-of-band rejection near the desired passband for closely spaced carriers. Achieving high rejection near the desired center wavelength can lead to narrow passbands. However, narrow passbands can be difficult to keep centered with aging and temperature variation. Narrow passbands also demand high-accuracy and high-stability transmitters.
- Nonadjacent carriers, which may be many nanometers distant from the filter passband, must also be rejected. This limits the usefulness of some interferometric filter technologies where wavelength periodic transmission occurs.

Because it can be difficult to achieve this performance with a single device, WDM demultiplex architectures often have staged filtering. For example, a $1:N$ wavelength-selective demultiplexer may "roughly" separate N carriers from a single fiber; the N receivers following the demux would then each have an individual channel filter. This concatenation allows the filter passbands to be multiplied, which reduces the demands on a single filter. Contemporary system designs employ eight 2.5-Gb/s carriers with 1- to 2-nm spacing and require out-of-band channel rejection of 25 dB.

Long dispersion-managed systems may also need per-carrier dispersion compensation, because delivering zero net dispersion for one carrier at the end of a system may result in a large net dispersion for a carrier several nanometers separate. On very long systems, this dispersion compensation can be so large as to require embedded gain to make up for compensation fiber losses. In these cases, the topology chosen for dispersion compensation can incorporate both wavelength demultiplexing and filtering functions, placing additional demands on the optical filter.

6. Provisions for Performance Monitoring in Long Amplified Systems

6.1 TERRESTRIAL SYSTEM MONITORING

Traditionally, systems that employ repeaters have used some form of remote performance monitoring for fault isolation and as an early warning of pending service affecting degradations. Systems that employ chains of fiber amplifiers are not immune to degradations, such as severed fibers or failed pump lasers. To detect remote failures, a method of sensing the failure locally must be devised, and a means to transmit that information to a remote terminal must be established. In some terrestrial systems a telemetry channel that uses the 1536-nm EDFA gain peak is used to transmit local EDFA parameters to the end terminals. See Chapter 9 in Volume IIIA for more details. In multichannel terrestrial systems, each optical channel has a unique low-frequency tone assigned to it that is monitored by each optical amplifier in the chain. The strength of this tone can be used to track the channel through the system (Hill *et al.* 1993). In addition, the tone can be used to estimate the SNR_O of each channel. Because the tone power is related to the actual optical signal power by a predetermined modulation index m_j, knowledge of all the tone powers, T_j, and the total optical signal level, P_T, including the noise, results in knowledge of the SNR_O when the following equation is used:

$$SNR_j = \alpha_j + 10 \log_{10} \left(\frac{T_j/m_j}{P_T - \sum_{i=1}^{i=N_{ch}} T_j/m_j} \right), \qquad (3.13)$$

where α_j relates the full EDFA noise bandwidth measured by the photodiode, and the actual 0.1-nm noise bandwidth typically used to measure

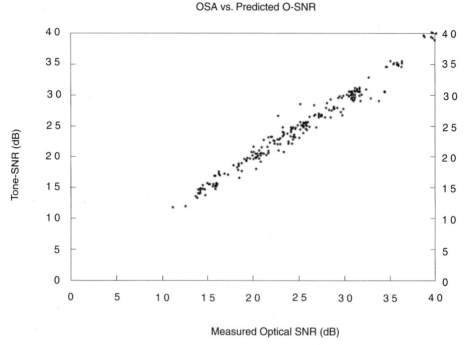

Fig. 3.10   Tone signal-to-noise ratio (SNR) versus optical SNR (SNR_O), showing a high degree of correlation.

the SNR_O. Depending on the channel, the value of α_j is 26–28 dB. Figure 3.10 shows the high degree of correlation between the measured SNR_O and those predicted on the basis of the tones method.

6.2   *UNDERSEA SYSTEM MONITORING*

Undersea system monitoring must

- Measure the system transmission margin — to provide confidence that high-capacity systems are functional and to provide an outage prediction capability should slow, unexpected degradations in transmission margin occur.
- Locate faults — to assist repair operations.

- Identify degradations in undersea hardware performance — so that system failure probabilities can be assessed to support backup, restoration, or repair planning.

Submarine systems employ several different methods to satisfy these needs.

6.2.1 System Margin

An overall system margin measurement capabilities exists at each cable station that terminates an undersea line. This function is supplied by high-speed receiver equipment that is specially modified to provide automatic measurement of the Q factor as described in Section 4.3.1.

6.2.2 Fault Location and Undersea Hardware Monitoring

There are three distinct monitoring systems employed in optically amplified undersea systems. The first uses a command and response channel reminiscent of that used in regenerative undersea systems, where the health of a repeater is assessed by commanding the repeater to respond with measurements of various parameters internal to the repeater. The second technique uses passive undersea optical loop-back paths and onshore signal processing electronics to infer the health of a system. The third technique uses coherent optical time-domain reflectometry (COTDR) and undersea loop-back paths to monitor a system.

6.2.2.1 Command and Response Systems

Command and response systems use an AM envelope, impressed upon the high-speed optical data signal, to command an undersea optical repeater. An optical receiver in the repeater detects this AM and converts it into a digital signal that is sent to processing electronics within the repeater. The processor interprets the command and performs an action, such as measuring optical output power, and encodes a digital response. This digital response is used to modulate the pump power of the optical amplifiers within the repeater. Modulating the pumps impresses an AM envelope on the high-speed optical data passing through the optical amplifier. The command channel modulation rate must be faster than the optical amplifier response, about 10 kHz, so that the signal is not tracked out. The response channel frequency must be low enough to modulate the gain of the optical amplifier, yet high enough to propagate with acceptable attenuation by the optical amplifier chain. Response channel frequencies between 6 and 8 kHz offer the best compromise (Lefranc *et al.* 1993). Shore electronics detect and decode the AM signal, which is then interpreted by monitoring system computers. From a transmission design perspective, one must trade off the command and response channel modulation amplitudes with the eye-closure impairments that they cause.

Command and response channels measure amplifier parameters such as input optical power, output optical power, and pump laser current. From these parameters, one can infer the health of the pump lasers, the loss of the fiber spans between repeaters, and the performance of some of the optical devices within the repeater. Command and response systems have advantages in their ability to measure an arbitrary number of amplifier parameters with an accuracy determined by the designer. The primary disadvantage is increased repeater complexity and the potential for reduced reliability.

6.2.2.2 Loop-back Systems

Loop-back systems couple energy from the outbound fiber path to the inbound fiber path in each repeater (Jensen *et al.* 1994). A portion of any outbound signal, or any modulation on any outbound signal, returns to the same end of the system from which the signal is launched, where it is then detected. Loop-back monitoring systems send a probe signal down the outbound fiber with the transmitted data. This probe signal either is an independent optical carrier (called a *sidetone*) multiplexed with the transmitted carrier or is an AM envelope impressed upon the high-speed transmitted data. The loop-back signal from any given repeater arrives back at the shore with a unique time delay that is dependent on the physical position of that repeater in the system. Through appropriate signal processing, shore electronics can measure a system signature, and from changes in this signature, changes in hardware performance can be inferred. From a transmission design perspective, one must trade off the modulation depth for AM systems, wavelength spacing and amplitude for sidetone systems, and loop-back loss against the impairments that they cause to transmission. Deeper modulation, higher sidetone amplitude, and lower loop-back losses improve the monitoring system SNR, but impair transmission by degrading the eye margin.

The topology for a typical repeater with a loop-back monitoring capability is shown in Fig. 3.11. The figure shows a repeater consisting of optical

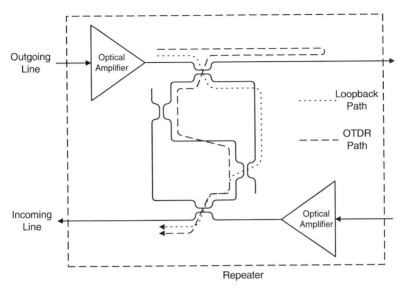

Fig. 3.11 Repeater showing loop-back and optical time-domain reflectometry (OTDR) paths.

amplifiers and an optical coupler network that supports the bidirectional amplification associated with a single fiber pair. If we trace the path followed by an optical signal in the outbound direction, we see that a small amount of the outbound signal is coupled to the inbound fiber (see the loop-back path [dotted line] in Fig. 3.11). Likewise, the inbound signal is coupled to the outbound direction, which allows system monitoring from both ends of the system.

Having suffered large losses when coupled into the opposite traffic direction, the looped-back probe signal received at the short is small with respect to the incoming data; SNRs of -90 dB are common. For this reason, the probe signal is modulated using a pseudo-random sequence. The shore electronics use correlation techniques and knowledge of the encoded sequence to extract the very low SNR probe from the inbound signal. The pseudo-random correlation technique also offers an easy way to determine the round-trip delay experienced by the probe signal. The amplitude of the cross correlation computed between the loop-back signal and a local pseudo-random sequence encodes the gain experienced by the probe signal at a given time delay. By shifting the local pseudo-random sequence an arbitrary number of bits, we can recompute the cross correlation and derive the relative gain experienced by a different time delay. If we repeat this process for many time delays, the correlation amplitude and round-trip delay produce a two-dimensional gain versus distance signature of the undersea system.

Changes in system gain (or loss) can be detected by recording the "baseline" gain versus time delay for a system and then monitoring for changes with time. Different failures have unique signatures. Figure 3.12 shows the signatures expected from an increased loss between two amplifiers and from a fiber break. In this figure, we see that a fiber break causes loss of the loop-back signal, and the loop-back signal drops to the noise floor of the detection equipment. For a finite loss, the loop-back signal gain is reduced in the first loop back following the loss (6 dB in this case). In subsequent amplifiers the lower input power reduces gain saturation, which increases signal gain. After a few amplifiers, the signal power recovers to its original level, and gains return to the baseline.

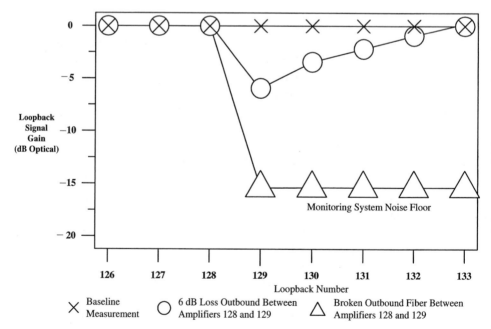

Fig. 3.12 Signatures from a fiber break and from 6-dB excess loss in an amplifier chain.

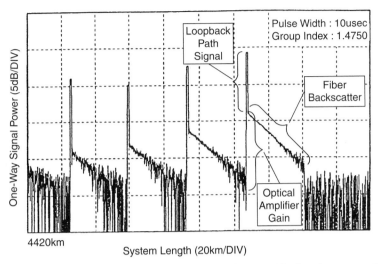

Fig. 3.13 Typical trace from a coherent OTDR (COTDR), showing an amplified system signature through several amplifiers at the end of approximately 4500 km.

A distinct advantage of loop-back monitoring systems lies in the simplicity of the undersea repeaters; there are few undersea components needed and they are all passive. However, designing a loop-back system capable of providing precise gain resolution while maintaining low system impairment is challenging.

6.2.2.3 COTDR Systems

COTDR can also be used to monitor the health of undersea systems. COTDR monitoring uses a very high-sensitivity OTDR that incorporates both coherent detection and pseudo-random sequence correlation to profile a system (Horiuchi, Yamamoto, and Akiba 1993). These systems are capable of measuring optical amplifier gain and distributed transmission fiber loss over 4500 km.

COTDR on amplified systems works much the same way as OTDR does on fiber, with one interesting exception: optical amplifiers typically employ isolators, so backscattered signals cannot travel backward down the launch fiber. The topology of the loop-back repeater solves this problem. Consider the OTDR path in Fig. 3.11. The forward-traveling probe signal from the OTDR backscatters off the outbound transmission fiber and then cross couples to the inbound fiber for return to the shore. By probing on the outbound fiber and detecting on the inbound fiber, the COTDR can derive the system loss profile. The COTDR must use its own carrier independent of the transmission signal: it is a sidetone system. COTDR is best suited as an out-of-service measurement, where interference with the signal carrier is acceptable. For in-service measurement, transmission designers must exercise care that the COTDR probe's proximity to the carrier and modulation amplitude do not impair transmission.

A typical COTDR trace is shown in Fig. 3.13. Note the optical gain present at the repeaters and the exponential signal decay in the transmission fiber. The chief advantages of COTDR are demonstrated by the figure: its ability to locate midspan fiber failure, and its easily interpreted output. Note that COTDR systems must have loop-back repeaters; COTDR can be used with loop-back monitoring and shares the same hardware advantages.

References

Agrawal, G. P. 1989. *Nonlinear fiber optics.* Boston: Academic Press.

Agrawal, G. P., P. J. Anthony, and T. M. Shen. 1988. Dispersion penalty for 1.3-μm lightwave systems with multimode semiconductor lasers. *J. Lightwave Tech.* 6:620–625.

Agrawal, G. P., and I. M. I. Habbab. 1990. Effect of four-wave mixing on multichannel amplification in semiconductor laser amplifiers. *IEEE J. Quantum Electron.* 26:501–505.

Andersson, P. O., and K. Akermark. 1992. Generation of BER floors from laser diode chirp noise. *Electron. Lett.* 28:472–474.

Aoki, M., M. Suzuki, H. Sano, T. Kawano, T. Ido, T. Taniwatatri, K. Uomi, and A. Takai. 1993. InGaAs/InGaAsP MQW electroabsorption modulator integrated with a DFB laser fabricated by bandgap control selective area MOCVD. *IEEE J. Quantum Electron.* 29:2088–2295.

Bell Telephone Laboratories. 1982. *Transmission systems for communications.*

Bergano, N. S., and C. R. Davidson. 1995. Polarization-scrambling-induced timing jitter in optical-amplifier systems. In *Technical Digest Optical Fiber Communications Conference, San Diego, CA.* Paper WG3, 122–123.

Bergano, N. S., C. R. Davidson, and F. Heismann. 1996. Bit synchronous polarization and phase modulation scheme for improving the performance of optical amplifier transmission systems. *Electron. Lett.* 32(1):52–54.

Bergano, N. S., F. W. Kerfoot, and C. R. Davidson. 1993. Margin measurements in optical amplifier systems. *IEEE Photon. Tech. Lett.* 5(3):304–306.

Bergano, N. S., V. J. Mazurczyk, and C. R. Davidson. 1994. Polarization hole-burning in erbium-doped fiber-amplifier transmission systems. In *Proceedings of ECOC'94, Florence, Italy, September,* 621–628.

Binder, J., and U. Kohn. 1994. 10 Gb/s dispersion optimized transmission at 1.55 μm wavelength on standard single mode fiber. *IEEE Photon. Tech. Lett.* 6:558–560.

Bungarzeanu, C. 1994. Limitations of dispersion supported transmission over standard single mode fiber. *IEEE Photon. Tech. Lett.* 6:858–859.

Chraplyvy, A. R., R. W. Tkach, L. L. Buhl, and R. C. Alferness. 1986. Phase modulation to amplitude modulation conversion of CW laser light in optical fibers. *Electron. Lett.* 22:409–410.

Chraplyvy, A. R., R. W. Tkach, A. H. Gbauck, B. L. Kasper, and R. M. Derosier. 1989. 8 Gb/s FSK modulation of DFB lasers with optical demodulation. In *Conference on Optical Fiber Communication (OFC'89), Houston, TX.* Paper PD17, PD17-1.

Chraplyvy, A. R., R. W. Tkach, K. C. Reichmann, P. D. Magill, and J. A. Nagel. 1993. End-to-end equilization experiments in amplified WDM lightwave systems. *IEEE Photon. Tech. Lett.* 4:428–429.

Chung, Y. C., R. M. Derosier, H. M. Presby, C. A. Burrus, Y. Akai, and N. Mauda. 1991. A 1.5 μm laser package frequency-locked with a novel miniature discharge lamp. *IEEE Photon. Tech. Lett.* 3:841–844.

Chung, Y. C., J. Jeong, and L. S. Cheng. 1994. Aging induced wavelength shifts in 1.5 μm DFB lasers. *IEEE Photon. Tech. Lett.* 6:792–795.

daSilva, V., and J. R. Simpson. 1994. Comparison of Raman efficiencies in optical fibers. In *Technical Digest Optical Fiber Communications Conference, San Jose, CA.* Paper WK13.

daSilva, V., D. L. Wilson, G. Nykolak, J. R. Simpson, P. F. Wysocki, P. B. Hansen, D. J. DiGiovanni, P. C. Becker, and S. G. Kosinski. 1995. Remotely pumped erbium-doped fiber amplifiers for repeaterless submarine systems. *IEEE Photon. Tech. Lett.* 7(9):1081–1083.

Dorgeuille, F., and F. Devaux. 1994. On the transmission performance and the chirp parameter of a multiple-quantum-well electroabsorption modulator. *IEEE J. Quantum Electron.* 30:2565–2572.

Duff, D. G. 1984. Computer-aided design of digital lightwave systems. *J. Select. Areas Commun.* 2:171–185.

Durhuus, T., C. Joergensen, B. Mikkelsen, R. J. S. Pedersen, and K. E. Stubkjaer. 1994. All optical wavelength conversion by SOA's in a Mach–Zehnder configuration. *IEEE Photon. Tech. Lett.* 6:53–55.

Edagawa, N., Y. Yoshida, H. Taga, S. Yamamoto, and H. Wakabayashi. 1990. 12300 ps/nm, 2.4 Gb/s nonregenerative optical fiber transmission experiment and effect of transmitter phase noise. *IEEE Photon. Tech. Lett.* 2:274–276.

Fells, J. A. J., M. A. Gibbon, G. H. B. Thompson, I. H. White, R. V. Penty, A. P. Wright, R. A. Saunders, C. J. Armistead, and E. M. Kimber. 1995a. Chirp and system performance of integrated laser modulators. *IEEE Photon. Tech. Lett.* 7(11):1279–1281.

Fells, J. A. J., M. A. Gibbon, G. H. B. Thompson, I. H. White, R. V. Penty, A. P. Wright, R. A. Saunders, C. J. Armistead, and E. M. Kimber. 1995b. Improving the system performance of integrated MQW laser modulators with negative chirp. In *Conference on Optical Fiber Communication (OFC'95), San Diego.* Paper TuF3, 23–24.

Fishman, D. A. 1990. Elusive bit-error-rate floors resulting from transient partitioning in 1.5 μm DFB lasers. *J. Lightwave Tech.* 8:634–641.

Fishman, D. A. 1991. Performance of single-electrode 1.5-μm DFB lasers in noncoherent FSK transmission. *J. Lightwave Tech.* 9:924–930.

Fishman, D. A. 1993. Design and performance of externally modulated 1.5-μm laser transmitter in the presence of chromatic dispersion. *J. Lightwave Tech.* 11:624–632.

Fishman, D. A., D. G. Duff, and J. A. Nagel. 1990. Measurement and simulation of multipath interference for 1.7 Gb/s lightwave transmission systems using single- and multifrequency lasers. *J. Lightwave Tech.* 8:894–905.

Fishman, D. A., S. Lumish, N. M. Denkin, R. R. Schulz, S. Y. Chai, and K. Ogawa. 1986. 1.7 Gb/s lightwave transmission field experiments. In *Conference on Optical Fiber Communication (OFC'86), Atlanta.* Paper PDP-11, PD11-1– PD11-5.

Fishman, D. A., and J. A. Nagel. 1993. Degradations due to stimulated Brillouin scattering in multigigabit intensity-modulated fiber-optic systems. *J. Lightwave Tech.* 11:1721–1728.

Giles, C. R., and E. Desurvire. 1991. Propagation of signal and noise in concatenated erbium-doped fiber optical amplifiers. *J. Lightwave Tech.* 9(2):147–154.

Gnauck, A. H., S. K. Korotky, J. J. Veselka, J. A. Nagel, C. T. Kemmerer, W. J. Minford, and D. T. Moser. 1991. Dispersion penalty reduction using an optical modulator with adjustable chirp. *IEEE Photon. Tech. Lett.* 3:916–918.

Hakki, B. W., F. Bosch, and S. Lumish. 1989. Dispersion and noise of 1.3-μm multimode lasers in microwave digital systems. *J. Lightwave Tech.* 7:804–812.

Hamide, J. P., P. Emplit, and J. M. Gabriagues. 1990. Limitations in long haul IM/DD optical fiber systems caused by chromatic dispersion and nonlinear Kerr effect. *Electron. Lett.* 26:1451–1453.

Hansen, P. B., L. Eskildsen, S. G. Grubb, A. M. Vengsarkar, S. K. Korotky, T. A. Strasser, J. E. J. Alphonsus, J. J. Veselka, D. J. DiGiovanni, D. W. Peckham, E. C. Beck, D. Truxal, W. Y. Cheung, S. G. Kosinski, D. Gasper, P. F. Wysocki, V. L. daSilva, and J. R. Simpson. 1995. 2.488 Gb/s unrepeatered transmission over 529 km using remotely pumped post- and pre-amplifiers, forward error correction, and dispersion compensation. *Electron. Lett.* 31(7):1460.

Heismann, F., D. A. Gray, B. H. Lee, and R. W. Smith. 1994. Electrooptic polarization scramblers for optically amplified long-haul transmission systems. *IEEE Photon. Tech. Lett.* 6(9):1156–1158.

Hill, G. R., P. J. Chidgey, F. Kaufhold, T. Lynch, O. Sahlen, M. Gustavsson, M. Janson, B. Lagerstrom, G. Grasso, F. Meli, S. Johansson, J. Ingers, L. Fernandez,

S. Rotolo, A. Antonielli, S. Tebaldini, E. Vezzoni, R. Caddeudu, N. Caponio, F. Testa, A. Scavennec, M. J. O'Mahony, J. Zhou, A. Yu, W. Sohler, U. Rust, and H. Herrmann. 1993. A transport network layer based on optical network elements. *J. Lightwave Tech.* 11:667–679.

Horiuchi, Y., S. Yamamoto, and S. Akiba. 1993. Highly accurate fault localization over 4500 km optical amplifier system using coherent Rayleigh backscatter reflectometry. In *Proceedings of ECOC'93, Montreux, Switzerland*, Paper MoC1.2, 5–8.

Humblet, P. A. 1991. On the bit error rate of lightwave systems with optical amplifiers. *J. Lightwave Tech.* 9(11):1576–1582.

Jensen, R. A., C. R. Davidson, D. L. Wilson, and J. K. Lyons. 1994. A novel technique for monitoring long haul undersea optical amplifier systems. In *Conference on Optical Fiber Communication (OFC'94)*. Paper ThR3, 256–257.

Johnson, J. E., T. Tanbun-Ek, Y. K. Chen, D. A. Fishman, R. A. Logan, P. A. Morton, S. N. G. Chu, A. Tate, A. M. Sergent, P. F. Sciortino, Jr., and K. W. Wecht. 1994. Low-chirp integrated ea-modulator/DFB laser grown by selective-area MOVPE. In *14th IEEE International Semiconductor Laser Conference, Maui, Hawaii, September.* Paper M4.7, 41–42.

Kasper, B. L. 1988. Receiver design. In *Optical fiber telecommunications II*, ed. S. E. Miller and I. P. Kaminow, 689–722. Boston: Academic Press.

Kataki, Y., and H. Soda. 1995. Time-resolved chirp measurement of modulator-integrated DFB LD by using a fiber interferometer. In *Conference on Optical Fiber Communication (OFC'95), San Diego.* Paper FC4, 310–311.

Kataoka, T., Y. Miyamoto, K. Hagimoto, K. Wakita, and I. Kotaka. 1992. Ultrahigh-speed driverless MQW intensity modulator, and 20 Gb/s 100 km transmission experiments. *Electron. Lett.* 28:897–898.

Koch, T. L., and J. E. Bowers. 1984. Nature of wavelength chirping in directly modulated semiconductor lasers. *Electron. Lett.* 20:1038–1040.

Korotky, S. K., P. D. Hansen, L. Eskildsen, and J. J. Veselka. 1995. Scheme for suppression of SBS. In *Technical Digest IOOC'95, Hong Kong.* Paper WD2-1, 110–111.

Kuo, C. Y., M. L. Kao, J. S. French, R. E. Tench, and T. W. Cline. 1990. 1.55 μm, 2.5 Gb/s direct detection repeaterless transmission of 160 km nondispersion shifted fiber. *IEEE Photon. Tech. Lett.* 2:911–913.

Langley, L. N., and K. A. Shore. 1992. The effect of external optical feedback on the turn-on delay statistics of laser diodes under pseudorandom modulation. *IEEE Photon. Tech. Lett.* 4:1207–1209.

Lefranc, E., B. Le Mouel, G. Bourret, B. Gherardi, and J. Chensnoy. 1993. Different supervisory systems in underseas equipments using optical amplification. In *Suboptic '93, Palais des Congres de Versailles, France.*

Lichtman, E. 1995. Limitations imposed by polarization-dependent gain and loss on all-optical ultra-long communication systems. *J. Lightwave Tech.* 13(5):906–913.

Linke, R. A. 1985. Modulation induced transient chirping in single frequency lasers. *IEEE J. Quantum Electron.* 21:593–597.

Marcuse, D. 1984. Computer simulation of laser photon fluctuations: Single-cavity laser results. *IEEE J. Quantum Electron.* 20:1148–1155.

Marcuse, D. 1990. Derivation of analytical expressions for the bit-error probability in lightwave systems with optical amplifiers. *J. Lightwave Tech.* 8(12):1816–1823.

Marcuse, D., A. R. Chraplyvy, and R. W. Tkach. 1991. Effect of fiber nonlinearity on long-distance transmission. *J. Lightwave Tech.* LT-9:356–361.

Marcuse, D., and T. H. Wood 1994. Time-dependent simulation of a laser–modulator combination. *IEEE J. Quantum Electron.* 30:2543–2555.

Mazurczyk, V. J., and D. G. Duff. 1995. Effect of intersymbol interference on signal-to-noise measurements. In *Technical Digest Optical Fiber Communications Conference, San Diego, CA.* Paper WQ1, 188–189.

Mazurczyk, V. J., and J. L. Zyskind. 1993. Polarization hole burning in erbium doped fiber amplifiers. In *CLEO'93, Baltimore.* Post deadline paper CPD26, 52–53. (Technical Digest Series, Vol. 11.)

McDonald, A. J., R. S. Fyath, and J. J. O'Reilly. 1989. Influence of extinction ratio on performance of optical receivers incorporating laser preamplifiers. *Electron. Lett.* 25:249–250.

Mikkelsen, B., T. Durhuus, C. Joergensen, S. L. Danielsen, R. J. S. Pedersen, and K. E. Stubkjaer. 1996. Wavelength conversion devices. In *Conference on Optical Fiber Communication (OFC'96), San Jose.* Paper WG1, 121–122.

Miller, S. E., and I. P. Kaminow. 1988. *Optical fiber telecommunications II.* Boston: Academic Press.

Ogata, T., S. Nakaya, Y. Aoki, T. Saito, and N. Henmi. 1992. Long-distance, repeat-erless transmission utilizing stimulated Brillouin scattering suppression and dispersion compensation. In *4th Optoelectronics Conference, Makuhari Messe, Japan.* Paper 16A4-3, 104–105.

Olsson, N. A. 1989. Lightwave systems with optical amplifiers. *J. Lightwave Tech.* 7:1071–1082.

Park, Y. K., and S. W. Granlund. 1994. *J. Opt. Fiber Technology: Materials, Devices, and Systems* 1(1):59–71.

Poole, C. D., and C. R. Giles. 1988. Polarization-dependent pulse compression and broadening due to polarization in dispersion-shifted fiber. *Opt. Lett.* 13:155–157.

Poole, C. D., R. W. Tkach, A. R. Chraplyvy, and D. A. Fishman. 1991. Fading in lightwave systems due to polarization-mode dispersion. *IEEE Photon. Tech. Lett.* 3:68–70.

Reichmann, K., P. D. Magill, U. Koren, B. I. Miller, M. Young, M. Newkirk, and M. D. Chien. 1993. 2.5 Gb/s transmission over 674 km at multiple wavelengths using a tunable DBR laser with an integrated electroabsorption modulator. *IEEE Photon. Tech. Lett.* 5:1098–1100.

Rühl, F. F., and R. W. Ayre. 1993. Explicit expressions for the receiver sensitivity and system penalties of optically preamplified direct-detection systems. *IEEE Photon. Tech. Lett.* 5(3):328–331.

Runge, P. K., and N. S. Bergano. 1988. Undersea cable transmission systems. In *Optical fiber telecommunications II,* ed. S. E. Miller and I. P. Kaminow, 879–909. Boston: Academic Press.

Sessa, W. B., and R. E. Wagner. 1992. Frequency stability of DFB lasers used in FDM multi-location networks. In *Conference on Optical Fiber Communication (OFC'92), San Jose.* Paper ThC3, 202.

Smith, R. G., and S. D. Personick. 1980. Receiver design for optical communications systems. In *Topics in applied physics.* Vol. 39, *Semiconductor devices for optical communications,* ed. H. Dressel, 89–160. New York: Springer-Verlag.

Suzuki, N., and Y. Hirayama. 1995. Comparison of effective α parameters for multiquantum-well electroabsorption modulators. *IEEE Photon. Tech. Lett.* 7:1007–1009.

Suzuki, N., and T. Ozeki, 1993. Simultaneous compensation of laser chirp, Kerr effect, and dispersion in 10 Gb/s long-haul transmission systems. *J. Lightwave Tech.* 11:1486–1494.

Taga, H., S. Yamamoto, N. Edagawa, Y. Yoshida, S. Akiba, and H. Wakabayashi. 1994. The experimental study of the effect of fiber chromatic dispersion upon IM-DD ultra-long distance optical communication systems with Er-doped fiber amplifiers using a 1000 km fiber loop. *J. Lightwave Tech.* 12(8):1455–1461.

Tiemeijer, L. F., P. I. Kuindersma, G. P. J. M. Cuijpers, P. J. A. Thijs, T. V. Dongen, J. J. M. Binsma, E. J. Jansen, and S. Walczyk. 1995. 102km 38.6 dB Budget 10 Gb/s NRZ repeaterless transmission at 1310 nm using a semiconductor booster amplifier module and a semiconductor preamplifier module. In *Proceedings of the 21st European Conference on Optical Communications (ECOC'95), Brussels.* Paper Tu.B2.4, 275–278.

Tiemeijer, L. F., P. J. A. Thijs, T. V. Dongen, J. J. M. Binsma, E. J. Jansen, and A. J. M. Verboven. 1994. 27-dB Gain unidirectional 1300-nm polarization-insensitive multiple quantum well laser amplifier module. *IEEE Photon. Tech. Lett.* 6:1430–1432.

Tkach, R. W., A. R. Chraplyvy, F. Forghiere, A. H. Gnauck, and R. M. Derosier. 1995. Four-photon mixing and high-speed WDM systems. *J. Lightwave Tech.* 13(5):889–897.

Trischitta, P. R., and E. L. Varma. 1989. In *Jitter in digital transmission systems,* chapter 2. Norwood, MA: Artech House.

Uomi, K., A. Murata, S. Sano, R. Takeyari, and A. Takai. 1992. Advantage of 1.55 InGaAs/InGaAsP MQW-DFB lasers for reducing waveform degradation and dispersion penalty for 2.5 Gb/s long-span normal fiber transmission. *IEEE Photon. Tech. Lett.* 4:657–660.

Velschow, B., L. D. Pedersen, C. G. Joergensen, M. Hogdal, and M. Joergensen. 1995. Comparison of electroabsorption and Mach–Zehnder modulators for more than 300 km of 2.488-Gb/s unrepeatered transmission. In *Conference on Optical Fiber Communication (OFC'95), San Diego.* Paper WL5, 151–152.

Verdiell, J-M., M. A. Newkirk, T. L. Koch, R. P. Gnall, U. Koren, B. I. Miller, and B. Tell. 1993. A frequency reference photonic integrated circuit for WDM with low polarization dependence. *IEEE Photon. Tech. Lett.* 4:451–456.

Vodhanel, R. S., M. Krain, R. E. Wagner, and W. B. Dessa. 1994. Long-term wavelength drift of the order of −0.01 nm/yr for 15 free-running DFB laser modules. In *Conference on Optical Fiber Communication (OFC'94), San Jose.* Paper WG5, 103–104.

Wedding, B. 1992. New method for optical transmission beyond the dispersion limit. *Electron. Lett.* 29:1298–1300.

Winters, J., and S. Kasturia. 1992. Adaptive nonlinear cancellation for high-speed fiber-optic systems. *J. Lightwave Tech.* 10:971–977.

Yamada, K., K. Nakamura, Y. Matsui, T. Kunii, and Y. Ogawa. 1995. Negative-chirp electroabsorption modulator using low-wavelength detuning. *IEEE Photon. Tech. Lett.* 10:1157–1158.

Yamamoto, S., M. Kuwazuru, H. Wakabayashi, and Y. Iwamoto. 1987. Analysis of chirp power penalty in 1.55-μm DFB-LD high speed optical fiber transmission systems. *J. Lightwave Tech.* 5:1518–1524.

Yanagita, H., K. Itoh, E. Ishikawa, H. Aoki, and H. Toratani, 1995. 26dB Amplification at 1.31 μm in a novel Pr^{3+}-doped InF_3/GaF_3-based fiber. In *Conference on Optical Fiber Communication (OFC'95), San Diego.* Paper PD-2.

Yanming, L., A. J. Antos, and A. Newhouse. 1996. Large effective area dispersion-shifted fibers with dual-ring index profiles. In *Conference On Optical Fiber Communication (OFC'96), San Jose.* Paper WK15, 165–166.

Chapter 4 | Laser Sources for Amplified and WDM Lightwave Systems

Thomas L. Koch

Lucent Technologies, Bell Laboratories, Holmdel, New Jersey

I. Introduction

The long transmission spans associated with amplified lightwave systems place severe requirements on the spectral and modulation characteristics of laser transmitters. Similarly, wavelength-division multiplexing (WDM) places an entirely new set of constraints on spectral stability associated with longitudinal mode selection and wavelength accuracy, wavelength stability with aging, and even the desire for advanced features such as electronic wavelength channel selection or simultaneous single-chip multi-channel transmission. These new requirements have led to a number of significant studies and refinements of existing laser technologies, and in several instances have provided a driver for the development and deployment of guided-wave integration technology known as *photonic integrated circuits* (*PICs*). This chapter briefly reviews the system requirements for sources used in long-haul amplified transmission and WDM transmission, and then examines the source technologies that have been explored or deployed to address these requirements.

II. Low-Chirp Transmission Sources

A. *SYSTEM REQUIREMENTS FOR AMPLIFIED TRANSMISSION SOURCES*

The advent of erbium-doped fiber amplifiers (EDFAs) has allowed for the analog boosting of unregenerated digital optical signals over extraordinary distances. In the case of transoceanic distances, fiber with carefully tailored

115

Copyright © 1997 by Lucent Technologies.
All rights of reproduction in any form reserved.
ISBN: 0-12-395171-2

dispersion characteristics is required to carefully control the pulse-distorting effects of linear dispersion and nonlinear propagation. In terrestrial systems with distances of 50–1000 km, a common situation involves the reuse or upgrade of existing installed cable that has a dispersion zero at a wavelength of 1.3 μm. With 1.5-μm sources mandated by the compelling combination of minimum fiber loss and Er-fiber amplification, dispersive distortion quickly becomes intolerable in both cases for sources that have a substantial optical bandwidth beyond the transform limit associated with the intensity-modulated envelope waveform of the encoded information. For a purely envelope-modulated carrier, simple numerical linear dispersive propagation modeling reveals that a 1-dB eye closure will occur at a limit of

$$B^2L \approx 6000 \cdot \left(\frac{17 \text{ ps/nm-km}}{D}\right) \text{km/s}^2, \tag{4.1}$$

where D is the actual dispersion parameter of the fiber of length L, and the bit rate is B. For 1.55-μm propagation in conventional fiber with $D = 17$ ps/nm-km, a 2.5-Gb/s signal will travel approximately 960 km before degrading the eye by 1 dB.

The ability to approach this limit is then strongly affected by transmitter *chirp*, which refers to any excursions of the carrier frequency during the digital bit stream, most commonly and problematically on time scales of one bit and less. These excursions are predominantly of a deterministic nature and have been well known in both directly modulated distributed feedback (DFB) lasers and externally modulated sources. The goal has therefore been to optimize the design of directly modulated lasers to provide for a low-cost, low-chirp source, or to engineer a low-chirp externally modulated product with simple packaging, good reliability, and reasonable cost. Both directly modulated lasers and externally modulated lasers have commonly been characterized by their *chirp parameter*, or *linewidth enhancement factor*, α, and simple modeling [1] suggests that the B^2L product in Eq. (4.1) will be reduced for a nonideal transmitter by a factor of approximately $\sqrt{1 + \alpha^2}$. However, in some cases external modulation has been used to intentionally chirp pulses in a specific manner so as to provide for a modest linear pulse compression that can *aid* in receiver sensitivity.

Technologies that have been brought to bear on low-chirp transmitters include DFB lasers optimized to reduce α from typical 1.55-μm bulk-active-layer laser values of about 6 to values of about 2 in properly engineered and detuned strained quantum well (QW) designs. Such sources may be suitable for distances approaching 200 km in standard fiber at synchronous optical network (SONET) or synchronous digital hierarchy (SDH) bit rates of 2.5 Gb/s (optical carrier [OC]-48). For applications requiring longer spans, external modulation is necessary. In such cases, LiNbO$_3$ interfero-metric designs can offer essentially perfect performance and may be suitable when package size and transmitter complexity are not serious constraints. Recently, semiconductor electroabsorption (EA) modulators have also proven to be relatively high performance, and they have the distinct advantage of being monolithically integratable with DFB lasers for a compact and robustly packaged low-chirp source. Both the external LiNbO$_3$ modulator and the integrated DFB–EA modulator have shown the capability of transmitting 2.5-Gb/s data over amplified spans of conventional fiber for distances extending beyond 600 km and approaching the transform-limited 1-dB penalty distances of about 1000 km. This chapter discusses these technologies in some detail. Although chirp is the primary new criterion, other factors influencing the desirability of a particular source for amplified transmission include output power and a requirement that the linewidth not be so excessively large as to cause phase-noise-induced stochastic dispersive effects. Typically this latter requirement is met by most DFB lasers with linewidths less than approximately 50 MHz.

B.   DIRECT MODULATION OF DFB LASERS

Although directly modulated DFB lasers are known to have inherent chirp associated with the transients that constitute the information coding, this chirp can be minimized by proper design. A particularly simple and useful model of chirping predicts that frequency excursions $\Delta\nu(t)$ are simply re-lated to the time-dependent output power $P(t)$ by the following relation [1]:

$$\Delta\nu(t) = \frac{\alpha}{4\pi}\left\{\frac{d}{dt}\ln[P(t)] + \kappa P(t)\right\}. \tag{4.2}$$

In this equation, the linewidth enhancement factor, α, is related to changes in the complex index of refraction of the gain medium $n = n_{real} + in_{imag}$, with the carrier density N as

$$\alpha = \left(\frac{\partial n_{real}}{\partial N}\right)\bigg/\left(\frac{\partial n_{imag}}{\partial N}\right), \tag{4.3}$$

and κ is related to nonlinear gain-compressing effects at a fixed carrier density [1]. This expression is effective at describing chirping, especially that associated with relaxation oscillation overshoots and the sharp turn-on and turn-off transients at digital pulse edges. However, it does ignore phenomena such as longitudinally nonuniform gain saturation (spatial hole burning) in DFB lasers that effectively detunes portions of the grating relative to others and thus actually alters the modal structure of the resonator dynamically. Although these effects can dominate the frequency-modulated (FM) response in the 10- to 100-MHz band and are important for analog transmission in dispersive systems, the fundamental transient chirping governed by Eq. (4.2) is of most importance for digital transmission.

From Eq. (4.2) it is clear that a smaller α reduces the largest contributions to chirp. Because $\partial n_{imag}/\partial N$ is the differential gain of the gain medium, a higher differential gain will generally produce lower chirp. The induced changes in real index tend to be smaller in magnitude and more distributed in optical frequency (as might be expected from the Kramers–Kronig relation between real and imaginary index (changes); thus, compensating effects from the denominator in Eq. (4.3) are usually not large enough to nullify any expected reductions in α.

The steplike density of states for QW gain media is well known to improve the differential gain in semiconductor lasers at low carrier densities [2]. This suggests a lower α for QW or multiple quantum well (MQW) lasers. Because MQW lasers typically operate with less state filling in each well, they have a large differential gain and thus a smaller α than those of single QW lasers [2]. This alone produces an α value for MQW DFB lasers at 1.55 μm of about 3, compared with approximately 5–6 for bulk 1.55-μm DFB lasers [3]. As pointed out in the system requirements, this would result in nearly a twofold improvement in transmission distance for MQW DFB lasers.

A further reduction in α can be realized by incorporating strained QWs. Compressive strain produces a lower effective mass in the valence band, which results in complete inversion at lower pump levels (see Chapter 5). In addition to producing very low I_{th} this further increases the differential gain.

Ketelson et al. [4] have explored these and other means for reducing α in 1.55-μm DFB lasers for digital transmission. The differential gain always decreases for increasing gain due to sublinearity in the gain versus carrier density relation. This suggests that a reduced cavity loss will generally result in an increased differential gain and thus a lower α. Therefore, longer cavity DFB lasers, where the distributed effective output coupling loss is decreased, should reduce chirping. Figure 4.1 shows a comparison of the

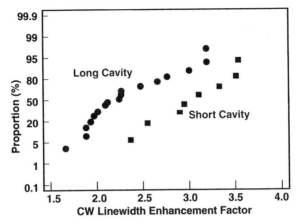

Fig. 4.1 Experimental distribution of measured α factors from two populations of distributed feedback (DFB) lasers that differ only in cavity length. CW, continuous wave.

experimentally evaluated α values for populations of longer and shorter DFB lasers, clearly verifying this trend.

The differential gain can also be improved by p-doping in the active layer, or δ-doping at the edge of the active layer, to provide high hole concentrations even without pumping. Such structures have been advanta-

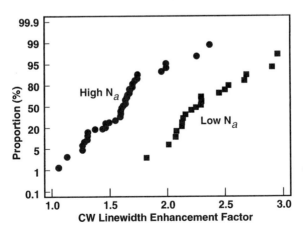

Fig. 4.2 Experimental distribution of measured α factors from two populations of DFB lasers that differ only in p-doping of the active layer.

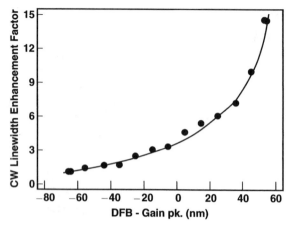

Fig. 4.3 Experimentally measured α factor of DFB lasers as a function of detuning of the lasing wavelength from the gain peak.

geously used to attain a high-speed modulation response by extending the relaxation oscillation frequency, which varies as $\sqrt{\partial n_{imag}/\partial N}$. Figure 4.2 displays the experimentally measured α factor for DFB lasers with different levels of p-doping in the active layers, again verifying the expected trend [4].

Finally, because lower energy states fill before higher energy states with increasing pumping, the differential gain will always be lower at lower energy, or longer wavelengths, than the gain peak. Conversely, at shorter wavelengths than the gain peak, the differential gain is higher, and α will be lower. Thus by intentionally detuning the grating pitch of DFB lasers to ensure lasing on the blue side of the gain peak, one can also achieve a lower α. Figure 4.3 displays the effect of detuning on the α value of DFB lasers, clearly illustrating the expected effect [4]. The challenge in this case is to provide sufficiently good antireflection coatings to suppress Fabry-Perot lasing at the gain peak rather than DFB lasing at the detuned, lower gain wavelength away from the gain peak.

C. EXTERNAL MODULATION

The previous section illustrates how careful optimization of epitaxy and laser design can reduce the α factor to allow transmission using direct modulation of DFB lasers over conventional 1.3-μm dispersion-zero fiber to distances of about 200 km, or even longer. However, it becomes exceedingly difficult to completely eliminate the index modulation that accompanies the temporal gain excursions that are inherent to direction intensity modulation. At 2.5 Gb/s, for distances of 150 km or longer, external modulation is advantageous.

External modulation affords the possibility of information encoding independent of the physics associated with the gain medium of a laser. The two common modulation technologies employed for high-speed digital telecommunications are $LiNbO_3$ traveling-wave Mach–Zehnder modulators and, more recently, semiconductor EA modulators.

Figure 4.4 shows a typical $LiNbO_3$ electrooptic Mach–Zehnder modulator, where a coplanar electrical transmission line allows for some increase in modulation bandwidth by partially matching the optical phase velocity to the electrical phase velocity of the information-encoding drive signal. In principle, if phase-velocity matching were perfect and there were no appreciable microwave drive or optical propagation losses, ever-longer devices would yield ever-smaller drive voltage rquirements. Typical phase-velocity mismatches, however, result in devices requiring about 5 V of drive for 2.5-Gb/s operation. Recent work has shown that very thick metallization (>50–100 μm) can improve the microwave phase-velocity matching and thus allow reduced drive voltages in high-speed modulators [5].

The principal attraction of external Mach–Zehnder modulators lies in their ability to produce nearly perfect intensity-modulated waveforms devoid of any excess phase modulation or chirp. In a typical low-chirp applica-

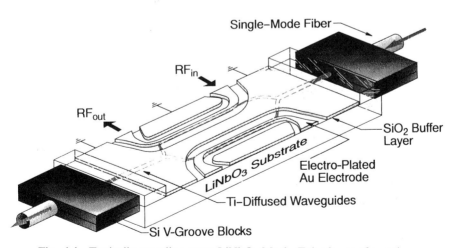

Fig. 4.4 Typically traveling-wave $LiNbO_3$ Mach–Zehnder configuration.

tion, the Mach–Zehnder is driven in a push–pull configuration, where one arm of the interferometer is driven to an increased index while the other arm is driven to a lower index; this yields an expression for the field of the form

$$E(t) \propto [e^{i\delta\varphi(t)} + e^{-i\delta\varphi(t)}] = \cos[\delta\varphi(t)], \tag{4.4}$$

where $\delta\varphi(t) \propto V(t)$, the applied digital voltage, through the electrooptic effect. For a modulator, the chirp parameter, α, can be defined in a manner analogous to that for the directly modulated laser, as

$$\alpha \equiv \frac{4\pi \, \Delta\nu(t)}{\dfrac{d}{dt} \ln P(t)} = \frac{2(d\phi dt)}{\dfrac{d}{dt} \ln P(t)}, \tag{4.5}$$

where $\phi(t)$ is the phase of the modulator's output optical field of power $P(t)$. Because the field represented by a pure envelope modulation has constant phase $\phi(t)$, the effective α by Eq. (4.5) is zero. However, if both arms are driven in phase rather than push–pull, the electric field in Eq. (4.4) becomes $E(t) \propto 2e^{i\delta\varphi(t)}$, which results in pure phase modulation, or $\alpha \to \infty$ according to Eq. (4.5). Because the relative magnitude and sign of the drive applied to each arm can be varied, appropriate choices can yield α factors of any value (at least in small signal), thus the Mach–Zehnder affords the possibility of an adjustable α-factor modulator. Extensive work has been carried out using LiNbO$_3$ Mach–Zehnder modulators with varying drive conditions to map out the impact of different chirp parameters (α) on high-speed, long-distance dispersive transmission. Figure 4.5 shows that a small negative (i.e., nonzero) chirp parameter can actually improve transmission by yielding modest pulse-compressing effects that tend to improve the eye margin of single ones bits and hence improve receiver sensitivity [6].

The principal drawbacks of LiNbO$_3$ modulators are the cost associated with an additional, separately packaged, fiber-coupled device, usually requiring polarization-maintaining fiber, and the sheer size of the additional device on a transmitter board. Some concerns also remain regarding potential long-term drifts of the bias voltages.

Semiconductor EA modulators rely on electric-field-induced absorption in guided-wave structures. In bulk absorbing layers this is termed the *Franz–Keldysh effect*, illustrated in Fig. 4.6 as a spatial quantum tunneling phenomenon that allows band-to-band absorption in a field-induced band-tilting situation, even with nominally insufficient energy to complete the transition at a fixed position in space. In QWs, enhanced confinement and excitonic effects alter this situation and yield significant enhancements in the respon-

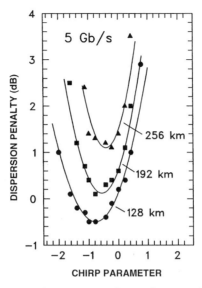

Fig. 4.5 Transmission penalty versus α factor for an adjustable α-factor Ti: LiNbO$_3$ modulator at 5 Gb/s.

sivity of the EA effects. EA in QWs is termed the *quantum-confined Stark effect* (*QCSE*). Closed-form analytical expressions for the Franz–Keldysh extinction have been evaluated [7], and Aspnes [8] has calculated the associated electrorefractive expressions. Combining these results allows an analyt-

Fig. 4.6 Below bandgap band-to-band absorption in the Franz–Keldysh effect by field-induced tunneling.

ical expression for the chirp parameter of a bulk Franz–Keldysh EA modulator, given by

$$\alpha = \left[\frac{\dfrac{\partial^2}{\partial\eta^2}\left(Ai(\eta)Bi(\eta)\right)}{\dfrac{\partial^2}{\partial\eta^2}\,Ai(\eta)^2)} \right], \tag{4.6}$$

where $Ai(\eta)$ and $Bi(\eta)$ are Airy functions, and

$$\eta = \left(\frac{E_g - \hbar\omega}{\hbar\theta}\right) \quad \text{and} \quad \hbar\theta = \left(\frac{e^2\hbar^2 F^2}{2\mu}\right)^{1/3} \tag{4.7}$$

for field strength F in volts per centimeter. Evaluation of this expression illustrates the point that the effective chirp parameter for EA is inherently a small-signal concept and does not fully represent the transmission performance of a device in a single number. Figure 4.7 shows the calculated extinction for bulk absorbing layers versus wavelength for various applied fields, in addition to the variation of α with wavelength for a particular field as evaluated by

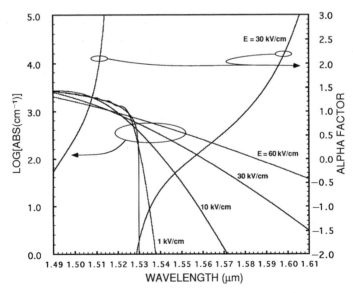

Fig. 4.7   Calculated extinction coefficient versus wavelength for a family of applied fields for 1.53-μm bandgap material. Also shown is the theoretical α factor for one field strength (30 kV/cm).

Eq. (4.6). Figure 4.8 shows the variation of α with applied field for various wavelengths. In this case, important observations are that small residual phase modulation is indeed possible, but that the effective α may be positive, may be negative, and can even change sign during the modulation cycle. Similar results occur for QCSE devices. Figure 4.9 shows the time-resolved wavelength of the output of a QCSE modulator for different biases at 10 Gb/s, clearly showing the magnitude of wavelength excursions reducing and eventually changing sign as the bias or zero-drive extinction is increased [9].

As a stand-alone device, semiconductor EA modulators offer compact size and low drive voltages. Typical lengths are less than 300 μm, and drive voltages are typically 2.0–4.0 V for bulk devices and 1.5–2.5 V for QCSE devices. The short lengths allow bulk-element electrical contacting with low enough capacitance to yield modulation bandwidths as high as 40 GHz [10]. The small-core semiconductor waveguides usually grown, and chosen to optimize the low drive voltages, require very large numerical aperture optics and

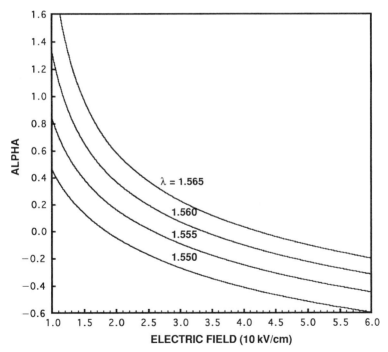

Fig. 4.8   Theoretical α factor versus applied field for a family of different wavelengths, again for 1.53-μm bandgap material. Note that α varies strongly with the applied field and can change sign during the modulation cycle.

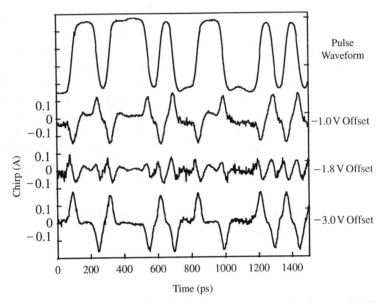

Fig. 4.9 Experimentally evaluated wavelength excursion during modulation at 10 Gb/s for various applied biases to a quantum well (QW) electroabsorption (EA) modulator. Note that the wavelength deviation changes sign as the bias is increased.

difficult alignment tolerances in packaging, as well as typical combined input and output fiber coupling losses of about 4–6 dB. Combined with the nonzero ON-state absorption loss, the net insertion loss is typically at least 6 dB. As in the case of LiNbO$_3$ devices, the stand-alone modulator requires an additional package, and the polarization sensitivity usually encountered requires some form of polarization maintenance. However, polarization-insensitive designs have been realized using combinations of compressive- and tensile-strained QWs that extinguish TE and TM light, respectively.

D. INTEGRATED LASER–EA MODULATORS

The most advantageous use of semiconductor EA modulators derives from their fabrication technology, which closely resembles that of semiconductor lasers. It has, therefore, proven realistic to fabricate PICs that combine on a single-chip a DFB laser with a bulk or QW EA modulator.

Such integrated laser–modulators totally eliminate the three principal drawbacks of discrete EA modulators: as an integrated device, the polarization state is inherently controlled at the modulator input; as an integrated device, there is no additional input and output fiber coupling insertion loss; and as an integrated device, there is no additional separate packaging cost for the modulator. The integrated laser–modulator concept thus offers the full operational simplicity of a directly modulated laser module, but the performance of an externally modulated transmitter system. The challenge in realizing this stems from the complexities of PIC fabrication technology.

Two general approaches have emerged for fabricating integrated laser modulators. They both address the complexity of generating two longitudinally distinct regions with two distinct bandgap energies along the same waveguide. The lower bandgap region, when forward biased, provides gain at a wavelength where the higher bandgap region is nominally transparent, but will absorb with a reverse bias as a result of EA. The typical difference in the photoluminescence wavelengths of the active and modulator layers is about 30–60 nm for 1.55-μm-range devices. The laser section, requiring a optical cavity independent of the EA region, is typically a DFB laser, but distributed Bragg reflector (DBR) lasers have also been integrated with EA modulators.

One approach to fabrication employs an abrupt transition from a gain medium to the EA medium. This can be realized by a "butt joint" where the gain waveguide layer is locally etched away and an EA waveguide layer is regrown, in its place, where the modulator will be. These steps are performed prior to the lateral definition of the waveguide, both for buried heterostructure guides and for ridge-guide designs. The goal is to form an appropriate longitudinal slab structure from which a conventional laser fabrication sequence can be executed. Such a structure [11] is shown in Fig. 4.10. Variants on this theme include a longitudinally uniform core waveguide with different upper layers in the gain and modulator regions, again formed by etch and regrowth techniques [12].

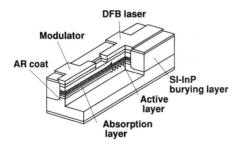

Fig. 4.10 Configuration of a bulk-active-layer integrated DFB laser–EA modulator fabricated using the butt-joint technique.

An increasingly prevalent approach is the application of selective area epitaxy, as discussed in detail in Chapter 5. This technique allows for the growth of longitudinally continuous QW active layers with varying bandgaps as controlled by laterally adjacent growth-inhibiting masks. This technique is especially simple in that the longitudinal slab is smooth, with no optical discontinuities to interfere with device operation and no physical layer discontinuities to interfere with the fabrication of the laser lateral stripe definition or current-blocking layers. This technique is inherently used with QWs because the bandgap variations result from quantum confinement shifts associated with the shifts in QW thickness. The selective area epitaxy is commonly performed with a wide enough region to allow nearly conventional laser waveguide fabrication, inlcuding lateral blocking structures, as if the device were a conventional laser. A typical selective area epitaxy integrated laser–EA modulator design [13] is shown in Fig. 4.11.

Both techniques have proven capable of generating high-performance devices, and the choice ultimately reduces to detailed manufacturing yield issues. Although the process is basically the same as combining a discrete DFB laser with a discrete modulator, there are two important additional fabrication requirements that strongly affect device performance. The first is electrical isolation between the laser and the modulator. The second is optical reflections from the modulator, which can disturb the operation of

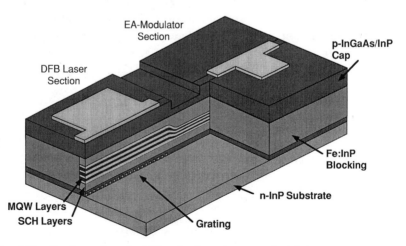

Fig. 4.11 Configuration of a QW active-layer integrated DFB laser–EA modulator fabricated using the selective area growth technique.

the laser, that would typically be avoided by using an isolated laser package in a discrete combination. A simple analysis reveals that these effects have dramatic consequences on the chirp performance, which is the principal reason for adopting external modulation. In a typical package without internal drives, the laser has a series 25- or 50-Ω matching resistor R_m for impedance matching, but otherwise electrically behaves like a 3- to 5-Ω resistor R_S when foward biased, typically with some parasitic contact and blocking layer capacitance. The modulation section, in the same typical package, is run in reverse bias, looks like a capacitor, and is usually shunted by a 25- or 50-Ω matching resistor R_m. Denoting the impedance between the laser and the modulator as Z, we have the simple equivalent circuit shown in Fig. 4.12.

Ignoring the laser capacitance for simplicity, and assuming that $R_S \ll \{R_m, Z\}$, we find that the fractional current leaking to the laser will be about R_m/Z. If we assume a typical FM efficiency of a laser to be approximately 1 GHz/mA, and a typical modulator drive to be 3 V, then a 1-kΩ isolation impedance would produce a 3-GHz frequency excursion. This is substantial compared with the information bandwidth limit of a 2.5-Gb/s signal, so more isolation is required. These effects are even more deleterious if the FM efficiency is enhanced in the region of the laser relaxation oscillation frequency. For this reason, isolation resistances of more than 10 kΩ are typically required, and values in excess of 100 kΩ can be achieved by

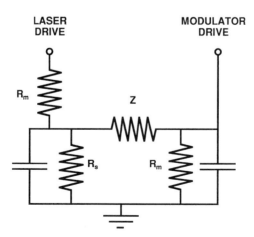

Fig. 4.12 Crude equivalent circuit to the estimate leakage current contributions to chirp in an integrated DFB laser–EA modulator photonic integrated circuit (PIC).

implantation or etching appropriate grooves in the upper contact and clad-ding layers between the laser and the modulator, as shown in Fig. 4.10 or Fig. 4.11. Clearly, the details of the device design, including doping and especially blocking layer design, will influence the electrical isolation.

Reflections from the modulator output facet can similarly influence chirp. Any residual output facet reflection produces a compound-cavity output mirror that varies in time as the modulator operates, because the modulator has an inherent chirp or phase modulation. Although phase variations in the residual reflection do change the resonant frequency of the laser, the dominant effects stem from the temporal *amplitude* variation of the com-pound mirror as the modulator goes from a transmitting state to an extin-guishing state. This is true because of the strong amplitude-phase coupling of the laser oscillator, governed by the α factor of the laser, not the modula-tor. Specifically, the extinction of the residual reflection lowers or raises the cavity Q and increases or decreases the clamped operating threshold gain or carrier density, which shifts the resonator frequency though the laser α.

An estimate of this effect can be realized by analyzing the modal fre-quency shifts of a Fabry-Perot laser with mirror reflectivity R under the influence of an external residual facet reflectivity R_1 for a compound mirror reflectivity of R_c, as shown in Fig. 4.13. Invoking a quasi-static analysis for modulation frequency components below the laser relaxation oscillation frequency, we can see that the frequency shift of the residual reflection

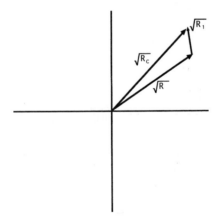

Fig. 4.13 Vector diagram showing the contribution of output coupler R and resid-ual facet reflection R_1 to effective composite output coupler R_c.

vanishes when the modulator is extinguished and reappears in the ON state. For α factors of the modulator of unity or less, the vector denoting the residual reflectivity will undergo less than a $\pi/2$ rotation before the facet reflectivity is extinguished by a factor of 10, so the net largest compound mirror reflectivity change can be approximated by $\Delta R \approx 2\sqrt{RR_1}$. The laser threshold gain change is then $\Delta g_{th} = (1/L)\sqrt{R_1/R}$. Because of the α factor of the laser gain α_G, this causes an effective index shift of $\Delta n = \alpha_G (\lambda/4\pi) \Delta g_{th}$, which gives rise to oscillation frequency shifts of

$$\Delta\nu = \frac{c\alpha_G}{4\pi n L} \sqrt{\frac{R_1}{R}}. \tag{4.8}$$

If we demand that the chirp be $\Delta\nu \ll B$, the bit rate, to achieve nearly transform-limited dispersion performance, we have

$$R_1 < \frac{16\pi^2 n^2 L^2 B^2 R}{c^2 \alpha_G^2}. \tag{4.9}$$

For application to an integrated DFB–EA modulator, we note that output coupling with $\kappa L \sim 2$ is comparable to that of a conventional cleaved-facet reflectivity $R \sim 0.3$. We then find that for typical laser parameters at 2.5 Gb/s, low reflections of $R_1 < 1 \times 10^{-3}$ are required for high-yield, nearly transform-limited operation. It is interesting to note that the required value of R_1 is a strong function of the α factor of the *laser*, not the modulator.

In addition to high-quality antireflection coatings, "window" structures have often been employed to achieve a low R_1 [14]. In this case, the modulator output waveguide terminates within the semiconductor. The coupling optics can still image and couple to the output waveguide through the uniform semiconductor material, but the freely expanding facet-reflected beam does not couple effectively into the waveguide. This high return loss can easily yield a 10-fold reduction in facet reflectivity below the artireflection-coated value.

Experimentally, the chirp of integrated laser–EA modulators has been measured by a number of techniques. The most complete measurement, as discussed in Chapter 3, is a full time-resolved frequency excursion determination under modulation. Examples of such measurements were shown in Fig. 4.9 for 10-Gb/s operation in an integrated MQW laser–EA modulator chip. This figure also displays the effective change in α factor as the modulator bias is varied, as discussed theoretically with regard to bulk devices. At 2.5 Gb/s, frequency excursions below 2 GHz are readily achieved, which allows transmission over distances approaching the transform limits of approximately 1000 km, as discussed in Section I.

III. WDM Sources

A. SYSTEM REQUIREMENTS FOR WDM SOURCES

WDM systems are initially proving economical for long-haul applications, where they allow not only lower cost repeaters using EDFAs without regeneration, but also the ability to upgrade gracefully at the terminals by adding incremental capacity with each added channel. Because these applications entail long transmission spans, the initial applications of WDM face, in many instances, all the requirements noted previously for amplified transmission, in addition to those related to WDM. However, the obvious requirements specific to WDM, regardless of distance, revolve around spectral stability.

Channel allocations for dense WDM systems are determined by complex engineering trade-offs relating both to transmission effects and to technology constraints at the transmitter and demultiplexers or receivers. Forces that encourage increased channel spacing include practical limits on source wavelength accuracy and long-term drifts, practical limits on the passband flatness and adjacent channel rejection in demultiplexers, nonlinear transmission effects that cause channel interactions over long fiber spans, and the larger inherent per-channel bandwidths required if time-division multiplexed (TDM) upgrades are anticipated. Forces that encourage narrow channel spacing include bandwidth efficiency of net WDM system capacity (larger number of channels), simplified amplifier gain equalization in amplified systems, and subtle advantages in source integration technology related to materials optoelectronic properties.

A balance among these forces is clearly application dependent, but the International Telecommunications Union (ITU) standards body is currently formulating a channel allocation grid with a 100-GHz channel spacing (~0.81 nm at a 1550-nm wavelength) on even 100-GHz intervals, counting nominally in both directions from 193.1 THz. Thus in the 1.55-μm band near the concatenated EDFA gain peak for long-haul systems, the grid appears as in Fig. 4.13. Some research consortia have explored systems configurations with channel spacings as narrow as 50 GHz (~0.4 nm), but a number of commercial applications have already adhered to the 100-GHz proposal, in some cases selecting out a 200-GHz comb of eight channels (spacing of ~1.61 nm), as highlighted in Fig. 4.14. For the remainder of this chapter, we examine sources with the expectation that they will be required to remain stable at the frequencies specified in Fig. 4.14.

With channel spacings of 200 GHz, the available demultiplexer technologies, such as those described in Chapter 8, require the sources to remain at the specified channel allocations with an accuracy of better than approximately ±25–50 GHz. This figure encompasses all sources of drift, including laser aging effects, effects associated with packaging and long- and short-term temperature stability of the laser chip, and effects related to any circuitry or optics that might be considered to stabilize or control the transmitter module.

Another criterion for WDM systems may become the ability of the source to address more than one WDM channel, or the ease with which a transmitter board can be provided for any system channel. Transmitter sources for WDM would ideally be able to select and transmit any WDM channel, or even simultaneously transmit all the channels of the WDM system. This suggests that a tunable laser or multifrequency laser would

ITU Standards Proposal: f = 193.1 ± m × 0.1 THz

FREQUENCY RELATIVE TO STANDARD (THz)	ABSOLUTE FREQUENCY (THz)	WAVELENGTH (nm)	STAND. CHANNEL
$f_{ref} + 5 \times 0.1$	193.600	1548.515	
$f_{ref} + 4 \times 0.1$	193.500	1549.315	*
$f_{ref} + 3 \times 0.1$	193.400	1550.116	
$f_{ref} + 2 \times 0.1$	193.300	1550.918	*
$f_{ref} + 1 \times 0.1$	193.200	1551.721	
f_{ref}	193.100	1552.524	*
$f_{ref} - 1 \times 0.1$	193.000	1553.329	
$f_{ref} - 2 \times 0.1$	192.900	1554.134	*
$f_{ref} - 3 \times 0.1$	192.800	1554.940	
$f_{ref} - 4 \times 0.1$	192.700	1555.747	*
$f_{ref} - 5 \times 0.1$	192.600	1556.555	
$f_{ref} - 6 \times 0.1$	192.500	1557.363	*
$f_{ref} - 7 \times 0.1$	192.400	1558.173	
$f_{ref} - 8 \times 0.1$	192.300	1558.983	*
$f_{ref} - 9 \times 0.1$	192.200	1559.794	
$f_{ref} - 10 \times 0.1$	192.100	1560.606	*

Fig. 4.14 Proposed International Telecommunications Unions (ITU) standard channel allocations for dense wavelength-division multiplexing (WDM).

be desirable for WDM. Although this is true in principle, the stability of tunable lasers with aging then becomes an issue, and means might be required to stabilize the frequency of the tunable laser against some external reference. This entails a more complex transmitter subsystem, and in this situation the detailed operating principle for achieving tuning becomes critical. These issues are discussed in more detail in the sections that follow.

B.   DISCRETE λ-SELECTED DFB LASERS

The initial deployments of WDM in long-haul networks are using DFB lasers, and this section briefly outlines the reasons for this choice. DFB lasers are well known in their ability to offer excellent spectral characteristics. These include very strong side-mode rejection (typically >40 dB) and excellent long-term stability of longitudinal mode selection. This fact stems from the relatively short resonator (typically <500 μm) that is entirely filled with a frequency-selective grating. In such short devices with near-optimum grating coupling ($\kappa L \sim 2$), the DFB modal selectivity is relatively robust against local small effective index excursions such as spatial hole-burning effects due to nonuniform intracavity intensity. Given the well-documented reliability of single-longitudinal mode operation that is essential for non-WDM applications, the only remaining concern is the long-term stability of the actual lasing wavelength. This stability could be affected by local temperature variations or carrier density variations that may arise from changes in leakage current or nonradiative recombination, or note increases in temperature from aging-induced drive current increases.

A critical feature of laser sources in this regard is the clamping of gain at the threshold value. The gain (imaginary index) and real index of refraction both are functions of carrier density. The frequency of the resonator is then fixed because clamping the gain at threshold also performs the function of clamping the effective index. Thus, despite unavoidable degradation from leakage or recombination, these demand only that a larger fraction of the drive current be required to maintain the lasing gain level and produce a reduction in output power for a given current. The frequency, however, is fixed.

This argument is only approximate because increases in drive current may increase chip temperature, either locally along the resonator or with respect to a temperature-controlled submount as a result of finite bonding thermal impedance. Also, although the average gain may be clamped, local reductions can be compensated for by increases in gain in longitudinally distinct regions of the DFB resonator. These effects can actually alter the internal mode structure of the chip by effectively introducing phase shifts along the grating length. Finally, a packaged device relies on the long-term integrity of the temperature control loop. At about 0.1 nm/°C, temperature stability well below 0.1 nm is advisable for dense WDM, demanding both a good control loop design and knowledge of the aging characteristics of the thermistor or other components in the loop.

The stability of DFB laser wavelength has proven to be extremely good, with drifts of less than 0.2 nm in 25 years expected. Figure 4.15 shows data acquired for a population of 100 lasers with no prior wavelength stability screening under different accelerated aging conditions [15]. Although the observed spread increases somewhat in conditions that most closely approach actual operation, the majority of the population remains within a ±0.2-nm spread over the extrapolated 25-year system life. It is also likely that rapidly drifting devices can be screened by other signatures. Additional studies have looked at similar populations of lasers and have observed average wavelength drift rates of less than 0.01 nm/year [16].

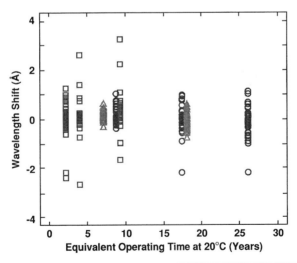

Fig. 4.15 Stability of 100 DFB lasers against wavelength drift under three accelerated aging conditions, normalized to equivalent aging under normal operation.

With these features, DFB lasers offer systems designers a viable choice to confidently proceed with WDM deployment.

C. FIBER-BASED LASERS

Recent work on UV-induced gratings in fibers has demonstrated that the filter characteristics of fiber gratings are likely to play a significant role in WDM technology. One application may lie in WDM source technology, where the attractive features of fiber gratings include the reproducibility and precision of the effective index of optical fiber, and the relative temperature insensitivity of filter elements made from silica ($\sim$0.01 nm/°C). The former characteristics allows for the placement of filter reflection bands at precise optical frequencies, which suggests that fiber gratings could be ideal for WDM channel definition.

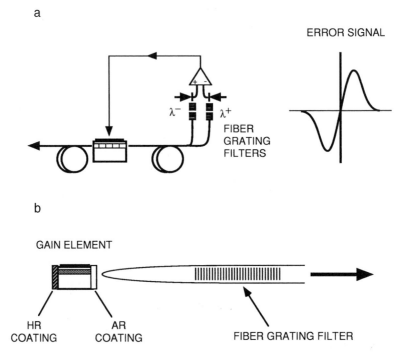

Fig. 4.16 Two configurations for using fiber gratings to stabilize laser wavelengths. (a) External reference filters with an electrical servo to lock DFB output to filter function. (b) Extended cavity using a fiber Bragg grating for optical feedback.

Two applications of fiber gratings are depicted schematically in Fig. 4.16. The first shows the use of grating filters to stabilize the frequency of a DFB by providing electrical feedback to control temperature, for example. In the event that channel spacings demand tighter control of frequency than aging drifts would allow, such a control servo may be advantageous. Another application might employ hybrid resonators where the feedback is optical. In this case, the highly reproducible center frequency of the fiber grating is coupled to a semiconductor gain element. In addition to accurate channel alignment maintenance, such a configuration may affect lower chirp for direct modulation at modest bit rates of 2.5 Gb/s and less. However, the longitudinal mode placement within the fiber Bragg envelope is reliant upon the thermal and mechanical stability of a hybrid package. Furthermore, the elimination of interface reflections is essential to stable operation. Again, the manufacturability of such a configuration has not been addressed.

Another application of fiber gratings would employ an actual fiber resonator with Er amplification. Spectral stability would demand very high gains and short cavities, and indeed such lasers have been fabricated [17]. As shown in Fig. 4.17, advantageous use can be made of the majority of the pump light that is not absorbed by the resonator in a follow-on booster EDFA. Such a master oscillator power amplifier (MOPA) configuration has reasonable output powers of more than 10 mW, but the manufacturability of such a configuration has not yet been addressed.

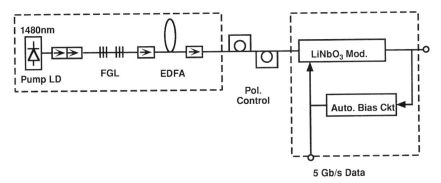

Fig. 4.17 Experimental demonstration of an Er-fiber laser with a fiber grating cavity and a master oscillator power amplifier (MOPA) configuration.

D. ACTIVE FILTER TUNABLE LASERS

The emergence of EDFAs and WDM transmission has evoked in many systems engineers a vision of tunable laser transmitters that are capable of accessing any of the frequency channels of the system. One approach that has been explored is the incorporation of a tunable active narrow band filter inside the laser resonator to force laser operation at the well-defined loss minimum of the filter. Sources are both hybrid, multielement resonators and monolithic semiconductor structures. The former includes a variety of designs incorporating $LiNbO_3$ acoustooptic and electrooptic filters, or numerous turnable bulk grating configurations. The latter includes monolithic tunable DBR lasers, vertical grating-assisted codirectional coupler filter lasers, and sampled-grating or superstructure grating (SSG) DBR lasers. In this section a few of these designs are discussed to illustrate both the strengths and the shortcomings of active filter tunable lasers.

Common to these designs is the insertion of a tunable filter function into a resonator that is characterized by a roughly evenly spaced longitudinal mode structure. The filter, tuned in wavelength, selects out a particular longitudinal mode. Most implementations of active filter tunable lasers are thus discretely tunable, with the mode spacing of the resonator given by $\Delta\lambda = \lambda^2/2n_g L$, where L is the resonator length with group index n_g. If we introduce phase shifting or adjustable optical path sections within the resonator capable of realizing at least $\Delta\phi = \pi$, or one-mode spacing, the longitudinal modes can be swept to allow so-called quasi-continuous access to any frequency within the tuning range of the active filter. Temperature tuning common to semiconductor resonators of about ~0.1 nm/°C can also be used to shift the discrete tuning comb for channel alignment at slower speeds.

The tunable DBR structure illustrated in Fig. 4.18 embodies these operational principles in a particularly simple fashion [18]. The Bragg section is composed of a double heterostructure or MQW active layer transparent to the propagating light but capable of negative index shifts with increasing carrier density resulting from both plasma and anomalous dispersion contributions. These phenomena have a combined magnitude as large as 2%, but reduced by modal confinement factors. The filter characteristic is constant in wave vector, $2\pi n_{eff}/\lambda$, so the maximum tuning range is simply governed by

$$\Delta\lambda/\lambda \approx \Delta n_{eff}/n_{g_{eff}}, \tag{4.10}$$

where n_{eff} and $n_{g_{eff}}$ are the effective and effective group indices, with $n_{g_{eff}} \equiv n_{eff} - \lambda\,(\partial n_{eff}/\partial\lambda)$. At a 1.55-$\mu$m wavelength, an approximately 10-nm tuning has been realized.

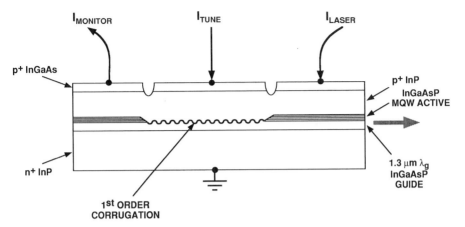

Fig. 4.18   Two-section tunable DER laser with an integrated back-facet monitor detector.

With precision fabrication of resonator lengths, the discrete tuning spacing can be chosen to match the WDM channel separation. Figure 4.19a shows the tuning characteristics of a DBR laser with a 50-GHz channel spacing, and Fig. 4.19b shows a detail of the errors from precise 50-GHz allocations as the device is tuned. This illustrates that precise, discrete channel selection can be achieved with a relatively simple structure. Such devices have been delivered in significant numbers to form the basis of WDM networking demonstrations [19].

The most serious practical consideration for such a tunable source is the long-term aging characteristics of the tuning calibration. The wavelength stability of DFB lasers discussed in the preceding section derives from the gain clamping of the DFB resonator. In the transparent section of the DBR reflector there is no gain clamping, and the mapping between index and current is then intimately tied to leakage currents and nonradiative recombination, which increase with aging.

If occasional recalibration cycles on a time scale of thousands of hours are permitted, such a source would be practical and cost-effective, capable of accessing many channels. The calibration could also be incorporated into the operation of a WDM multiaccess system. However, the initial WDM deployments require that sources be maintained at their respective channels with no provisions for in-service calibration. The problem lies in the fact that the externally observable signatures of intracavity filter drifts with aging can be observed only with sophisticated spectral or active moni-

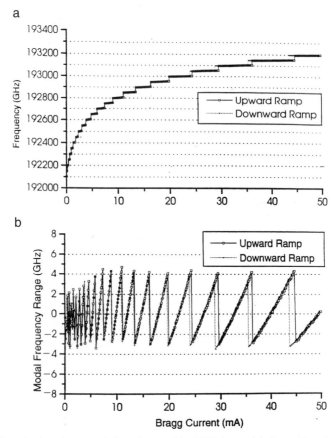

Fig. 4.19 Tuning characteristics of a tunable DBR laser. (a) Overall tuning characteristics with discrete steps every 50 GHz. (b) Detail of the errors from precise 50-GHz channel allocations versus tuning current.

toring equipment that is likely to add significantly to the cost of implementation. As the filter becomes misaligned from the selected mode, the laser will continue to operate on that mode until the misalignment becomes severe enough to hop to the next mode. This event is likely to cause a channel dropout, to interfere with the next adjacent channel, or, at best, to introduce unacceptable relative intensity noise (RIN) due to the hopping instability. Although some relatively simple maintenance schemes have been demonstrated, none have been pursued to a point of practical implementation.

Numerous research programs have focused on the issue of increasing the tuning range of active filter tunable lasers. Figure 4.20 shows a vertically grating-assisted coupler filter (VGF) laser [20] that employs asynchronous waveguides in proximity that can forward couple only by virtue of coarse grating-assisted phase matching $n_1/\lambda = n_2/\lambda + 1/\Lambda_G$, where n_1, n_2, and Λ_G are the effective phase indices of waveguides 1 and 2, and the grating pitch, respectively. Effecting an index change in n_1, for example, gives tuning of the amount

$$\Delta\lambda/\lambda \approx \Delta n_1/(n_{g_1} - n_{g_2}), \tag{4.11}$$

which represents a dramatic enhancement over the values in Eq. (4.10). The tuning characteristic of such a device is shown in Fig. 4.21, where a total range of 70 nm is reported. Such devices have also been fabricated to include phase shifters for continuous tuning. One additional complexity of such a device stems from the inherently closer mode spacing of the longer cavity and the inherently broader filter response of the wider tuning design. The inherent selectivity of the selected mode over the next adjacent mode is small and is marginally capable of achieving the greater than 30-

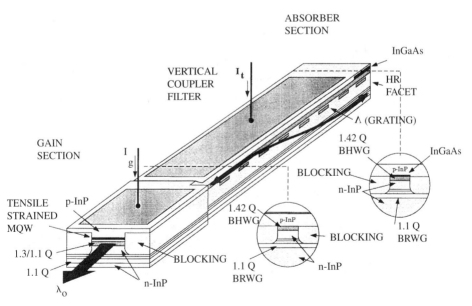

Fig. 4.20   Configuration of a monolithic vertical grating-assisted codirectional coupler filter (VGF) extended tuning laser.

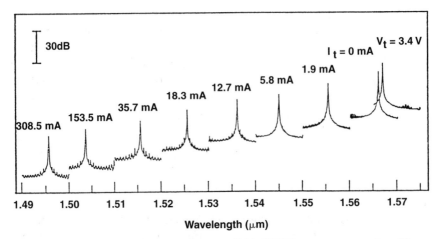

Fig. 4.21 Tuning characteristics of a monolithic VGF laser over more than 70 nm.

dB side-mode rejection required for most WDM systems. This is compounded by the fact that unintentional fixed intracavity reflections as small as −40 dB, arising from layer structure discontinuities or waveguide imperfections, can introduce interferometric ripple in the cavity mode losses of the same magnitude as the filter response adjacent mode rejection. Tuning of such a device to sequentially select out successive modes can thus be challenging. This is then compounded by the aging characteristics discussed with respect to the simpler tunable DBR laser, which had much stronger modal selectivity and a coarser mode structure.

SSG lasers employ either blanked-out sampling of the Bragg reflection [21], or a periodic chirp in the grating [22], to introduce an additional, longer periodicity to the Bragg reflector. This produces "ghosts" or additional replicas of the primary Bragg peak in frequency to produce a comb of narrow reflection bands that can extend for more than 100 nm. The breadth of the comb is determined by the reflection bandwidth of each sampling or chirped subsection, whereas the narrow width of each comb element is determined by the length of the overall reflector structure.

The tuning concept relies on different comb spacings for each end of the resonator, so low-loss lasing can occur for only the particular comb elements that are spectrally aligned. As one comb is tuned with respect to the other, the lasing will jump, or "strobe," to the next aligned comb element, as shown in Fig. 4.22. This discrete tuning with very large jumps can be achieved over huge ranges limited by the bandwidth of the gain

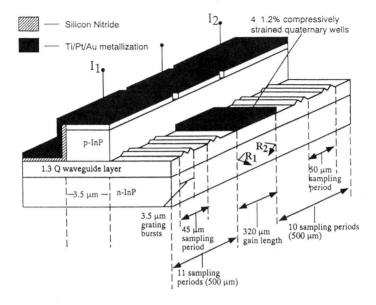

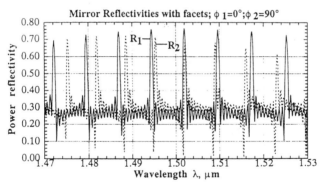

Fig. 4.22   Sample grating DBR laser showing the device and the alignment of end mirror reflectivity combs.

element. Because the narrow comb elements have resolution corresponding to reflective gratings over the entire length, the modal selectivity of the SSG laser is better than that of the VGF device. If phase shifting sections are added and both ends tuned, all frequencies can be accessed. Figure 4.23 shows a result where more than 34 nm is continuously accessed by appropriate electrode control [22]. These devices, however, still suffer from aging-induced calibration drifts. The longer cavities and considerably more

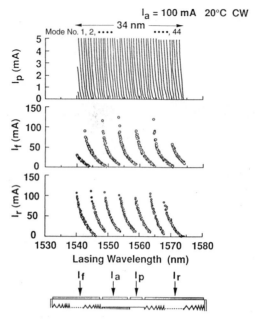

Fig. 4.23 Tuning currents applied to various sections in a multisection superstructure grating (SSG) DBR to obtain quasi-continuous tuning over 34 nm.

complex mode-selection mechanism and control make the logistics of long-term longitudinal mode stability considerably more problematic than that of the simple DBR, which itself has not been satisfactorily controlled for commercial deployment.

E.   GEOMETRIC λ-SELECTION LASERS

Another scheme to achieve WDM sources involves tuning, not by the analog control of an intracavity quantity, but by the geometric cavity design and activation of appropriate gain elements. These designs also offer the exciting promise of simultaneous multichannel transmission from a single laser chip. Figure 4.24 illustrates the concept in a particularly simple fashion [23]. A transparent slab planar waveguide is terminated on one side by a series of individually activated stripe waveguide gain elements. The other end of the freely diffracting planar waveguide region incorporates an etched-facet focusing grating that images the waveguide gain elements back on themselves in a spectrally selective manner. Activating the central gain

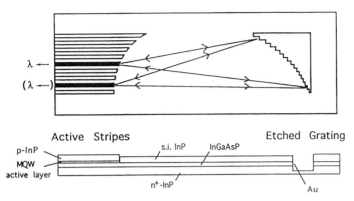

Fig. 4.24 Monolithic multifrequency WDM laser source using an etched-facet Rowland circle grating configuration.

element together with another gain element will produce lasing at that particular wavelength aligned to image back onto the gain element that is activated. Fiber coupling can be made at one output port, and numerous channels can be sequentially or simultaneously activated and modulated. Although it has proven challenging to achieve low losses because of the complicated etched-facet grating, impressive channel alignment and large numbers of channels (>60) have been demonstrated [23].

More recently, substantial effort has been invested in variants of the structure shown in Fig. 4.25 [24]. This class of structures incorporates a planar waveguide grating router (WGR) rather than the diffraction grating discussed previously. This router is discussed in detail in Chapter 8 in the context of Si–SiO$_2$ passive demultiplexers. These devices function in transmission in a manner similar to a higher order diffraction grating in reflection. The individual reflections of the grating teeth are replaced instead by the light traversing the individual curved waveguides of the router. Each arm of the interferometer is incrementally longer than the adjacent arm by nominally an integral number of wavelengths, typically about 20–100. If we trace a path through the resonator, light enters the first free-space region reflected from the single output port and diffracts to illuminate all the curved waveguides of the router. At the far end, if the path length differences are exactly multiples of λ, the outputs are phased to focus the light onto the central waveguide gain element. For slightly different wavelengths, the deviation from exact multiples of λ introduces a phase tilt across the waveguides entering the second free-space region, which

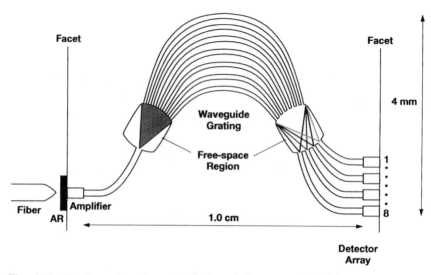

Fig. 4.25 Configuration of a monolithic multifrequency WDM laser source using a waveguide grating router (WGR) as an intracavity wavelength-selective filter element.

then focuses the light onto a different waveguide gain element. The activation of a particular gain element thus defines a wavelength path for which a low-loss cavity exists, and different wavelengths are generated by activating different gain elements, either in succession or simultaneously if desired.

The wavelength selectivity in this case is not governed by the analog setting of some intracavity element, but rather by the geometric layout of the filter and a discrete choice of gain elements, just as in the case of the reflective grating device discussed previously. Hence, tuning in such resonators is termed *geometric selection* and is particularly well suited to precise and deterministic maintenance of channel spacing where large numbers of channels are needed, spanning a large range of wavelength.

Figure 4.26 shows a typical output spectrum taken by the successive activation of each gain element in a device designed for eight-channel operation at a 200-GHz (~1.61-nm) channel separation. Devices have also been demonstrated that successfully generate larger numbers of channels, up to 24, which may be critical for networking and distribution applications of WDM [25].

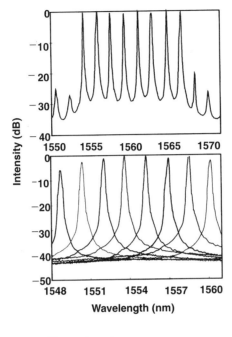

Channelspacing
Designed for 1.62 nm
Av. measured 1.633 nm

Fig. 4.26 Spectrum of a WGR laser under sequential and simultaneous multichannel operation with a 200-GHz channel spacing.

As with the tunable DBR laser, the geometric selection WDM sources have strengths and weaknesses. One important and enabling advantage is the ability to simultaneously generate more than one wavelength channel. Also important is the ability to span large tuning ranges limited in principle only by the spectral breadth of the gain elements employed, the spectral loss characteristics of the passive regions, and the loss of the router itself when large numbers of channels are designed over a large spectral range. This requires a large number of router arms, just as high-resolution gratings require the illumination of many grating teeth. Further desirable features, as in the case of the tunable DBR, is channel spacing defined by geometry. Although not as precisely uniform as the longitudinal mode spacing of a DBR, the interferometric multichannel passband characteristic of the router is "frozen" in the chip and not expected to experience significant aging effects.

The alignment of longitudinal modes with respect to the router passbands in this class of lasers, however, is random. Furthermore, the typically large size (~5- to 15-mm chip size) results in dense longitudinal modes (typically ~2–5 GHz), thus the stability of selection of a particular longitudinal mode is not guaranteed. Also, whereas alignment and stability against mode hops may be good for one channel (i.e., one gain element selection), the next channel will have a different random longitudinal mode alignment with respect to the selection filter. It is thus expected that the light-current characteristics of successive channels will display mode hops in some cases, and this is indeed observed. As gain elements age, the thermal loading is likely to shift the current operating range where mode hops occur. One problem, though, is the likelihood of lost bits of information during mode hops, particularly if the intermodal beat frequency is within the information band of transmission.

Recent work has illustrated that nonlinear mode-stabilizing effects are dramatically enhanced in these long-cavity structures; these effects result in better single-mode stability or side-mode suppression than would be expected on the basis of the mode spacing and filter bandwidth alone [26]. This feature stems from intermodal beat frequencies in the longer structures that lie in a range where the carrier density can be substantially modulated. These results suggest that low RIN single-longitudinal-mode operation is readily achieved, but they do not address the likelihood of mode hops in the operating range of the device. In some applications, such infrequent mode hopping may not impair system operation, particularly in shorter distance, lower data rate, dense WDM scenarios.

Another difficulty of the WGR laser is the ability to modulate at high speed. Because of the long cavity, the relaxation oscillation limited bandwidth is typically less than 2 GHz and then 2.5-Gb/s operation is problematic. Innovative solutions to this include the addition of higher order WGR output ports with an integrated EA modulator [27]. However, this configuration precludes the operation of the device as a simultaneous multichannel transmitter. Finally, the large size of the chip (~1-cm scale) raises serious questions about the cost of manufacture.

With these strengths and weaknesses, the geometric selection devices are attractive in the context of large numbers of simultaneous WDM channels at lower (~1-Gb/s) data rates and short enough distances where mode hops may not introduce errors. If the manufacturing costs of centimeter-scale

InP chips can be properly managed, local networking and distribution systems such as fiber to the home may be such an application.

F. DBF ARRAY WDM SOURCES

Section III.B described in detail the reasons why DFB lasers were a viable source for WDM transmission and have been the choice for initial WDM deployments. Primary among these reasons is the excellent spectral stability, both against aging drifts and in longitudinal mode selectivity. The latter derives from the highly spectrally selective intracavity grating combined with the short cavity and its resulting coarse longitudinal mode spectrum. Although temperature tuning provides adjustments at the 1-nm scale, accessing large spectral ranges without analog intracavity elements or multi-section designs suggests the application of arrays of DFB lasers.

For simultaneous operation, one DFB laser can be allocated for each channel where transmission is desired. For "tunable" single-channel operation, the only requirement is that the wavelength spacing of successive DFB lasers be within a temperature tuning range, nominally less than 2 nm for ± 1-nm tuning ($\pm 10°C$ temperature adjustment).

DFB arrays can either be fiber coupled as an array or employ on-chip waveguide combining networks. Furthermore, they can be directly modulated, or on-chip externally modulated with EA modulators. The latter can be provided for each laser for simultaneous low-chirp transmission, or as a single modulator for encoding any activated DFB as a single-channel "tunable" or "channel-selectable" laser. These two configurations are shown in Fig. 4.27.

The advantage of using a DFB array lies in the adoption of all the desirable spectral stability features of discrete DFB lasers, but with the added provision of selectable or simultaneous multichannel operation over a wide spectrum. The primary challenge in implementing DFB arrays is the ability to fabricate individual DFB laser elements at precisely defined frequencies. Assuming that a $\lambda/4$-shifted or gain-coupled design is incorporated to eliminate modal degeneracy, for a given pitch the frequency of a DFB laser is governed by the effective index of the waveguide.

Physical phenomena that can alter the waveguide effective index $n_{eff}(\lambda)$ include variations in the thickness or composition of the epitaxial layers that compose the waveguide, variations in the carrier density required for threshold, and variations in the width of the waveguide. If β_{lase} is fixed by the spatial boundary conditions of the resonator (i.e., the corrugation period

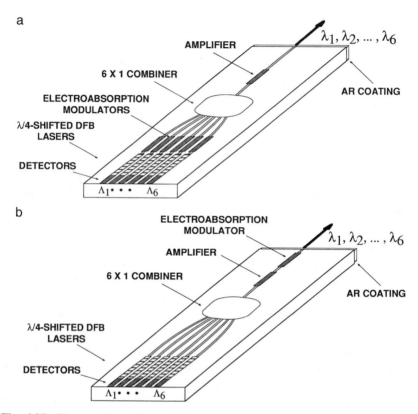

Fig. 4.27 Two configurations of DFB array WDM laser sources. (a) Each laser can be separately modulated at high speed for simultaneous multichannel transmission. (b) All lasers combine into one modulator for single-channel, wavelength-selectable operation.

and strength, phase-shift locations, relative facet locations, and resonator physical length), then for any uniform variation Δn_{eff}, the wavelength deviation is the same as that given by Eq. (4.10):

$$\Delta\lambda = \lambda \cdot \frac{\Delta n_{eff}}{n_{g_{eff}}}, \qquad (4.12)$$

which can be expanded to

$$\Delta\lambda = \frac{\lambda}{n_{eff_{group}}} \cdot \left\{ \frac{\partial n_{eff}}{\partial t} \cdot \Delta t + \frac{\partial n_{eff}}{\partial \lambda_{PL}} \cdot \Delta\lambda_{PL} + \frac{\partial n_{eff}}{\partial W} \cdot \Delta W - \frac{\lambda\alpha}{4\pi} \cdot \Delta g_{threshold} \right\}.$$

$$(4.13)$$

In Eq. (4.13), the first term refers to changes in thickness, the second assesses deviations in epitaxial composition characterized by the photoluminescence wavelength λ_{PL}, and the third assesses deviations in the waveguide width. The last term arises from carrier-induced index changes that would occur in devices with different threshold gains.

The impact of these terms can be evaluated using the effective index method for a typical bulk DFB structure and a typical MQW structure [28]. The first term is characterized for convenience in terms of wavelength shift (in nm) per percentage change in epitaxial layer thickness, assuming that all epitaxial layers shift in thickness by the same percentage as a result of fluctuation of growth rates. The derivative for this term,

$$\frac{\partial \lambda}{\partial(\% \, \Delta t)} = \frac{\lambda}{n_{eff_{group}}} \cdot \frac{\partial n_{eff}}{\partial(\% \, \Delta t)}, \tag{4.14}$$

is shown in Fig. 4.28. Also shown on each curve is the width at which cutoff occurs for the first higher order lateral mode in the waveguide. In this figure, the increase in sensitivity to epitaxial thickness variations with increasing guide width is merely a consequence of the increasing lateral confinement factor as the guide gets wider. This plot indicates that a typical standard deviation of $\sigma_t \sim \pm 2\%$ will lead to a wavelength standard deviation

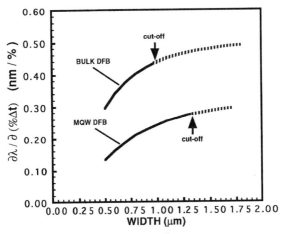

Fig. 4.28 Theoretical evaluation of wavelength deviation as a function of the percentage change in epitaxial layer thickness for a typical bulk and QW buried heterostructure DFB laser design.

of $\sigma_\lambda \sim \pm 0.4$ nm for a modern MQW DFB. The next term arises from variations in epitaxial composition:

$$\frac{\partial \lambda}{\partial \lambda_{PL}} = \frac{\lambda}{n_{eff_{group}}} \cdot \frac{\partial n_{eff}}{\partial \lambda_{PL}}. \tag{4.15}$$

This derivative is shown in Fig. 4.29, and in this case the units are dimensionless, expressing lasing wavelength shift per unit photoluminescence wavelength shift of the constituent layers. For simplicity, all nonbinary layers are assumed to shift by the same amount, and the increase with increasing width again results simply from an increasing lateral confinement factor. We see for the MQW structure that an epitaxial photoluminescence standard deviation of $\sigma_{\lambda_{PL}} \sim \pm 5$ nm will give rise to a σ_λ of $\sim \pm 0.4$ nm. The third term constitutes the contributions arising from waveguide width fluctuations:

$$\frac{\partial \lambda}{\partial W} = \frac{\lambda}{n_{eff_{group}}} \cdot \frac{\partial n_{eff}}{\partial W}. \tag{4.16}$$

Again, the results are dimensionless, and they are shown in Fig. 4.30 for a range of waveguide widths. In this case the trend is different, with the sensitivity to width fluctuations diminishing rapidly as the waveguide width

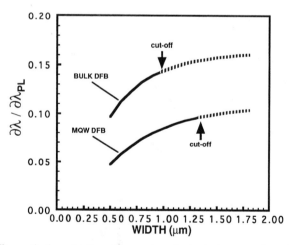

Fig. 4.29 Theoretical evaluation of wavelength deviation as a function of change in compositional change, characterized by photoluminescence wavelength, for a typical bulk and QW buried heterostructure DFB laser design.

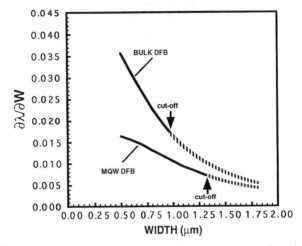

Fig. 4.30   Theoretical evaluation of wavelength deviation as a function of wave-guide width change for a typical bulk and QW buried heterostructure DFB laser design.

increases. This finding suggests designing for as wide a waveguide as practical while still staying safely below the higher lateral mode cutoff guide width. At a typical design of about 1 μm for the MQW, a standard deviation in mesa width of $\sigma_W \sim \pm 0.1$ μm gives rise to $\sigma_\lambda \sim \pm 0.8$ nm. Although reactive ion etching techniques may provide improvements in mesa width reproducibility, wavelength fluctuations due to waveguide width control are the largest contribution and are likely to remain the most challenging.

Figure 4.31 shows the raw frequency data from 14 randomly selected 6-element $\lambda/4$-shifted MQW DFB arrays from 2 wafers of the design shown in Figs. 4-27b. A photolithographic printing process was used employing electron-beam-generated near-field holographic photomasks to define the $\lambda/4$-shifted gratings [29]. Although direct electron-beam exposure has also been demonstrated for array fabrication [30], the photomask printing technique allows for the simultaneous printing of all the required pitches in a high-throughput, low-cost process. These data reveal a spread in laser frequency at each channel that is comparable to the channel spacing of 200 GHz, or 1.61 nm. However, careful inspection of the data reveals that the *spacing* in the arrays is much more precise than the overall spread observed in Fig. 4.31 at each channel frequency. Stated alternatively, the processing uniformity within the wafer area of a particular array is such that the various contributions to σ_λ discussed previously are relatively small.

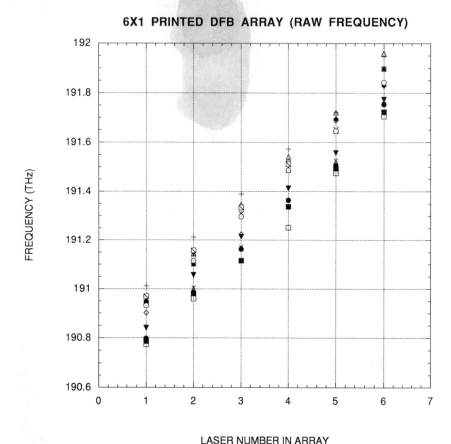

Fig. 4.31   Raw frequency data for 14 randomly selected 6-element arrays of phase-mark-printed λ/4-shifted DFB lasers.

Figure 4.32 shows the same data, where each array has undergone a simulated uniform temperature adjustment to align the combs together as closely as possible, as would be done operationally. In this circumstance, the histogram of deviations from perfect channel alignment for the 84 lasers in the population is shown in Fig. 4.33, with a standard deviation of 0.18 nm, an order of magnitude smaller than the values commonly observed for the distribution of wavelengths in waferwide measurements of DFB lasers.

Figure 4.34 shows the calculated yield reduction due to wavelength inaccuracy, assuming a Gaussian distribution. The result is plotted for various array sizes, as a function of the allowed tolerance in wavelength accuracy normalized to the σ of the fabrication process, as discussed

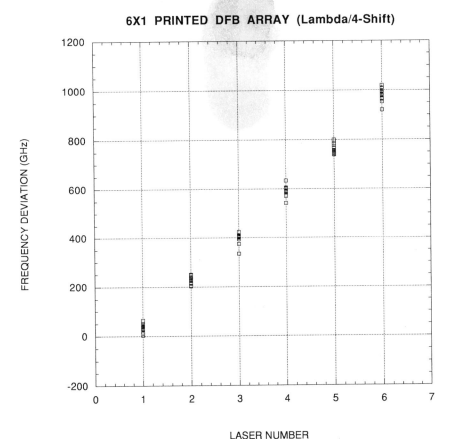

Fig. 4.32 Compressed data shown in Fig. 4.14. A simulated temperature adjustment of each array has been used to bring the arrays as closely into alignment as possible.

previously. As an example, for a wavelength accuracy requirement of ±0.2 nm, the observed σ of 0.18 nm would lead to a yield of about 25 to 6% for a four- to eight-element array. This level of wavelength precision may be required for an array in which all lasers are operating simultaneously. This is challenging, but sufficient for prototype development. Further uniformity in growth and processing, as well as design of lasers for reduced fabricational sensitivity, should improve this.

For single-channel λ-selectable operation, there is no requirement that all channels be simultaneously aligned. The accuracy of each channel is then extended to the full allowable temperature tuning of any element,

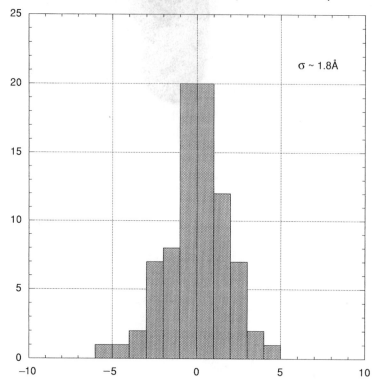

6X1 PRINTED DFB ARRAY (Lambda/4-Shift)

WAVELENGTH DEVIATION FROM UNIFORM SLOPE (Å)

Fig. 4.33 Histogram of frequency deviations from uniform channel spacing for the lasers described in Figs. 4.30 and 4.31.

which may be ±0.8 nm. In this circumstance, the array yield rapidly approaches unity even for large numbers of elements, as depicted in Fig. 4.34. Such a λ-selectable WDM transmitter PIC was shown in Fig. 4.27b, comprising a six-channel laser array with an integrated combiner, amplifier, modulator, and back-face monitor. This PIC has been demonstrated experimentally [31] and designed to operate in a standard-size 14-pin butterfly package. The output optical spectrum is shown vertically displaced in Fig. 4.35 successively for each laser, and the 3-dB bandwidth of the unpackaged device is approximately 3.8 GHz, which is adequate for high-quality operation at 2.5 Gb/s. The eye diagram at 2.5 Gb/s with a 2^{23}-1 pseudo-random

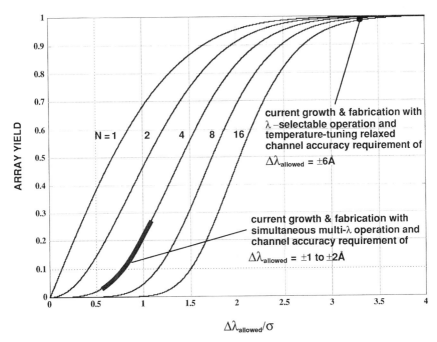

Fig. 4.34 Calculated yield reductions due to wavelength inaccuracy for DFB laser arrays, assuming a Gaussian distribution in wavelength. The tests are a normalized function of the ratio of the allowed wavelength error to the standard deviation of errors from uniform channel spacing.

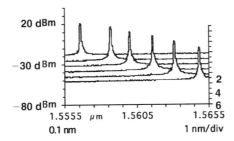

Fig. 4.35 Spectrum of a packaged 6-element phase-mark-printed $\lambda/4$-shifted DFB laser array. The device is in a standard 14-pin butterfly package with a standard single high-speed coplanar drive.

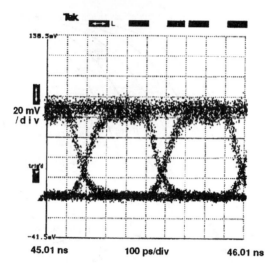

Fig. 4.36   Eye diagram of a WDM array PIC chip under 2.5-Gb/s modulation.

bit stream is shown in Fig. 4.36. Neither the small-signal nor the large-signal response showed any significant sensitivity to laser or amplifier bias conditions.

The chirp was also evaluated for this PIC using an automated time-resolved spectral measurement system based on a fiber-coupled scanning monochromator and high-speed temporal waveform recording. Under 2.5-Gb/s operation, the peak-to-peak wavelength excursion for all six selectable wavelengths ranged from 0.09 to 0.13 Å, with an average of 0.1 Å and a typical result shown in Figure 4.37. Such a packaged module has been evaluated in a transmission experiment to determine bit error rate performance over 600 km of conventional single-mode fiber, and the channel-independent low-chirp characteristics of the PIC module were verified by successfully switching channels without substantial variations in transmission characteristics [31].

IV.   Conclusions

A combination of technology advances and unique market pressures has resulted in the rapid development and deployment of amplified transmission and WDM systems. Amplified transmission has resulted in the adoption of external modulation instead of direct laser modulation for most of these

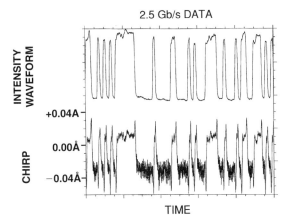

Fig. 4.37 Time-resolved wavelength excursions under 2.5-Gb/s modulation. The peak-to-peak wavelength deviation is about 0.1 nm, which is sufficient for excellent long-haul transmission over conventional fiber (>600 km).

systems. This represents a remarkable departure from the simplicity of directly modulated lasers prevalent up to this point in optical fiber communications.

In the case of WDM, the only laser source technology that meets system requirements at the time of this publication is individually selected DFB lasers or DFB-based integrated EA modulator sources. These can be either free running or supplemented with external frequency references. However, advances in WDM source technology are likely to provide high functionality λ-selectable or simultaneous multi-λ transmitters.

In many instances, the same 14-pin butterfly package has evolved first to provide nearly chirp-free sources to nearly chirp-free λ-selectable sources. The trend appears to be ever-higher functionality from the same module, and this trend is likely to continue.

References

1. Koch, T. L., and R. A. Linke. 1986. Effect of nonlinear gain reduction on semiconductor laser wavelength chirping. *Appl. Phys. Lett.* 48:613. *See also* Koch, T. L., and J. E. Bowers. 1984. Nature of wavelength chirping in directly modulated semiconductor lasers. *Electron. Lett.* 20:1038.
2. Yariv, A. 1989. *Quantum electronics.* 3d ed. New York: Wiley, 264–276.

3. Kitamura, K., S. Takano, N. Henmi, T. Sasaki, H. Yamada, Y. Shinohara, H. Hasumi, and I. Mito. 1988. 1.5 μm Multiple quantum well distributed feedback laser diodes grown on corrugated InP by MOVPE. *Electron. Lett.* 24: 1045–1046.

4. Ketelson, L. J. P., I. Kim, L. E. Eng, J. A. Grenko, D. L. Coblentz, T. H. Wessel, U. K. Chakrabarti, and R. L. Hartman. 1995. Directly modulated low dispersion penalty lasers at 1.55 μm for 2.5 Gb/s applications. In *LEOS '95, San Francisco*. Paper SCL5.1.

5. Seino, M. 1996. Recent progress in high-speed and highly reliable LiNbO$_3$ optical modulators. In *Technical Digest of OECC '96, Chiba*. Invited paper 17D2.

6. Gnauch, A. H., S. K. Korotky, J. Veselka, J. Nagel, C. T. Kemmerer, W. J. Minford, and D. T. Moser. 1991. Dispersion penalty reduction using an optical modulator with adjustable chirp. *IEEE Photon. Tech. Lett.* 3:916–918.

7. Tharmalingam, K. 1963. Optical absorption in the presence of a uniform field. *Phys. Rev.* 130:2204–2206.

8. Aspnes, D. E. 1966. Electric field effects on the dielectric constants of solids. *Phys. Rev.* 153:972–982.

9. Morton, P. A. 1996. High-speed integrated DFB/electroabsorption modulated lasers. In *Technical Digest of CLEO '96, Anaheim*, 314. Invited paper CWL1.

10. Wakita, K., I. Kotaka, O. Mitomi, H. Asai, and U. Kawamura. 1991. Observation of low-chirp modulation in InGaAs-InAlAs multiple quantum-well optical modulators under 30 GHz. *IEEE Photon. Tech. Lett.* 3:138–140.

11. Soda, H., M. Furutsu, K. Sato, N. Okazaki, Y. Yamazaki, H. Nishimoto, and H. Ishikawa. 1990. High-power and high-speed semi-insulating BH structure monolithic electroabsorption modulator/DFB laser light source. *Electron. Lett.* 26:9–10.

12. Reichmann, K. C., P. D. Magill, U. Koren, B. I. Miller, M. Young, M. Newkirk, and M. D. Chien. 1993. 2.5 Gb/s transmission over 674 km at multiple wavelengths using a tunable DBR laser with an integrated electroabsorption modulator. *Photon. Tech. Lett.* 5:1098–1100.

13. Johnson J. E., T. Tanbun-Ek, Y. K. Chen, D. A. Fishman, R. A. Logan, P. A. Morton, S. G. N. Chu, A. Tate, A. M. Sargent, P. F. Sciortino, Jr., and K. W. Wecht. 1994. Low-chirp integrated EA-modulator/DFB laser grown by selective-area MOVPE. In *Technical Digest of 14th IEEE International Semiconductor Laser Conference, Maui*, 41–42. Paper M4.7.

14. Akiba, S., M. Usami, and K. Utaka. 1987. 1.5 μm λ/4-Shifted InGaAsP/InP DFB lasers. *IEEE J. Lightwave Tech.* LT-5:1564–1573.

15. Chung, Y. C., and J. Jeong. 1994. Aging-induced wavelength shifts in 1.5 μm DFB lasers. In *Technical Digest of OFC'94, San Jose*, 104–105. Paper WG6.

16. Vodhanel, R. S., M. Krain, R. E. Wagner, and W. B. Sessa. 1994. Long-term wavelength drift of the order of -0.01 nm/yr for 15 free-running DFB laser modules. In *Technical Digest of OFC'94, San Jose*, 103–104. Paper WG5.

17. Mizrahi, V., D. J. DiGiovanni, R. M. Atkins, S. G. Grubb, Y. K. Park, and J.-M. P. Delavaux. 1993. *IEEE J. Lightwave Tech.* 11.

18. Koch, T. L., and U. Koren. 1990. Semiconductor lasers for coherent optical fiber communications. *IEEE J. Lightwave Tech.* 8:274.

19. Kaminow, I. P., C. R. Doerr, C. Dragone, T. Koch, U. Koren, A. A. M. Saleh, A. J. Kirby, C. M. Ozveren, B. Schofield, R. E. Thomas, R. A. Barry, D. M. Castagnozzi, V. W. S. Chan, B. R. Hemenway, Jr., D. Marquis, S. A. Parikh, M. L. Stevens, E. A. Swanson, S. G. Finn, and R. G. Gallager. 1996. A wideband all-optical WDM network. *IEEE J. Select. Areas Commun.* 14:780–799.

20. Kim, I., R. C. Alferness, U. Koren, L. L. Buhl, B. I. Miller, M. G. Young, M. D. Chien, T. L. Koch, H. M. Presby, G. Raybon, and C. A. Burrus. 1993. Broadly tunable vertical-coupler filtered tensile-strained InGaAs/InGaAsP multiple-quantum-well laser with 74.4nm tuning range. *Appl. Phys. Lett.*

21. Jayaraman, V., A. Mathur, L. A. Coldren, and P. D. Dapkus. 1993. Extended tuning range in sampled grating DBR lasers. *IEEE Photon. Tech. Lett.* 5:489–491.

22. Tohmori, Y., Y. Yoshikuni, H. Ishii, F. Kano, T. Tamamura, and Y. Kondo. Over 100 nm wavelength tuning in superstructure grating (SSG) DBR lasers. *Electron. Lett.* 29:352–354.

23. Soole, J. B. D., K. R. Poguntke, A. Schere, H. P. LeBlanc, C. Chang-Hasnain, J. R. Hayes, C. Caneau, R. Bhat, and M. A. Koza. 1992. Wavelength-selectable laser emission from a multistripe array grating integrated cavity laser. *Appl. Phys. Lett.* 61:2750–2752.

24. M. Zirngibl, and C. H. Joyner. 1994. A 12-frequency WDM laser source based on a transmissive waveguide grating router. *Electron. Lett.* 30:700–701.

25. Zirngibl, M., C. H. Joyner, C. R. Doerr, L. W. Stultz, and H. M. Presby. 1996. An 18-channel multifrequency laser. *IEEE Photon. Tech. Lett.* 8:870–872.

26. Doerr, C. R., M. Zirngibl, and C. H. Joyner. 1995. Single longitudinal-mode stability via wave mixing in long-cavity semiconductor lasers. *IEEE Photon. Tech. Lett.* 7:962–964.

27. Joyner, C. H., M. Zirngibl, and J. C. Centanni. 1995. An 8-channel digitally tunable transmitter with electroabsorption modulated output by selective-area epitaxy. *IEEE Photon. Tech. Lett.* 7:1013–1015.

28. Tennant, D. M., and T. L. Koch. 1996. Fabrication and uniformity issues in $\lambda/4$-shifted DFB laser arrays using e-beam generated contact grating masks. *Microelectron. Eng. Special Issue on Nanotechnologies.*

29. Young, M. G., T. L. Koch, U. Koren, D. M. Tennant, B. I. Miller, M. Chien, and K. Feder. 1996. Wavelength uniformity in $\lambda/4$-shifted DFB laser array WDM transmitters. In *Technical Digest of OSA/IEEE topical meeting on semiconductor lasers: Advanced devices and applications, Keystone, Co, August.* Paper WA6.

30. Zah, C. E., F. J. Favire, B. Pathak, R. Bhat, C. Caneau, P. S. D. Lin, A. S. Gozdz, N. C. Andreadakis, M. A. Koza, and T. P. Lee. 1992. Monolithic

integration of multiwavelength compressive strained multiquantum-well distributed-feedback laser array with star coupler and optical amplifiers. *Electron. Lett. 28:2361–2362.*

31. Young, M. G., T. L. Koch, U. Koren, G. Raybon, A. H. Gnauck, B. I. Miller, M. Chien, K. Dreyer, R. E. Behringer, D. M. Tennant, and K. Feder. 1995. Six-channel WDM transmitter module with ultra-low chirp and stable λ selection. In *Technical Digest of ECOC '95, Brussels,* vol. 3, 1019–1022. Postdeadline paper Th.B.3.4.

Chapter 5 | Advances in Semiconductor Laser Growth and Fabrication Technology

Charles H. Joyner

Lucent Technologies, Bell Laboratories, Holmdel, New Jersey

I. Introduction

In the 1988 edition of this text, *Optical Fiber Telecommunications II,* an excellent discussion of semiconductor laser history and theory of operation was given by J. E. Bowers and M. A. Pollack in Chapter 13. Since that time, many advances in design as well as fabrication techniques have taken place. Using the foundation laid by Bowers and Pollack as a starting point, we attempt to present the most striking of these advances in laser performance that have a strong impact on optical communications.

II. New Sources and Growth Apparatus

Although the principles of laser fabrication remain the same, advances in technique and starting materials have had a significant impact since 1990. InP substrates have improved in quality so that wafers with etch pit densities less than $500/cm^2$ have become common. Despite a 15-year lead for GaAs, these now rival the best GaAs substrates. In addition, 3-in.-diameter InP substrates are already available from some suppliers. Wafers come chemically etched and sealed in nitrogen-filled packages in a "growth-ready" form that removes uncertainty as to surface preparation and reproducibility.

The availability of high-quality metal-organic sources has revolutionized growth processes. These are chemical compounds using organic ligands attached to metal hosts to improve the volatility of the metal species. For a long time, indium was the difficult element to deliver reproducibly because its large mass and poor electron affinity rendered even its trimethylated

OPTICAL FIBER TELECOMMUNICATIONS,
VOLUME IIIB

Copyright © 1997 by Lucent Technologies.
All rights of reproduction in any form reserved.
ISBN: 0-12-395171-2

form a solid. Recently, several suppliers discovered techniques to finely divide the solid and support it on inert beads or suspend it in an inert liquid to allow reproducible gas-phase entrainment when hydrogen bubbles through or over the increased surface area that the small grain size provides. All other commonly required elemental components — Ga, As, P, Zn, Si, Cd, Fe, and Be — are readily available in extremely pure form as liquid or gaseous precursors with low cracking temperatures. As a result, liquid-phase epitaxy and other solid-source growth techniques have vanished from both research and industry. Most popular are techniques employing source material pyrolysis at or on the substrate surface (650°C) under ambient pressures between 50 and 800 torr. This growth method is called *metal-organic vapor-phase epitaxy (MOVPE)*. Another highly competitive technique precracks the precursors in heated nozzles and projects elemental species at a relatively low-temperature substrate (~500°C) at pressures between 1×10^{-2} and 10^{-5} torr. This is most descriptively called *metal-organic molecular beam epitaxy (MOMBE)*. For a detailed review of growth techniques, see Razeghi (1989); Long, Logan, and Karlicek (1988); or Pearsall (1982).

Currently, both techniques provide monolayer abruptness at interfaces. Efforts are now centered on achieving layer composition and thickness uniformity over as wide an area as possible. To this end, vertical reactor designs with rotating susceptors have achieved the best results. Commercially available reactors growing quantum well structures routinely achieve one-sigma variations of 3 nm, with total wavelength variations less than 9 nm from center to edge. For bulk layers — e.g., InGaAs by Bhat, Koza, and Hennessey (1995) — thickness uniformities vary as little as ±0.65% from center to edge of a 2-in. wafer. Similarly, lattice constant variations are as low as ±150 arcsec with standard deviations on Ga and As compositions of 0.12 and 0.41%, respectively. This outstanding uniformity and reproducibility is the enabling technology that makes possible the advances discussed in the next sections and Chapter 4 (Volume IIIB).

III. Band Structure Engineering by Means of Strained Multiple Quantum Wells

Historically it was believed that in order to have good long-term reliability, every epitaxial layer in a compound semiconductor structure should be lattice-matched to the substrate host. In 1986, Adams, as well as Yablonovitch and Kane, proposed that strain could be used to both lower the thresholds and improve the efficiency of multiple quantum well (MQW) structures. It was widely assumed that strain would lead to dislocation defects and shorter laser lifetimes, especially for high-power lasers where high current densities would provide the driving force to disorder atomic structures under high stress. However, since this bold proposal, it has been amply shown that strained-layer MQW structures retain the reliability benefits of lattice-matched MQW structures, while allowing the engineer to take advantage of new physical effects and materials combinations. In this chapter, we see that strain allows all of Adams's predictions to come true, in addition to other beneficial effects. Strain has proven to be a practical tool as well as a research probe into the basic physics of semiconductor band structure. An excellent detailed technical review of these developments can be found in the February 1994 special issue of the *IEEE Journal of Quantum Electronics*. We particularly recommend the work of O'Reilly and Adams, and Thijs and coworkers. Unless otherwise noted, references to these authors refer to this February 1994 publication. In the next sections, we summarize and update the high points of improving laser performance by use of strain.

A. ELASTIC PROPERTIES OF CRYSTALS AND THE EFFECT OF STRAIN ON BANDGAP

Let us first consider some of the problems that exist for bulk semiconductor lasers. Although our primary concern is with quantum well structures, the physics of bulk semiconductors provides a good comparative reference. Lattice-matched direct bandgap bulk semiconductors have broad energy bands that make current injection and carrier transport possible. However, of the wide range of energy and momentum values these carriers (holes and electrons) can assume, only a very small fraction have the correct values to contribute to laser output with a precisely defined energy. Part of the problem in dealing with the energy spread of the carriers is that the effective mass of the electrons (m_c) in the conduction band is less than the effective mass of the holes (m_h) in the valence band. Therefore, as the carrier density increases just prior to achieving transparency, the quasi-Fermi level for the electrons is well into the conduction band before the quasi-Fermi level for the holes has reached the valence band edge. This causes a large valence band density of states, increasing the threshold current required to initiate lasing.

Another problem arises from the electronic symmetry of the valence band structure. In a bulk semiconductor, the valence band state for the holes is equally composed from electronic orbitals of p_x, p_y, and p_z character. Although all these become equally occupied at population inversion, only the p_y states (TE mode, $\sim\frac{1}{3}$) have the correct optical symmetry to contribute significantly to laser output. This reduces quantum efficiency.

In an unstrained semiconductor the light-hole (LH) and heavy-hole (HH) valence bands are degenerate at the valence band maximum. Because of its smaller width, the LH band has fewer carriers, but they are shorter lived, which causes both bands to contribute almost equally to the radiative current density. However, only the electron-HH transition contributes to stimulated emission.

To make matters even worse, all the carriers can be consumed by nonradiative loss processes and spontaneous emission, both of which decrease laser efficiency. We now consider how strain affects these situations.

Mathews and Blakeslee (1974) have shown that if a lattice-mismatched epitaxial layer is below a certain critical thickness for a given strain, the resulting biaxial in-plane strain causes tetragonal deformation of the cubic symmetry of the crystal lattice without inducing any dislocation defects. Simply put, the strained layer expands or contracts in the normal direction to the host crystal surface while the relative distances between atoms in the plane parallel to the host plane remain the same as those of the host. Using the coordinate system of Fig. 5.1, Thijs and coworkers showed that the total strain can be resolved into a purely hydrostatic component,

$$\varepsilon_{hy} = \Delta V/V = \varepsilon_{xx} + \varepsilon_{yy} + \varepsilon_{zz} = 2\varepsilon_{\|} + \varepsilon\perp, \tag{5.1}$$

and a purely axial component,

$$\varepsilon_{ax} = \varepsilon_{zz} - \varepsilon_{xx} = \varepsilon\perp - \varepsilon_{\|}. \tag{5.2}$$

The hydrostatic component changes the mean bandgap energy, E_g, whereas the axial component causes a splitting of the LH and HH states at the valence band maximum. This is schematically illustrated for the simple case of a bulk semiconductor bandgap in Fig. 5.2. Note that the crystal distortion has the effect of modifying the valence sub-band structure, while having little effect on the conduction band. With strain introduced, the valence band structure becomes highly anisotropic. For compressive strain, the highest valence band of Fig. 5.2b is heavy along the strain axis, $k\perp$, and light in the growth plane, $k_{\|}$. The reverse is true for tensile strain in Fig. 5.2c.

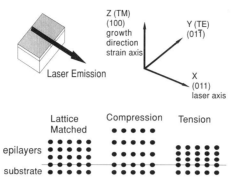

Fig. 5.1 Schematic laser structure showing the coordinate geometry for this chapter. Note that TM gain is polarized along z, TE gain is polarized along y, and spontaneous emission is polarized along the laser axis, x. We also adopt the convention that a plus sign indicates compressive strain and a minus sign tensile strain in reference to epilayers.

To fully evaluate the effect of this modified band structure, we also need to recall that for quantum well structures the differential gain (g) is given by

$$dg/dN \propto (m_h m_c)^{1/2}/(m_h + m_c), \qquad (5.3)$$

where N is the transparency carrier density per quantum well. Note that decreasing the hole effective mass (m_h) increases the differential gain, and that the transparency and threshold currents will be reduced.

Now refer to Fig. 5.2 and reconsider the problems mentioned at the beginning of this section. First, consider the compressively strained case — Fig. 5.2b. In this situation, the heavy hole has shifted upward at the valence band maximum. For quantum well structures, the confinement energy is determined by the effective mass along the growth direction (z = strain axis, $k\perp$) and here has HH character. The density of states is determined by the effective mass in the plane of the quantum well, $k_\parallel$, which is now light. Thus, as m_h approaches m_c, there is a marked reduction in the density of valence band states at population inversion and hence a reduction of the carrier density as well as a reduction in the spread in electron energies at threshold. The practical benefit is an increase in the differential gain (see Eq. [5.3]) and a reduction of the threshold current.

Another beneficial consequence of compressive strain arises from the fact that the HH state at the center of the Brillouin zone has no p_z character (see Ghiti and O'Reilly [1993] or Corzine, Yan, and Coldren [1993]), which implies that TM emission is strongly suppressed. The remaining valence

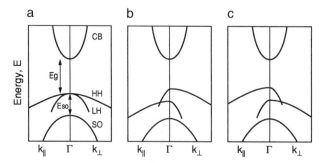

Fig. 5.2 Schematic representation of the band structure of an unstrained direct-gap tetrahedral semiconductor. (a) The light-hole (LH) and heavy-hole (HH) bands are degenerate at the Brillouin zone center Γ, and the spin–split-off (SO) band lies E_{SO} lower in energy. The lowest conduction band (CB) is separated by the bandgap energy, E_g, from the valence bands. (b) Under biaxial compression, the hydrostatic component of the stress increases the mean bandgap, whereas the axial component splits the degeneracy of the valence band maximum and introduces an anisotropic valence band structure. Note that this introduces an ambiguity in terminology, because the highest band is now *heavy* along $k_\perp$, the strain axis (= growth direction), but *light* along $k_\parallel$ (in the growth plane). We label the bands in this chapter by their mass along the growth direction, so that a band that is referred to as heavy hole may in fact have a low in-plane mass, as here. (c) Under biaxial tension, the mean bandgap decreases and the valence band splitting is reversed, so that the highest band is now *light* along $k_\perp$, the strain axis and comparatively *heavy* along $k_\parallel$. [Reprinted with permission from O'Reilly, E. P., and A. R. Adams. 1994. Band structure engineering in strained semiconductor lasers. *IEEE J. Quantum Electron.* 30(2):366–379. Copyright © 1994 IEEE.]

band character is equally divided between p_x (propagating transverse to the laser cavity and hence lost) and p_y, which provides TE gain for stimulated emission along the laser axis. Thus, from arguments of polarization symmetry we obtain approximately 50% conversion efficiency compared with about 33% for the bulk laser.

The case for tensile strain is a little more complicated. Recall that for a lattice-matched quantum well, the quantum-confined Stark effect brings the heavy hole to the top of the valence band. A small amount of tensile strain can then cause the LH and HH bands to coincide. Sub-band mixing effects then give the highest valence band a large effective mass. However, if the tensile strain is large enough and the wells are thick enough, Zah *et al.* (1992) (closely followed by Krijn *et al.* [1992]) have shown that the highest LH band shifts above the HH band and the effective mass as well as the density of states is again reduced (see Silver and O'Reilly [1994] for detailed theory). This once again lowers the laser threshold and improves quantum efficiency as for the compressive case. For biaxial tension, the valence band state now has $\frac{2}{3}p_z$-like (TM) character with $\frac{1}{6}p_y$ (TE), and $\frac{1}{6}p_x$ (transverse/lost). This causes enormous enhancement of the TM polarization, with approximately two-thirds of the carriers contributing to the dominant gain mechanism.

Note that the manner in which the material is grown is critical to achieving this desired result. Techniques such as MOMBE, with a relatively low growth temperature, have less thermal energy available at the growth interface to promote dislocations than other techniques do. Growth rate is another important factor. The longer a highly strained layer remains exposed without being covered by a lattice-matched or strain-compensating layer, the greater the chance that dislocations are generated. Therefore, fast growth allows higher degrees of strain than those of slow-growth techniques. Andersson *et al.* (1987) demonstrated a (% lattice-mismatch) $\times$ (layer thickness) product of 200 Å% for InGaAs on GaAs by molecular beam epitaxy (MBE) (540°C at 1 μm/h). For the tensile-strained InGaAs/ InGaAsP on InP material system, Bhat (1992) (1.6% $\times$ 125 Å $=$ 200 Å%) and Krijn *et al.* (1992) (1.5% $\times$ 160 Å $=$ 240 Å%) have also demonstrated very impressive dislocation free results. Miller *et al.* (1991) have shown that by straining the barriers in a quantum well structure in the opposite way from the wells, it is possible to strain-compensate the entire superlattice so that it is overall strain neutral. By minimizing the quantity

$$\varepsilon_{\text{ave}} = (\varepsilon_b t_b + \varepsilon_w t_w)/(t_b + t_w), \qquad (5.4)$$

where ε is percentage strain ($-$ for tension, $+$ for compression), t is thickness, w is well, and b is barrier, Miller *et al.* found marked improvement ($\sim$45%) in photoluminescence full width at half maximum (FWHM) over structures without compensation. The other desired properties of strain in the wells remain largely unaffected, and it is possible to incorporate an arbitrary number of highly strained wells into a device structure. Figure 5.3 plots quantum well thickness as a function of InAs mole fraction for 1.5-μm MQW lasers made from InGaAs/InGaAsP. This is an excellent map delineating the narrow path to avoiding the pitfalls of exceeding critical layer thicknesses, creating a type II semiconductor, or landing on the HH-LH valence band crossing in the case of tensile strain, as mentioned previously.

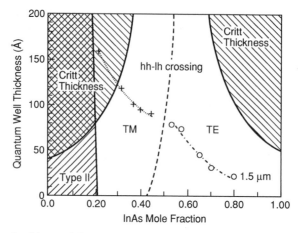

Fig. 5.3 Strained-layer 1.5-μm-wavelength In$_x$Ga$_{1-x}$As/InGaAsP quantum well laser map. The *shaded areas* are unfavorable due to the critical thickness limitation and the type II quantum wells. The remaining area is divided by another unfavorable area due to the HH-LH sub-band crossing. The InAs mole fractions and the quantum well widths required for 1.5-μm-emission-wavelength quantum well lasers are indicated. The *open circles* represent TE-polarized compressively strained and the *crosses* TM-polarized tensile-strained quantum well lasers. [Reprinted with permission from Thijs, P. A. J., L. F. Tiemeijer, J. J. M. Binsma, and T. van Dongen. 1994. Progress in long-wavelength strained-layer InGaAs(P) quantum-well semiconductor lasers and amplifiers. *IEEE J. Quantum Electron.* 30(2):477–499. Copyright © 1994 IEEE.]

Another potential problem with the use of tensile strain is the reduction of the conduction band discontinuity. As tensile strain reduces the effective bandgap of the well, the difference in energy between the well and barrier in the conduction band becomes smaller. When this differential is small enough, carrier spillover into the barrier and separate confinement layers will detrimentally increase the temperature sensitivity of the threshold current. Note that if the technique of strain compensation is used, the barrier height in the conduction band is further reduced, which further increases carrier spillover. Thus, a judicious choice of initial well and barrier bandgaps is required for the optimum tensile-strained structure.

Figure 5.4 shows a plot of threshold reduction as a function of strain, using experimental results from many contributors. Here again, excellent experimental confirmation of the previously mentioned benefits of strain is provided. Note that for the tensile case a minimum threshold is achieved

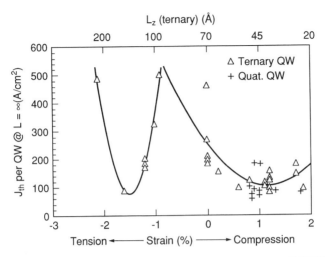

Fig. 5.4 Summary of threshold current densities per quantum well (QW) deduced for infinite cavity length 1.5-μm lasers versus the strain in the InGaAs(P) quantum wells, using data referenced in Thijs *et al.* (1994). The *solid lines* represent fits through the data points. J_{th}, current threshold at a given temperature. [Reprinted with permission from Thijs, P. A. J., L. F. Tiemeijer, J. J. M. Binsma, and T. van Dongen. 1994. Progress in long-wavelength strained-layer InGaAs(P) quantum-well semiconductor lasers and amplifiers. *IEEE J. Quantum Electron.* 30(2):477–499. Copyright © 1994 IEEE.]

over a narrower range of strain than that in the compressive case. Also, more strain is needed with tension ($\sim -1.6\%$) than with compression ($\sim +1.2\%$) for a similar threshold. This can cause a problem in exceeding the critical well thickness for tensile structures, particularly because the wells must be thick to avoid the LH-HH crossing.

Thijs, Osinski, *et al.* (1992); Temkin *et al.* (1990); and Zah *et al.* (1990) have all succeeded in creating sub-mA threshold lasers of one or both strain types, with the current absolute records being narrowly held by Thijs and coworkers. One structure (September 1991) used two 1.2% compressively strained $In_{0.7}Ga_{0.3}As$ quantum wells, whereas the other (1992) used a single 1.6% tensile-strained $In_{0.32}Ga_{0.68}As$ quantum well. At 10°C with reflective coatings of 92 and 98%, they both had 0.8-mA thresholds and delivered 1 mW at a 10-mA drive current. These lasers are very important to low-power systems where direct modulation with no prebias current is desired, or systems where thermally induced cross talk is a problem.

B. INTRINSIC LOSS MECHANISMS VERSUS STRAIN

So far, we ascribed the beneficial effects of strain shown in Fig. 5.4 to the reduction of the hole effective mass caused by the modification of the valence band structure. Another possibility is that strain has in some way reduced the internal nonradiative losses in the cavity caused by processes such as Auger recombination (AR) and intervalence band absorption (IVBA). Figure 5.5a illustrates the IVBA process using solid arrows. A photon emitted during the lasing process may be reabsorbed by a transition that lifts an electron from the spin–split-off band (lowest curve) into a vacancy in the HH band. This unwanted absorption increases the gain required at threshold. The dotted arrows of 5.5a show what happens as the bandgap, E_g, is increased (as it is for the hydrostatic component of compressive strain). Now the IVBA process must move to a larger wave vector, which makes the process less likely. The degree to which IVBA contributes to loss depends directly on the density of holes at k_{IVBA}. Assum-

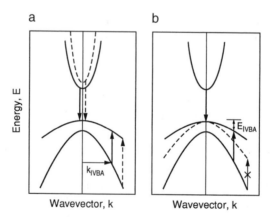

Fig. 5.5 (a) In the intervalence band absorption (IVBA) process, a photon emitted in the lasing process is reabsorbed by exciting an electron from the spin–split-off band into a hole state in the heavy-hole band (*solid upward-pointing vertical arrow*). The *dotted arrows* indicate how IVBA moves to larger wave vector k_{IVBA} and is thereby reduced with increasing bandgap brought about, for example, by hydrostatic pressure. (b) Reducing the hole effective mass (*dotted curve*) also increases k_{IVBA} and leads to the effective elimination of IVBA in strained structures as a result of the large value of E_{IVBA}. [Reprinted with permission from O'Reilly, E. P., and A. R. Adams. 1994. Band structure engineering in strained semiconductor lasers. *IEEE J. Quantum Electron.* 30(2):366–379. Copyright © 1994 IEEE.]

ing a parabolic band structure, O'Reilly and Adams gave the kinetic energy of the holes as

$$E_{\text{IVBA}} = m_s(E_g - E_{SO})/(m_h - m_s), \qquad (5.5)$$

where m_s is the spin–split-off effective mass. Thus we see that as the HH mass, m_h, decreases toward that of m_s, the energy required for IVBA increases, as illustrated in Fig. 5.5b. In fact, if m_h equals m_s, Eq. (5.5) shows that IVBA becomes impossible.

Ring *et al.* (1992) and Adams *et al.* (1993) have performed a clever experiment to test the detrimental contribution of IVBA on lasing threshold independent of the other loss mechanism, AR, which causes an increase in the current needed to achieve threshold but has little effect on differential quantum efficiency above threshold. These researchers placed lasers of many types in a pressure cell and measured their light-current characteristics up to 1.5 times threshold with short pulses to avoid heating. They then measured normalized efficiency over pressure ranges from 0 to 6 kbar. Both the bulk and unstrained lasers showed a marked increase in differential quantum efficiency at pressures up to 4 kbar; however, the highly compressive and tensile-strained quantum well lasers showed no improvement. This indicates that the hydrostatic pressure from the ingrown strain has already removed IVBA from these devices.

Consider now AR as it is depicted in Fig. 5.6. For the band-to-band transitions of Figs. 5.6a and 5.6b, the Auger coefficient is given by

$$C(T) = C_0 \exp(-E_a/kT), \qquad (5.6)$$

with the activation energies E_a for each process given by

$$E_a(\text{CHCC}) = m_c E_g/(m_c + m_h)$$
$$E_a(\text{CHSH}) = m_s(E_g - E_{SO})/(2m_h + m_c - m_s). \qquad (5.7)$$

Reducing m_h for these processes decreases their contribution to AR by several orders of magnitude as E_a appears in the exponent. The phonon-assisted process in Fig. 5.6 is shown by O'Reilly and Adams to be proportional to m_h^2. Therefore, this component of AR is also reduced by decreasing the hole mass by means of strain.

Using the same equipment described previously, Adams and coworkers (1993) studied the variation of threshold current with hydrostatic pressure. In this case, all the devices showed a considerable decrease in threshold

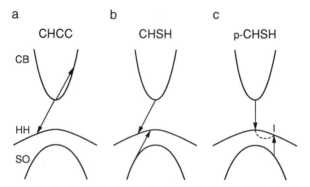

Fig. 5.6 (a) In the direct CHCC Auger recombination process, the energy and momentum released when a **C**onduction band electron (CB) and a **H**eavy hole (HH) recombine across the bandgap is used to excite a **C**onduction electron to a higher **C**onduction band state. (b) In the CHSH process, the energy and momentum released excite an electron from the **S**pin–split-off (SO) band into a state in the **H**eavy-hole band. (c) An example of a phonon-assisted CHSH Auger transition. The electron excited from the split-off band passes through a forbidden intermediate (I) state and is then scattered with the absorption or emission of a phonon to the final state. The phonon allows conservation of energy and momentum in the overall process. [Reprinted with permission from O'Reilly, E. P., and A. R. Adams. 1994. Band structure engineering in strained semiconductor lasers. *IEEE J. Quantum Electron.* 30(2):366–379. Copyright © 1994 IEEE.]

current, including the strongly strained MQW lasers of both types. This implies that the dominant mechanism controlling threshold current is AR and that the hydrostatic strain component is key to reducing thresholds. However, even in structures strained at 1.8% there is still a significant AR loss contribution. As further confirmation, Adams, Heasman, and O'Reilly (1989) pointed out that for GaAs-based lasers, where AR is not significant, thresholds actually increase with hydrostatic pressure due to increasing E_g.

Another important figure of merit to consider is characteristic temperature, T_0, given by

$$T_0 = (T_2 - T_1)\{\ln[J_{th}(T_2)/J_{th}(T_1)]\}^{-1}, \tag{5.8}$$

where J_{th} refers to the current threshold at a given temperature. For long-wavelength lasers, T_0 is typically between 35 and 75°K. The more insensitive to temperature a laser's threshold, the larger the value of T_0. This is a very important number for the optical systems designer. If a laser's performance did not depend on its temperature, it would not be necessary to cool it externally for most applications. This would represent a significant cost savings in both packaging and power budget. With compressive or tensile strain, T_0 values have achieved maxima around 100°K.

This is a significant improvement but not as large as if AR were eliminated completely. With their narrow wells and consequent smaller optical confinement, a compressive structure will degrade more rapidly than a tensile one with the same number of wells, because the temperature dependency of the transparency carrier density is smaller. This shows only that strain is no cure for poor design. Well width and number, in addition to barrier height and optical confinement cladding, still play a major role.

Tiemeijer *et al.* (1992) have shown a strong wavelength dependence of the transparency current for both (+) and (−) strained InGaAs/InGaAsP laser amplifiers. Bhat (1992) has shown a monotonic decrease in threshold current with increasing cavity length for AlGaInAs on InP (+ and −) strained lasers. Both results indicate that with optimal strain the threshold current density is now loss limited rather than transparency limited.

C. OUTPUT POWER AND RELIABILITY VERSUS STRAIN

For InP-based lasers used for optical communications, the maximum power output is limited ultimately by the heating of the device as more current is applied. In previous sections, we showed how strain reduces the threshold current as well as its temperature dependence. Strain also improves differential quantum efficiency and reduces internal loss. In short, more of the carriers that previously produced heat are now allowed to produce light. The result is improved output power. This is especially important to the fabrication of erbium-doped fiber amplifiers (EDFAs) (See Chapter 2 in Volume IIIB). To our best knowledge, the current record for continuous-wave (CW) output power at 1.5 μm is 325 mW (Thijs, Binsma *et al.* 1991), obtained from a 1000-μm-long cavity ($R_f = 4\%$, $R_r = 98\%$), compressively strained, $In_{0.7}Ga_{0.3}As/InGaAsP$ MQW-SIPBH laser. For tensile strain, a record 220 mW (Thijs *et al.* 1992) has been obtained from single quantum well (SQW) $In_{0.32}Ga_{0.68}As/InGaAsP$ lasers. Thijs *et al.* (1992) have also shown that when strain compensation (see Miller *et al.* [1991]) is used, lasers show virtually no change in threshold current after they are stressed by running at elevated temperatures (100°C) and currents (150 mA) for 20 h. Uncompensated structures typically showed J_{th} increases of 0.5%. Long-term reliability studies on strained-layer InGaAs/InP 1480-nm pump lasers were also performed. These were operated continuously at 80- to

90-mW output power with a heat sink temperature of 70°C (junction, ~100°C) for 9500 h. Even at this elevated temperature, the data show a projected median lifetime of 37 years. As a practical note, it has been observed that small amounts of unwanted strain are introduced into the active regions of devices by processing, mounting, and packaging. For lattice-matched structures, this small variable strain can greatly affect performance and reproducibility. For highly strained MQW structures, these small stresses are relatively insignificant and have little effect.

D. LASER LINEWIDTH, CHIRP, AND MODULATION LIMITS VERSUS STRAIN, DOPING, AND DETUNING

For optical communications networks employing wavelength-division multiplexing (WDM; see Chapter 15 in Volume IIIA), the linewidth of the laser output and the degree to which wavelength changes upon modulation of the drive current are critical to determining the optical channel spacings as well as allowed transmission distances because of dispersion effects. Henry (1982) has shown that the spectral linewidth $\Delta v \propto (1 + \alpha_H^2)/P$, and Koch and Link (1986) have shown that the rate at which wavelength changes with modulation current $\Delta v_{\text{chirp}} \propto (1 + \alpha_H^2)^{1/2}$, where P is power and the linewidth enhancement factor, α_H, is as follows:

$$\alpha_H = (-4\,\pi/\lambda)[(dn/dN)/(dg/dN)], \tag{5.9}$$

where n is the refractive index, N is the carrier concentration, and g is the gain. The goal is obviously to reduce α_H to improve both figures of merit, Δv and Δv_{chirp}. Recall that changing N will cause a change of n. Strain first helps by reducing the N needed to obtain a given power and thereby lessening α_H. Strain helps again by increasing gain as a function of current, dg/dN, which further reduces α_H. After tabulating comparative studies of strained versus lattice-matched structures, O'Reilly and Adams found a general reduction of α_H by a factor of 2 with strain.

For many applications, increasing the rate at which a laser can be directly modulated is the lowest cost solution to maximizing the data handling capacity of an optical network. Maintaining the previous symbol definitions, Olshansky et al. (1987) have shown that the maximum possible intrinsic −3-dB modulation bandwidth is

$$f_{-3\text{dB}} = (v_g \Gamma \alpha_{\text{loss}} P\, dg/dN)/\pi h v\, V_{\text{act}}\, \alpha_m (\gamma - 1/\tau_s)\sqrt{2}, \tag{5.10}$$

where v_g is the group velocity in the cavity, Γ is the optical confinement factor, α_{loss} is the total cavity loss, hv is the photon energy, V_{act} is the active layer volume, α_m is the mirror loss, γ is the damping rate due to nonlinear gain, and τ_s is the carrier lifetime.

Assuming that the cavity design is already optimized for small active volume and large optical confinement, Morton, Temkin, et al. (1992) have shown that the higher gain saturation of MQW lasers can be almost exactly compensated for by increased dg/dN. This leaves carrier transport across the separate confinement heterostructure (SCH) layers and the capture and emission times from the quantum wells as the dominant limitations to bandwidth. By Zn doping the entire MQW stack as well as the upper SCH region at 1×10^{18}, Morton et al. (1992, 1994) were able to achieve a record direct modulation bandwidth of 25 GHz at 1.55 μm for a bias current of 180 mA from a Fabry–Perot laser.

Kano et al. (1994) have studied compressively strained MQW lasers as a function of wavelength detuning the distributed feedback (DFB) grating wavelength from the gain peak of the active region, with and without modulation doping (p-type 1.5×10^{18}) of the barriers and upper SCH. Modulation doping with $+1.4\%$ strain produced α_H's near 1 for 1200-μm-long cavities. Optically detuning the grating wavelength -50 nm from the gain peak in addition to modulation doping produced spectral linewidths of 100 kHz for output powers of about 50 mW with 1500-μm-long cavities. Morton et al. used -20 nm DFB detuning with 300-μm-long cavities to achieve a record 25-GHz bandwidth for only 50 mA of bias in a fully packaged device.

E. POLARIZATION INSENSITIVE AMPLIFIERS BY MEANS OF STRAIN

Amplifiers are an essential component of most optical communications systems. For WDM networks, EDFAs are strongly preferred because of the low level of optical cross talk imposed on signals of different wavelength passing through the amplifier simultaneously. They are expensive, however. Work by Darcie, Jopson, and Saleh (1988) and Doerr et al. (1995) has shown that optical cross talk in semiconductor optical amplifiers (SOAs) can be effectively eliminated by electrically prebiasing the common amplifier with the sum of the electrical drive currents for the optical signals. Semiconductor amplifiers are essential for photonic circuit integration, and they are the low-cost method of choice for single-signal amplification. To compete effectively, however, an SOA must be polarization independent or risk loosing the signal intensity if the random polarization vector of the incoming light does not match the preferred polarization of the amplifier. Strain again comes to the rescue.

Recall from Section III.A that compressively strained wells allowed the exclusive emission of TE polarized light while ($\sim -1\%$) tensile wells preferred TM emission by a ratio of 4:1. Tiemeijer et al. (1993) reported a 1.3-μm SOA in which the barriers were lattice-matched, but 110-Å $In_{0.5}Ga_{0.5}As_{0.78}P_{0.22}$ tensile (T) wells alternated with 45-Å $In_{0.83}Ga_{0.17}As_{0.67}P_{0.33}$ compressive (C) wells to produce the structure schematically diagrammed in Fig. 5.7a. With both types of wells adjusted to the same bandgap, the number of wells of each type is adjusted to achieve the same gain for TE versus TM polarized light. A 3T–4C structure was found to be polarization independent to within 1 dB over the 3-dB bandwidth (1280–1330 nm) of the amplifier, with drive currents from 25 to 200 mA. A fiber-to-fiber gain of 16 dB with a coupled noise figure of 6.5 dB was demonstrated for 200 mA of drive current. Newkirk et al. demonstrated a similar structure for 1.55-μm amplification in 1993, achieving nearly identical figures of merit.

The novel structure of Fig. 5.7b was first proposed by Magari et al. in 1994 using lattice-matched InGaAs wells and was later modified to

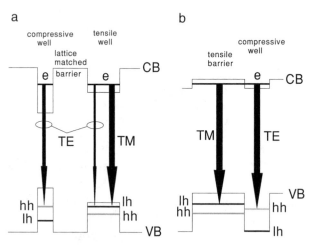

Fig. 5.7 Schematic (not to scale) band diagram of two strained layer MQW schemes to achieve polarization independent optical amplification. (a) In an MQW stack, the TE:TM emission ratio is adjusted by the amount of strain and the number of compressive versus tensile wells having identical effective bandgaps. (b) Compressive wells and tensile barriers produce an overall strain-balanced structure in which the electron-LH transition within the barrier (mostly TM) is slightly favored over the ($\sim$20-meV) larger energy electron-HH TE well transition.

incorporate compressive InGaAsP wells for superior performance by Gode-froy *et al.* and Ougazzaden *et al.* in 1995. In this structure, compressively strained wells alternate with tensile-strained barriers to produce an overall strain-neutral lattice. The small confinement energy in the conduction band ensures carrier spillover to fill the barrier states as well. Again, the well electron-HH transition contributes TE emission, but now a transition elec-tron-LH *within the tensile barrier* contributes dominantly TM radiation. Because of the slightly lower (~20-meV) energy of the electron-LH transi-tion in the barrier, it is possible to balance the TE:TM ratio within each well–barrier couple, by choosing the correct layer compositions, thick-nesses, and strain. With this approach, the optical confinement factor can be adjusted independently by changing the number of well–barrier couples without affecting the overall strain or the polarization ratio. This is a very powerful tool for photonic integration. The hero result of Ougazzaden *et al.* was a 27-dB gain at 240 mA through a 700-μm-long cavity, with less than 1 dB of polarization sensitivity over an 85-nm (1500- to 1585-nm) bandwidth completely independent of drive current. They used 16 couples of (+1.1%) compressively strained 80-Å InGaAsP wells and (−0.9%) tensile 70-Å InGaAs barriers.

IV. Selective Area Growth

Improved functionality for optical subsystems requires the integration of active (light-emitting, -modulating, and -detecting) devices with passive (waveguiding, splitting, and filtering) components. Photonic integrated cir-cuits (PICs) offer reduced loss from device to device on chip as a result of the small distances involved and the tight optical confinement of semiconductor waveguide structures. It is possible to accomplish integration by sequentially growing a base structure, etching away sections, and regrowing new struc-tures. This etch-and-regrow technique experiences problems with reflective loss at the butt-coupled interfaces and becomes much more difficult as the number of different devices needed increases. Selective area growth (SAG) is a technique that allows the engineer to selectively determine the local bandgap of many different devices within a single plane simultaneously. It is especially applicable to MQW structures.

Growth on partially masked GaAs substrates was begun by Gale *et al.* as early as 1982. Murata *et al.* began the first study of InGaAs growth in small mask openings on totally covered InP to produce integrated pin photodiodes for optoelectronic circuits in 1989. The first detailed SAG study of the InGaAs/InP material system for optical emitters was conducted by Galeuchet, Roentgen, and Graf in 1990. The original intent was to devise a technique whereby an entire diode device could be produced in a single growth through an opening, with the mask remaining to contribute to the current confinement. These studies quickly led to the concept of local bandgap variation and photonic integration using quantum wells with numerous contributors. An interesting variation also emerged in which the InP surface is not masked but etched in a pattern prior to growth. In this case, the difficult to nucleate (111)B crystal face serves as a "mask" relative to the (100) or (111)A face, and similar principles apply. For comprehensive reviews on selective and nonplanar growth, see Bhat (1992) and Thrush *et al.* (1993).

The concept of bandgap control by SAG through masks is illustrated in Fig. 5.8. We assume metal-organic vapor-phase epitaxy with generic precursors given in the upper part of the figure, but the same principles apply to many growth techniques and starting materials. Source material arriving from the gas phase will grow epitaxially in regions where there is no mask. Where source material lands on the dielectric mask, it will not

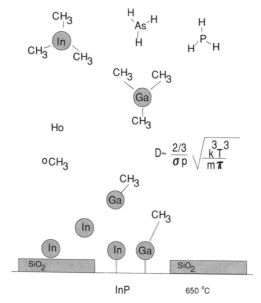

Fig. 5.8   Schematic (not to scale) of selective area growth through a mask.

readily nucleate. On the mask surface, it will travel for a distance that depends mainly on the surface temperature and ambient pressure. If it arrives at the mask edge, it will nucleate on the semiconductor surface; if not, it will return to the gas phase and diffuse, because of local concentration gradients, to find an unmasked area. Thus two processes, *surface* and *gas-phase diffusion*, contribute to epitaxy in the mask openings. For the majority of mask patterns commonly used (generally twin-stripe covered regions with openings of 10 μm or more), the dominant process is gas-phase diffusion, with the diffusion constant for each species given by

$$E = \tfrac{2}{3}(1/\sigma p)(k^3 T^3/m\pi)^{1/2}, \tag{5.11}$$

where σ is the collision cross section (πd^2), p is the pressure, k is Boltzmann's constant, T is absolute temperature in degrees Kelvin, and m is the mass of the species. At first glance it looks as though mass should be the dominant factor. Thus, in the diffusion race between indium (114.8 amu) and gallium (69.8 amu), Ga should be about 13% faster than In. The fact that this does not happen prompted research by Caneau *et al.* (1992), Caneau *et al.* (1993), Jensen and Coronell (1991), and Eckel *et al.* (1994) in which they studied growth rate and layer composition as a function of precursor molecule. It was determined that at the ambient growth temperature of about 600°C, most In-containing metal-organics decompose completely above the surface so that In transports as an atomic species. Most Ga precursors, however, because of the relative strength of the Ga−C bond, do not decompose completely in the gas phase. Gallium travels with a single methyl (−CH$_3$) group attached and is therefore stearically hindered. The methyl group is released as Ga begins to bond to the semiconductor lattice. Another look at Eq. (5.11) reveals the collision cross section in the denominator. Indium, with a σ nearly equal to 6.5 Å^2, compared with Ga−CH$_3$, which has a σ nearly equal to 38 Å^2, gets a diffusive head start by a factor greater than 5. There are other mitigating factors, such as the incorporation efficiency for each species, but the overall result is an increase in In content for epilayers on masked versus unmasked surfaces. This effect occurs in addition to the fact that the epilayer will be thicker. The growth rate in the neighborhood of the mask is enhanced as a result of the increased concentration of source species migrating over from the covered surface. Thus, from both the quantum size effect and the change in alloy composition, a quantum well structure is shifted to a lower energy bandgap (longer wavelength) near a masked surface.

There are many parameters that affect the specific bandgap change for SAG. Choice of group III and group IV precursors, III/V ratio, growth temperature, growth rate, pressure, carrier gas flow rate, mask composition, thickness, width, and distance of the point in question from the mask all contribute. In most cases, variations in individual reactor geometry, in addition to the aforementioned, prevent the formulation of any universally applicable equation to quantify the SAG process. Mathematical expressions have been proposed by Thrush et al. (1993), Itagaki et al. (1994), and Fujii, Ekawa, and Yamazaki (1995b); however, all require experimentally determined variables (Fujii et al. 1995a). In practice, it is quite simple to grow quantum wells over a test mask with varied parameters, and construct empirical calibrations that are quite reproducible if the total surface area and filling factor (ratio of masked to unmasked surface) are maintained. Figure 5.9 (Joyner et al. 1992) shows a typical plot of bandgap shift versus mask parameters. Under the proper conditions, near-linear behavior is obtained and extrapolation to predict shifts for untried structures is relatively simple. Suzuki et al. (1994) have shown that in the InGaAs/InGaAsP/ InP material system it is possible to construct MQW structures that are low-loss waveguides (λ_g = 1.24 μm) far from masked regions but shift so far as to be high-quality emitters and detectors (λ_g = 1.66 μm) near masked

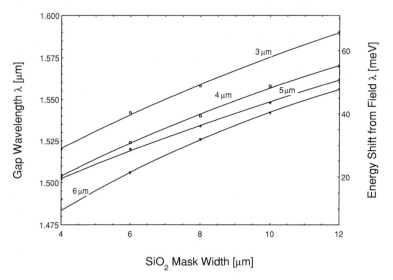

Fig. 5.9 Gap wavelength versus mask width for 100-torr InGaAs/InGaAsP (1.31-μm) MQWs. Field wavelength is 1.465 μm. Gap width is labeled above each curve.

regions. They used an opening of 10 μm with twin-stripe masks as wide as 300 μm to achieve this record bandgap shift of 253 meV with no degradation of photoluminescence intensity relative to unmasked growth. Also note that even though a mask ends abruptly in the direction of light propagation, the nature of diffusion is such that the MQW bandgap thickness changes over tens of microns. There is no abrupt index change to cause unwanted reflections, and the active and passive elements all form in the same optical plane. This is demonstrated in Fig. 5.10. A twin-stripe mask (of dimensions given) was overgrown with InGaAs/InP MQWs and subsequently cleaved along the propagation direction (dotted line). A scanning electron micrograph reveals the end of the oxide mask as a dark depression in the upper right of the large image. Detailed studies of the material inside the gap and far from the mask in the propagation direction show a smooth variation

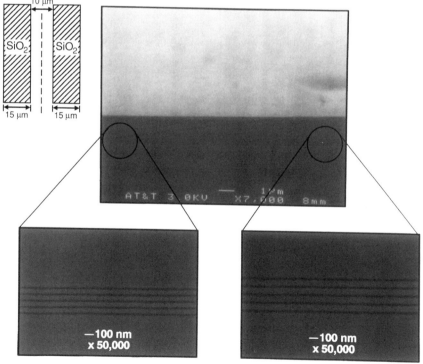

Fig. 5.10 A scanning electron micrograph showing the well and barrier thickness variation with distance along the propagation direction of a twin-stripe mask structure.

of well and barrier thickness until the effect of the mask is no longer present at approximately 14 μm from the mask edge.

There are problems with SAG. Having a mask present before quantum well growth implies that the surface must have been photolithographically processed first. Meticulous cleaning procedures or postmask etching is then necessary to ensure good growth nucleation. If the masked growth is too thick, the mask height will be exceeded and runaway growth will occur at the mask edge, making the surface nonplanar and difficult to process further. Indeed, unless the mask is very thick or undercut, in most cases the growth within a few microns of the mask edge is not usable for devices. If the mask is too large or the growth pressure is too high, polycrystalline nucleation will occur on the mask. This growth can also "run away" and deplete its surroundings of source material. There is also a limit to how closely adjacent devices may be spaced due to the surface area that masking requires. SAG's most distressing feature is the inability to completely characterize small areas near masks in nondestructive ways. The most powerful tool, X-ray diffraction, is not feasible to apply for practical device dimensions. This leaves microphotoluminescence, photoreflectance, and (thanks to Finders *et al.* [1991]) possibly Raman spectroscopy to uniquely determine compositions. The destructive tests of precision cleaving followed by scanning electron microscopy, as well as micro-Auger spectroscopy (Suzuki), are essential on test structures to quantify the nature of the growth at critical points.

Fortunately, the natural tendency of SAG is to strain both wells and barriers into compression. Simply starting with tensile barriers and letting nature take its course to produce (+) strain can have big advantages, as we saw in the previous sections. Additionally, because the highly strained material covers such a small surface and the growth rate is accelerated, it is possible to exceed the critical thicknesses defined by Mathews and Blakeslee (1974). Suzuki (1994) has demonstrated growth rate increases of a factor of 4 and strains in excess of +1% starting from lattice-matched compositions with no dislocations.

The first demonstration of an integrated photonic device using SAG was provided by Kato *et al.* in 1991. In the regime employing both surface and gas-phase diffusion, they used a 2-μm-wide opening with mask widths of 3 and 6 μm to create a modulator (1.48 μm) and laser (1.54 μm), respectively, from 1.3-μm InGaAs/InGaAsP MQWs. They achieved 100% coupling efficiency between the laser and the modulator. The modulator extinction ratio was 20 dB at 2 V. The integrated source transmitted error free at 2.5 Gb/s. True to Galeuchet's original hope, the mask was retained to increase current confinement, which left the final structure a nonplanar ridge. This demonstration captured the imagination of device makers, and a host of PICs ensued. There are, in fact, an infinite number of mask designs to achieve a given bandgap shift. The twin-stripe configuration is still most popular; however, the need for a planar final surface has led most fabricators away from Kato and colleagues' process. A planar surface makes subsequent photolithography steps easier and facilitates p-side down-bonding. Figure 5.11 gives a generic recipe for constructing planar channel mesa buried heterostructure (CMBH) devices using SAG. For the most complex PIC there are generally five main growth–processing steps: (1) a base wafer with an underlying waveguide or grating structure is created, (2) a SAG mask is applied and MQW growth is performed, (3) the SAG mask is removed and the surface is planarized with p-doped InP, (4) stripe masks

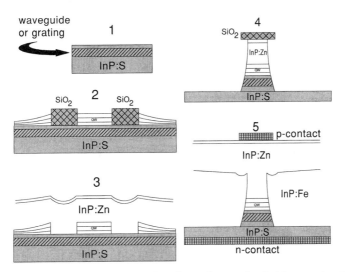

Fig. 5.11 Generic processing steps for channel mesa buried heterostructure devices using SAG (not to scale). (1) A base epitaxial layer is deposited if an underlying waveguide or grating is necessary. (2) A dielectric mask is lithographically defined and quantum wells are grown with varying bandgap by means of SAG. (3) The SAG mask is removed and the MQW structure is buried. (4) A new strip mask is defined and the active mesa is etched from the center point of the SAG mask. (5) Semi-insulating material (e.g., InP:Fe) is grown with the stripe mask in place. Subsequently the mask is removed and a final heavily p-doped growth completes the diode structure.

are applied and mesas are etched to define the active and passive waveguide regions, and (5) semi-insulating growth is performed with the mesa-oxide in place. Finally, with the oxide removed, a heavily P-doped layer followed by metallization completes the diode structure. Using this "planar" technique, Aoki *et al.* (1992) demonstrated a greater than 10 Gb/s integrated laser–modulator with low drive voltage (1 V peak to peak) and 13-dB extinction. Aoki *et al.* subsequently (1994) fabricated a five-channel MQW-DFB array of lasers with wavelengths from 1545.4 to 1555.5 nm at (313 GHz) 2.5 ± 0.2-nm intervals. They used a constant pitch grating (240.4 nm) and SAG masks of constant 15-μm openings, with widths that changed in 5-μm increments from 10 to 30 μm. The lasing wavelength was controlled by increasing the thickness of the optical guiding layer over the grating and thus changing the effective refractive index. The same masks also modified the thickness and composition of the quantum wells to shift the gain peak wavelength to match that of the grating–waveguide structure. Stable single-mode operation was observed for all channels with side mode suppression ratios greater than 35 dB. Output powers of 20 mW at 80 mA attest to the quality of the material grown in the region of the mask.

Using the nonplanar process of Kato *et al.* (1991) with a constant pitch grating and similar parameters to those used by Aoki *et al.* (1994), Sasaki, Yamaguchi, and Kitamura (1994) constructed a 10-channel WDM-DFB laser array with 2.5-nm spacings spanning 1537.9–1560.1 nm. Again, the performance of each laser was comparable to the best produced by un-masked growth. Both Hitachi and Lucent provide SAG electroabsorption modulated lasers as commercial products for 2.5-Gb/s transmission.

A novel class of light source has been created by Osowski *et al.* (1995) and Osowski *et al.* (1994) using SAG. In this case, the mask width is tapered throughout the active region so that the MQW stack changes bandgap along the laser axis. Because the carrier diffusion length is many times shorter than the active region is long, emission occurs at a continuum of wavelengths. With this method, InGaAs/GaAs/AlGaAs light-emitting diodes have been created with 3-dB spectral breadths of 165 nm centered at 960 nm. Such sources are highly valued to improve the spatial resolution of optical time domain reflectometry and to reduce the Rayleigh backscattering noise of optical fiber gyroscopes.

Since its first InP-based components in 1991, SAG has become a manufacturing platform technology that is expandable to produce a large number of PICs (Joyner *et al.* 1994; Joyner *et al.* 1995).

V. Selective Area Etching

We have seen that SAG allows one to change epilayer thickness as a function of length along a semiconductor plane by using a mask during growth. Brenner (1994) has shown that an exactly analogous processing technique exists whereby a semiconductor surface is masked during wet chemical etching. Under conditions where transport of the etchant species is the limiting factor for removal of the semiconductor layer, Brenner derived the following equation for the unmasked etch rate (nm/s):

$$r = M_{sc}D_cC_b/p\rho_{sc} \; \delta 10^4, \tag{5.12}$$

where M_{sc} (g/mol) is the molecular weight of the semiconductor, D_c (cm^2/s) is the diffusion coefficient of the etchant species, C_b (mol/liter) is the etchant concentration in the bulk solution, p is the number of etchant species necessary to remove one III/V pair, ρ_{sc} (g/cm^3) is the density of the semiconductor, δ (cm) is the diffusion layer thickness, and 10^4 is the conversion factor for centimeters to nanometers. Note that the materials parameters M_{sc} and ρ_{sc} are similar for all III/V compounds so that in the diffusion-limited regime etch rate is relatively independent of epilayer composition. This is not true of the case where the supply of etchant exceeds the rate at which it is consumed at the semiconductor surface. Under these conditions, anisotropic etching occurs to expose well-defined crystallographic planes.

Choosing the diffusion-limited regime with unstirred solutions of dilute bromine–methanol (0.2–0.8% by volume), Brenner and Melchoir (1994) performed a series of experiments as illustrated in Fig. 5.12. It was determined that the relative etch rate, r_e, of masked versus unmasked InP or InGaAsP is given by

$$r_e = \Phi^{-2/3}, \tag{5.13}$$

where Φ is defined as the ratio of the width of the mask opening to the period of the repeated mask opening, and the repetition period is large relative to the diffusion constant of the etchant. The $1/e$ diffusion length for bromine was found to be 65 μm. It was found that the relative etch rates for masks oriented along the (011) and (01$\bar{1}$) are equal, and that r_e remained constant to temperature variation from -9 to 25°C. Using this technique, Brenner, Bachman, and Melchior (1994) fabricated vertically tapered optical mode shape adapters. The polarization insensitive fiber-

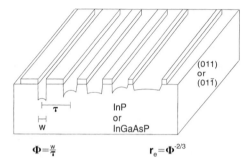

$$\Phi=\frac{w}{\tau} \qquad\qquad r_e=\Phi^{-2/3}$$

Fig. 5.12 Selective area etching, where r_e is defined as the relative etch rate of masked versus unmasked material. The formula is appropriate for cases where the etchant diffusion rate is small compared with the mask opening repetition rate (Brenner *et al.* 1994).

to-waveguide coupling loss was less than -1 dB with relaxed alignment tolerances of ± 2.5 μm in both horizontal and vertical directions. The longitudinal separation from flat-end single-mode fiber for a -1-db penalty was 28 μm!

The selective area etching technique is as powerful for longitudinal index control as SAG is for bandgap control and provides an added flexibility to the fabrication of complex PICs.

VI.   Beam Expanded Lasers

For fiber to the home to become a reality, every component to be used on the customer's premises must be inexpensive, must be conservative in its use of electrical power, and must involve as low a packaging cost as possible. For semiconductor emitters to be useful in optical telecommunications, it is necessary for light to couple efficiently from a semiconductor chip facet into an optical fiber or a silica-based planar lightwave circuit (PLC). The far-field emission angle (3 dB/FWHM) for generic CMBH active semiconductor devices is about 25°, implying mode field diameters (MFDs) of approximately 2 μm. However, for fibers and PLCs the acceptance angle (at 3 dB) is about 7°, with an average MFD of 8 μm. This factor of approximately 4 mismatch in mode size causes direct coupling losses on the order of -8 dB. Historically, lenses or lensed fibers have been used to improve coupling, despite the added expense and packaging complexity that this implies. More recently, many clever designs have been demonstrated to expand the optical mode at the output of semiconductor lasers. As the far-field emission angles drop to resemble those of fiber, the coupling efficiencies to flat "as cleaved" fiber approach 50% and the alignment tolerances improve tremendously. The horizontal and vertical translation tolerances for −1 dB of added loss change from 0.5 to more than 3 μm, whereas axial distances approach 18 μm of separation before a −1-dB penalty is imposed. This allows whole subsystems to be assembled by means of automation, and may eliminate the need for "active" (i.e., powered-device) coupling optimization, thus greatly reducing packaging cost. We examine four of these beam expansion schemes, as illustrated in Fig. 5.13. We keep a tabulation of some important figures of merit in Table 5.1. These are only representative of general principles and not meant as an exhaustive treatment of all work in this field.

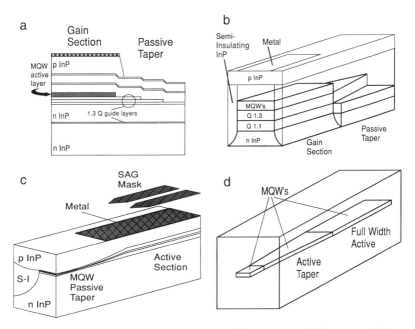

Fig. 5.13 Designs for beam expanded lasers (not to scale). (a) Waveguides are feathered by etching to cause mode expansion at the output facet. (b) A waveguide layer under the active layer is tapered horizontally. (c) The active region is grown through an SAG mask so that the active layer changes in index, thickness, and bandgap at the output facet. (d) The active layer itself is tapered at the output facet and electrically pumped throughout.

Table 5.1   Expanded Beam Laser Comparison

Technique/Source	λ (μm)	dQE (%)	Coupling Loss (dB)	I_{th} at 25°C (mA)	$-1\text{-}dB$ Vert./ Horiz. Tolerance ($\pm \mu m$)	Vert./ Horiz. Far Field
Fig. 5.13a/Koch et al. (1990)	1.48	30	−4.2	45	2.1/3.8	12/12
Fig. 5.13b/Ben-Michael et al. (1994)	1.55	50	−3.5	12	2.6/3.1	7/5
Fig. 5.13b/Ben-Michael et al. (1994)	1.3	51		22	2.5/2.8	8/6
Fig. 5.13c/Takemoto et al. (1995)	1.3		−2.6	22[a]		12/11
Fig. 5.13c/ Yamamoto et al. (1995)	1.3	35	−3.8	6.5	2/2	10.8/9.2
Fig. 5.13d/Fukano et al. (1995)	1.3	54	−2.3[b]	4.7	3.2/2.8	
Fig. 5.13d/Doussiere et al. (1994)	1.48	52	−4.9	25		15/13
SAE[c]/Brenner, Hess, and Melchior (1995)	1.3		−3.0	54	1.8/1.8	

[a] Uncoated facets.
[b] Dispersion-shifted 4-μm core fiber.
[c] SAE, selective area etching.

Pioneering work in this field was done by Koch and coworkers in 1990 using InGaAs/InGaAsP at 1.48 μm. According to their initial publication, expansion was accomplished by a multilayered structure in which the MQW active layer sat atop a stack of three separate 1.3-μm waveguide layers, as shown in Fig. 5.13a. Prior to mesa formation, the active region was defined and successive waveguiding layers were etched away along the axis of the laser cavity so that only the thinnest remained at the output facet. As light emerged from the MQW active section toward the right, it experienced weaker and weaker confinement as the effective index of the structure dropped, which caused the optical mode to expand at the right facet. Far-field angles of 12° were achieved in both horizontal and vertical directions. These researchers achieved a 30% differential quantum efficiency despite the −1.4-dB added cavity loss introduced by the beam expansion section. The output mode butt-coupled into a single-mode fiber with only −4.2-dB loss, and had a −1-dB alignment tolerance of ±3.8 μm parallel to the laser junction and ±2.1 μm perpendicular to the laser junction. Although the index steps at the end of each waveguide truncation were extremely small, the vertical far field was displaced by several degrees at an angle toward the substrate; presumably, this displacement was due to an admixture of the noncoupled light from the taper steps. Nevertheless, it was a powerful proof of principle.

In subsequent work, Ben-Michael, of the same group, and colleagues (1994) devised the scheme depicted in Fig. 5.13b for a 1.55-μm laser. In this scheme, there are no abrupt index steps in the beam expansion section. The 1.3-μm InGaAsP quaternary (Q) layer below the active layer is etched to form a laterally tapered adiabatic mode expander. As light approaches the output facet, it is finally supported by only an 800-Å-thick 1.1-μm InGaAsP layer. With this technique, lasers were fabricated at both 1.3- and 1.55-μm wavelengths (with both waveguide layers being Q1.1 for the 1.3-μm lasers). At 1.55 μm the lasers showed extremely narrow far-field patterns with FWHM of 5° laterally and 7° vertically (6° × 8° for the 1.3-μm devices). The −1-dB excess loss point due to misalignment was ±2.6 μm (±2.5 μm) and ±3.1 μm (±2.8 μm) laterally for the λ = 1.55 (1.3) μm laser. This is a further improvement in performance. However, both the aforementioned schemes are rather complex lithographically and might suffer lower production yields in consequence. This inspired the search for simpler structures with similar performance.

Yamamoto et al. (1995) of Fujitsu, as well as Takemoto et al. (1995) of Nippon Telephone and Telegraph (NTT), have produced beam expanded lasers using SAG. We saw in the previous section that the use of a mask during epitaxy could cause thicker growth near the mask boundary. Taken to its extreme (wide masks and narrow openings), SAG can produce growth enhancements of factors greater than 4, with laser-quality material produced in these narrow openings. Thus, starting with MQW material of effective bandgap 1.1 μm, the bandgap is shifted to 1.3 μm in a mask opening approximately 10 μm across flanked by masked regions about 100 μm wide. This is shown schematically in Fig. 5.13c. The mask shown floating above the laser would have been used at the time of MQW growth, resulting in the continuous vertical tapering of the thickness of the highest index (MQW) layer. The tapers on the mask itself may be used to cause an even more gradual vertical growth taper. The processed result was a standard CMBH structure as seen at the output cross section in Fig. 5.13c. Lithographically this is quite simple. Some uncertainty exists as to how far past the masked (active) region current must be injected to achieve transparency as the bandgap changes along the laser axis. By varying electrode lengths, Yamamoto *et al.* have rigorously shown that for this mask design this distance is 50 μm past the end of the mask. The results for both groups give outstanding figures of merit, as seen in Table 5.1. In addition, Yamamoto *et al.* achieved excellent high-temperature performance with thresholds of 22.2 mA at 85°C. One remaining problem for this technique is the fact that SAG still requires an extra lithography step compared with a standard CMBH process. In addition, there is p-doped InP over the passive taper, which could cause unwanted free-carrier absorption loss as the optical mode expands to fill the output end of the laser. Finally, note that all the previously mentioned techniques have essentially taken the best approximately 300-μm-long laser available and added an approximately 300-μm-long passive beam expander, thus doubling the laser length and cutting the normal production yield in half.

In 1994, Doussiere *et al.* proposed a simple solution to the aforementioned problems. In a standard CMBH process with 1.48-μm MQW material, they created a mesa definition mask with a taper at one end, as shown in Fig. 5.13d. By pumping the entire cavity, they obtained high differential quantum efficiency and reduced far-field angles. Fukano *et al.* (1995), working at 1.3 μm, have shown even more impressive results with the same structure, as seen in Table 5.1 (note 4.7-mA I_{th} at 25°C). Most significant, they have now reduced the length of the laser cavity to a more conventional 225 μm, improving the yield per wafer, in this respect alone, to the standard values without tapers. In addition to meeting the other desirable criteria of beam expansion, Fukano *et al.* demonstrated a lasing threshold at 85°C of only 18 mA and produced a 20-mW output for 90 mA at 85°C. These are exactly the figures of merit necessary to providing robust uncooled sources that are inexpensive to package for local access. The only drawback of the Fig. 5.13d technique is the small dimensions required. Whereas the back side of the mesa at the active layer is about 1 μm wide (at the limit common to industrial production), the output side is only 0.4 μm wide. Both Doussiere *et al.* (1994) and Fukano *et al.* used reactive ion etching rather than wet chemical etching to achieve this result. It may be a challenge to reproduce this 0.4-μm dimension accurately in large numbers.

Brenner, Hess, and Melchior (1995) have taken yet another approach using selective area etching. First, a base wafer consisting of a 4.5-μm-thick stack of widely separated quantum wells capped by a Q1.1 waveguide layer was grown. With selective area etching the structure was feathered away toward the output facet. The active MQWs and upper P-doped cladding were grown over this. The mesa formation mask was then made to flare out at the output facet from 3 to 7 μm and left as a freestanding ridge. The final chip length was only 300 μm, and as with Fukano *et al.*'s work, the entire mesa was electrically pumped. These researchers achieved a coupling efficiency to flat-end fiber of -3 dB and vertical–horizontal alignment tolerances of ±1.8 μm.

Finally, it is obvious that the data in Table 5.1 are taken from laboratories all over the world, thus there may be slight differences in calibration. However, it is interesting to note that the first three entries were all analyzed with the same far-field stage and coupled to the same cleaved fiber. There is clear evidence that reducing the beam divergence past an optimal angle produces poorer coupling. For lowest butt-coupling loss, it is necessary to *match* the acceptance angle of the component receiving the light.

VII. Conclusion

The 1990s have been an extremely exciting time for semiconductor laser physics. Advances in growth techniques have allowed monolayer abrupt structures with high degrees of uniformity to become routine. The introduction of strain has taught us a great deal about the band structure of semiconductors and has enabled us to greatly improve laser thresholds, power, reliability, efficiency, and high-temperature performance. The techniques of SAG and selective area etching have provided a platform for photonic integration that has already yielded commercial products. Clever designs using all of the aforementioned are finally creating easily packaged components and subsystems, which can lower the cost of photonics to a degree that makes them candidates for local access as well as long-distance networks.

References

Adams, A. R. 1986. Band structure engineering for low-threshold high efficiency semiconductor lasers,. *Electron. Lett.* 22:249–250.

Adams, A. R., M. J. Hawley, E. P. O'Reilly, and W. S. Ring. 1993. The success of strained layer lasers elucidated by high pressure experiments. *Japn. J. Appl. Phys.* 32(suppl. 32-1):358–360.

Adams, A. R., K. C. Heasman, and E. P. O'Reilly. 1989. In *Band structure engineering in semiconductor microstructures*, ed. R. A. Abrams and M. Jaros, 279–301. New York: Plenum.

Andersson, T. J., Z. G. Chen, V. D. Kulakovski, A. Uddin, and J. T. Vallin. 1987. Variation of the critical layer thickness with In content in strained InGaAs-GaAs quantum wells grown by MBE. *Appl. Phys. Lett.* 51:752–754.

Aoki, M., M. Suzuki, M. Takahashi, H. Sano, T. Ido, T. Kawano, and A. Takai. 1992. High-speed (10 Gbit/s) and low-drive-voltage (1V peak to peak) InGaAs/InGaAsP MQW electroabsorption-modulator integrated DFB laser with semi-insulating buried heterostructure. *Electron. Lett.* 28:1157–1158.

Aoki, M., T. Taniwatari, M. Suzuki, and T. Tsutsui. 1994. Detuning adjustable multiwavelength MWQ-DFB laser array grown by effective index/quantum energy control selective area MOVPE. *Photon. Tech. Lett.* 6:789–791.

Ben-Michael, R., U. Koren, B. I. Miller, M. G. Young, M. D. Chien, and G. Raybon. 1994. InP-based multiple quantum well lasers with an integrated tapered beam expander waveguide. *IEEE Photon. Tech. Lett.* 6:1412–1414.

Bhat, R. 1992. Current status of selective area epitaxy by MOCVD. *J. Cryst. Growth* 120:362–368.

Bhat, R., M. A. Koza, J. Hennessey. 1995. Growth studies in a vertical rotating-disk reactor. In *EW-MOVPE VI, Belgium*. Paper E9.

Bhat, R., C. E. Zah, M. A. Koza, D-M. D. Hwang, F. J. Favire, and B. Pathak. 1992. OMCVD growth of strained $Al_xGa_yIn_{1-x-y}As$ for low threshold 1.3 μm and 1.55 μm quantum well lasers (paper presented at the fourth international conference on InP and related materials, Newport, RI, April), Paper THD2.

Bowers, J. E., and M. A. Pollack. 1988. Semiconductor lasers for telecommunications. In *Optical fiber telecommunications II*, ed. S. E. Miller and I. P. Kaminow, 509–561. New York: Academic Press.

Brenner, T. 1994. Monolithically integrated semiconductor optical amplifiers and optical mode adapters. Ph.D. diss., Swiss Federal Institute of Technology ETH no. 10961, 54–63.

Brenner, T., M. Bachman, and H. Melchior. 1994. Vertically tapered InGaAsP/InP waveguides for highly efficient coupling to flat end single mode fibers. *Appl. Phys. Lett.* 65:798–800.

Brenner, T., R. Hess, and H. Melchior. 1995. Compact InGaAsP/InP laser diodes with integrated mode expander for efficient coupling to flat-ended singlemode fibers. *Electron. Lett.* 31:1443–1445.

Brenner, T., and H. Melchior. 1994. Local etch-rate control of masked InP/InGaAsP by diffusion limited etching. *J. Electrochem. Soc.* 141:1954–1956.

Caneau, C., R. Bhat, C. C. Chang, K. Kash, and M. A. Koza. 1993. Selective organometallic vapor phase epitaxy of Ga and In compounds: A comparison of TMIn and TEGa. *J. Cryst. Growth* 132:364–370.

Caneau, C., R. Bhat, M. R. Frei, R. J. Chang, R. J. Deri, and M. A. Koza. 1992. Studies on the selective OMVPE of (Ga,In)/(As,P). *J. Cryst. Growth* 124:243–248.

Corzine, S. W., R-H. Yan, and L. A. Coldren. 1993. Optical gain in III-V bulk and quantum well semiconductors. In *Quantum well lasers*, ed. P. S. Zory, 17–96. New York: Academic Press.

Darcie, T. E., R. M. Jopson, and A. A. Saleh. 1988. Electronic compensation of saturation induced crosstalk in optical amplifiers. *Electron. Lett.* 24:1154–1155.

Doerr, C. R., C. H. Joyner, M. Zirngibl, L. W. Stulz, and H. Presby. 1995. Elimination of signal distortion and crosstalk from carrier density changes in the shared amplifier of multifrequency signal sources. *IEEE Photon. Tech. Lett.* 7:1131–1133.

Doussiere, P., P. Garabedian, C. Graver, E. Derouin, E. Gaumont-Goarin, G. Michaud, and R. Meilleur. 1994. Tapered active stripe for 1.5 μm InGaAsP/InP strained multiple quantum well lasers with reduced beam divergence. *Appl. Phys. Lett.* 64:539–541.

Eckel, M., D. Ottenwalder, F. Scholz, G. Frankowsky, T. Wacker, and A. Hangleiter. 1994. Improved composition homogeneity during selective area epitaxy of GaInAs using a novel In precursor. *Appl. Phys. Lett.* 64:854–856.

Finders, J., J. Geurts, A. Kohl, M. Weyers, B. Opitz, O. Kayser, and P. Balk. 1991. Composition of selectively grown InGaAs structures from locally resolved Raman spectroscopy. *J. Cryst. Growth* 107:151–155.

Fujii, T., M. Ekawa, and S. Yamazaki. 1995a. Growth pressure dependence of selective area metalorganic vapor phase epitaxial growth on planar patterned substrates. *J. Cryst. Growth* 156:59–66.

Fujii, T., M. Ekawa, and S. Yamazaki. 1995b. A theory for metalorganic vapor phase epitaxial growth on planar patterned substrates. *J. Cryst. Growth* 146:475–481.

Fukano, H., Y. Katoda, Y. Kondo, M. Ueiki, Y. Sakai, K. Kasaya, K. Yokoyama, and Y. Tohmori. 1995. 1.3 μm large spot-size laser diodes with laterally tapered active layer. *Electron. Lett.* 31:1439–1440.

Gale, R. P., R. W. McClelland, J. C. C. Fan, and C. O. Bozler. 1982. Lateral epitaxial overgrowth of GaAs by organo-metallic vapor-deposition. *Appl. Phys. Lett.* 41:545–547.

Galeuchet, Y. D., P. Roentgen, and V. Graf. 1990. InGaAs/InP selective area metalorganic vapor phase epitaxy for one-step-grown buried low-dimensional structures. *J. Appl. Phys.* 68:560–568.

Ghiti, A., And E. P. O'Reilly. 1993. Valence band engineering in quantum well lasers. In *Quantum well lasers*, ed. P. S. Zory, 329–366. New York: Academic Press.

Godefroy, A., A. Le Corre, F. Clerot, S. Salaun, S. Loualiche, J. C. Simon, L. Henry, C. Vaudry, J. C. Keromenes, G. Jouile, and P. Lamouler. 1995. 1.55-μm polarization-insensitive optical amplifier with strain-balanced superlattice active layer. *IEEE Photon. Tech. Lett.* 7:473–475.

Henry, C. H. 1982. Theory of the linewidth of semiconductor lasers. *IEEE J. Quantum Electron.* QE-18:259–264.

Itagaki, T., T. Kimura, Y. Goto, Y. Mihashi, S. Takamiya, and S. Mitsui. 1994. Analysis of the in-plane bandgap distribution in selectively grown InGaAs/

InGaAsP multiple quantum well by low pressure metalorganic chemical vapor deposition. *J. Cryst. Growth* 145:256–262.

Jensen, K., and D. Coronell. 1991. Analysis of MOCVD of GaAs on patterned substrates. *J. Cryst. Growth* 114:581–592.

Joyner, C. H., S. Chandrasekhar, J. W. Sullhoff, and A. G. Dentai, 1992. Extremely large band gap shifts for MQW structures by selective area epitaxy on SiO$_2$ masked substrates. *IEEE Photon. Tech. Lett.* 4:1006–1009.

Joyner, C. H., M. Zirngibl, and J. C. Centanni. 1995. An 8-channel digitally tunable transmitter with an electroabsorption modulated output by selective area epitaxy. *IEEE Photon. Tech. Lett.* 7:1013–1015.

Joyner, C. H., M. Zirngibl, and J. P. Meester. 1994. A multifrequency waveguide grating laser by selective area epitaxy. *IEEE Photon. Tech. Lett.* 6:1277–1279.

Kano, F., T. Yamanaka, N. Yamamoto, H. Mawatari, Y. Thomori, and Y. Yoshikuni. 1994. Linewidth enhancement factor in InGaAsP/InP modulation-doped strained multiple-quantum-well lasers. *IEEE J. Quantum Electron.* 30:533–537.

Kato, T., T. Sasaki, N. Kida, K. Komatsu, and I. Mito. 1991. Novel MQW DFB laser diode/modulator integrated light source using bandgap energy control epitaxial growth technique. In *ECOC/IOOC '91 Proceedings,* part 2, 429–432 Paper WeB7-1.

Koch, T. L., U. Koren, G. Eisenstein, M. G. Young, M. Oron, C. R. Giles, and B. I. Miller. 1990. Tapered waveguide InGaAs/InGaAsP multiple-quantum-well lasers. *IEEE Photon. Tech. Lett.* 2:88–90.

Koch, T. L., and R. A. Linke. 1986. Effect of nonlinear gain on semiconductor laser wavelength chirping. *Appl. Phys. Lett.* 48:613–615.

Krijn, M. P. C. M., G. W. t'Hooft, M. J. B. Boermans, P. J. A. Thijs, J. J. M. Binsma, L. F. Tiemeijer, and C. J. van der Poel. 1992. Improved performances of compressively as well as tensile-strained quantum well lasers. *Appl. Phys. Lett.* 61:1772–1774.

Long, J. A., R. A. Logan, and R. F. Karlicek, Jr. 1988. Epitaxial growth methods for lightwave devices. In *Optical fiber telecommunications II,* ed. S. E. Miller and I. P. Kaminow, 631–663. New York: Academic Press.

Marari, K., M. Okamoto, K. S. Suzuki, Y. Noguchi, and O. Mikami. 1994. Polarization-insensitive optical amplifier with tensile-strained-barrier MQW structure. *IEEE J. Quantum Electron.* 30:695–701.

Mathews, J. W., and A. E. Blakeslee. 1974. Defects in epitaxial multilayers. *J. Cryst. Growth* 27:118–125.

Miller, B. I., U. Koren, M. G. Young, and M. D. Chien. 1991. Strain-compensated strained-layer superlattices for 1.5 μm wavelength lasers. *Appl. Phys. Lett.* 58:1952–1954.

Morton, P. A., R. A. Logan, T. Tanbun-Ek, P. F. Sciortino, Jr., A. M. Sergent, R. K. Montgomery, and B. T. Lee. 1992. 25 Ghz bandwidth 1.55 μm GaInAsP p-doped strained multiquantum-well lasers. *Electron. Lett.* 28:2156–2157.

Morton, P. A., T. Tanbun-Ek, R. A. Logan, N. Chand, K. W. Wecht, A. M. Sergent, and P. F. Sciortino, Jr. 1994. Packaged 1.55 μm DFB laser with 25 GHz modulation bandwidth. *Electron. Lett.* 30:2044–2045.

Morton, P. A., H. Temkin, D. Coblentz, R. A. Logan, and T. Tanbun-Ek. 1992. Enhanced modulation bandwidth of strained MQW lasers. *Appl. Phys. Lett.* 60:1812–1814.

Murata, M., T. Morita, K. Koike, T. Katsuyama, and K. Ono. 1989. Planar imbedded growth of GaInAs by MOVPE. In *International Symposium on GaAs and Related Compounds, Karuizawa, Japan,* Physics Conference Series No. 106, 87–92.

Newkirk, M. A., B. A. Miller, U. Koren, M. G. Young, M. Chien, R. M. Jopson, and C. E. Burrus. 1993. 1.5 μm multiquantum-well semiconductor optical amplifier with tensile and compressively strained wells for polarization independent gain. *IEEE Photon. Tech. Lett.* 5:406–408.

Olshansky, R., P. Hill, V. Lanzisera, and W. Powazinik. 1987. Frequency response of 1.3 μm InGaAsP high speed semiconductor lasers. *IEEE J. Quantum Electron.* QE-23:1410–1418.

O'Reilly, E. P., and A. R. Adams. 1994. Band structure engineering in strained semiconductor lasers. *IEEE J. Quantum Electron.* 30(2):366–379.

Osinski, J. S., P. Grodzinski, Y. Zou, P. D. Dapkus, Z. Karim, and A. R. Tanguary, Jr. 1992. Low threshold current 1.5 μm buried heterostructure lasers using strained quaternary quantum wells. *IEEE Photon. Tech. Lett.* 4:1313–1317.

Osowski, M. L., R. M. Cockeril, R. M. Lammert, D. V. Forbes, D. E. Ackley, and J. J. Coleman. 1994. A strained-layer InGaAs-GaAs-AlGaAs single quantum well broad spectrum LED by selective-area metalorganic chemical vapor deposition. *IEEE Photon. Tech. Lett.* 6:1289–1292.

Osowski, M. L., R. M. Lammert, D. V. Forbes, D. E. Ackley, and J. J. Coleman. 1995. Broadband emission from InGaAs-GaAs-AlGaAs LED with integrated absorber by selective-area MOCVD. *Electron. Lett.* 31:1498–1499.

Ougazzaden, A., D. Sigogne, A. Mircea, E. V. K. Rao, A. Ramdane, and L. Silvestre. 1955. Atmospheric pressure MOVPE growth of high performance polarisation insensitive strain compensated MQW InGaAsP/InGaAs optical amplifier. *Electron. Lett.* 31:1242–1244.

Pearsall, T. P. 1982. *InGaAsP alloy semiconductors.* New York: Wiley.

Razeghi, M. 1989. *The MOCVD challenge.* Philadelphia: Adam Hilger.

Ring, W. S., A. R. Adams, P. J. A. Thijs, and T. van Dongen. 1992. Elimination of the intervalence band absorption in compressively strained InGaAs/InP 1.5 μm MQW lasers observed by hydrostatic pressure measurements. *Electron. Lett.* 28:569–570.

Sasaki, T., M. Yamaguchi, and M. Kitamura. 1994. 10 wavelength MQW-DBR lasers fabricated by selective MOVPE growth. *Electron. Lett.* 30:785–786.

Silver, M., and E. P. O'Reilly. 1994. Gain and radiative current density in InGaAs/InGaAsP lasers with electrostatically confined electron states. *IEEE J. Quantum Electron.* 30(2):547–553.

Suzuki, M., M. Aoki, T. Tsuchiya, and T. Taniwatari. 1994. 1.24–1.66 μm quantum energy tuning for simultaneously grown InGaAs/InP quantum wells by selective-area metalorganic vapor phase epitaxy. *J. Cryst. Growth* 145:249–255.

Takemoto, A., Y. Miyazaki, K. Shibata, Y. Hisa, K. Goto, T. Itagaki, T. Takiguchi, and E. Omura. 1995. High-speed and narrow-beam 1.3 μm diode lasers with monolithically integrated waveguide lens formed by selective-area epitaxial growth. In *IOOC '95 Proceedings*. Paper PD1-4.

Temkin, H., N. K. Dutta, T. Tanbun-Ek, R. A. Logan, and A. M. Sergent. 1990. InGaAs/InP quantum well lasers with sub-mA threshold current. *Appl. Phys. Lett.* 57:1610–1612.

Thijs, P. A. J., J. J. M. Binsma, L. F. Tiemeijer, and T. van Dongen. 1991. Improved performance 1.5 μm wavelength tensile and compressively strained InGaAs-InGaAsP quantum well lasers. In *Technical Digest ECOC/IOOC Proceedings, Paris, France, September*, vol. 2, 31–38.

Thijs, P. A. J., J. J. M. Binsma, L. F. Tiemeijer, and T. van Dongen. 1992. Sub-mA threshold current (0.62 mA) and high power (220 mW) 1.5μm tensile strained InGaAs single quantum well lasers. *Electron. Lett.* 28:829–830.

Thijs, P. A. J., L. F. Tiemeijer, J. J. M. Binsma, and T. van Dongen. 1994. Progress in long-wavelength strained-layer InGaAs(P) quantum-well semiconductor lasers and amplifiers. *IEEE J. Quantum Electron.* 30(2):477–499.

Thijs, P. A. J., L. F. Tiemeijer, P. I. Kuindersma, J. J. M. Binsma, and T. van Dongen. 1991. High performance 1.5 μm wavelength InGaAs/InGaAsP strained quantum well lasers and amplifiers. *IEEE J. Quantum Electron.* 27:1426–1439.

Thrush, E. J., J. P. Stagg, M. A. Gibbon, R. E. Mallard, B. Hamilton, J. M. Jowett, and E. M. Allen. 1993. Selective and non-planar epitaxy of InP/GaInAs(P) by MOCVD. *Mater. Sci. Eng.* B21:130–146.

Tiemeijer, L. F., P. J. A. Thijs, J. J. M. Binsma, and T. van Dongen. 1992. Direct measurement of the transparency current and valence band effective masses in tensile and compressively strained InGaAs/InP multiple quantum-well laser amplifiers. *Appl. Phys. Lett.* 60:554–556.

Tiemeijer, L. F., P. J. A. Thijs, T. van Dongen, R. M. W. Slootweg, J. M. M. van der Heijden, J. J. M. Binsma, and M. P. C. M. Krijn. 1993. Polarization insensitive multiple quantum well laser amplifiers for the 1300 μm window. *Appl. Phys. Lett.* 62:826–828.

Yablonovitch, E., and E. O. Kane. 1986. Reduction of the lasing threshold current density by lowering the valence band effective mass. *J. Lightwave Tech.* LT-4(May):504–506, with correction in LT-4(July):961.

Yamamoto, T., H. Kobayashi, M. Ekawa, T. Fujii, H. Soda, and M. Kobayashi. 1995. High temperature operation of 1.3 μm narrow beam divergence tapered-thickness waveguide BH MQW lasers. *Electron. Lett.* 31:2178–2179.

Zah, C. E., R. Bhat, F. J. Favire, B. Pathak, C. Caneau, N. C. Andreadakis, P. S. D. Lin, A. S. Gozda, and T. P. Lee. 1992. Long wavelength strained-layer quantum well lasers. *SPIE, Quantum Well and Superlattice Physics IV* 1675: 172–179.

Zah, C. E., F. J. Favire, R. Bhat, S. G. Menocal, N. C. Andreadakis, D. M. Hwang, M. Koza, and T. P. Lee. 1990. Submilliampre-threshold 1.5 μm strained-layer multiple quantum well lasers. *IEEE Photon. Tech. Lett.* 2:852–853.

Chapter 6 | Vertical-Cavity Surface-Emitting Lasers

L. A. Coldren
B. J. Thibeault

Department of Electrical and Computer Engineering, University of California, Santa Barbara, California

I. Introduction

The inclusion of a vertical-cavity surface-emitting laser (VCSEL) chapter within this relatively practical book indicates that VCSELs are now considered to be viable candidates for many real communications applications. This is a remarkable turn of events because at the beginning of the 1990s these devices were considered little more than laboratory novelties. Their rise in credibility has largely been forced by the rapid evolution of their performance as well as the more widespread recognition of their compatibility with low-cost wafer-scale fabrication and characterization technologies. They are especially interesting for array applications because specific arrays can be formed with no change in the fabrication procedure. In this chapter, the design, fabrication, and performance of a variety of VCSEL structures are reviewed, and their use in a number of applications is introduced.

Most characteristics of GaAs-based VCSELs in the 0.8- to 1.0-μm wavelength range are now comparable to or better than those of edge-emitters in the lower power ($\sim$1-mW) regime where many short-haul data communications applications fall. For these applications, fiber loss and dispersion are generally not significant factors. These VCSELs are also proposed for many other applications, ranging from printing to optical switching. As we learn in the following sections, their improved characteristics are a natural consequence of scaling the active and modal volumes of these diode lasers. The appeal of the VCSEL structure is that it enables this scaling in a simpler way than with edge-emitters. Besides the manufacturability feature, additional attractive characteristics of VCSELs include their circularly shaped, low-numerical-aperture output beams for easy coupling to fibers or free-space optics; their single-axial-mode spectra (although lateral modes still need to be dealt with) for potential wavelength-division multiplexing

OPTICAL FIBER TELECOMMUNICATIONS,
VOLUME IIIB

Copyright © 1997 by Lucent Technologies.
All rights of reproduction in any form reserved.
ISBN: 0-12-395171-2

(WDM) or wavelength addressing schemes; their high power conversion efficiency in the low power range for reduced heating in highly integrated circuits; and their natural vertical emission for array applications.

Efforts in InP-based longer wavelength VCSELs, favored for long-haul fiber-based telecommunications (1.3–1.6 μm), have met with slower progress because of inherent difficulties in constructing highly reflecting mirrors as well as high-gain active regions. Nevertheless, recent work with wafer-bonded GaAs mirrors and strained quantum well active regions suggests that viable devices may be forthcoming. Similarly, shorter wavelength, visible VCSELs have been somewhat more difficult to engineer, but again, recent progress suggests that viable devices will shortly emerge. These are desired for plastic-fiber data links (~650 nm) as well as storage and display applications.

II. Structures

A. ETCHED MESA

Figure 6.1 illustrates the three most popular structures that have emerged in the GaAs-based devices. The first, termed the *etched-mesa structure* [1–4], is very analogous to the edge-emitting ridge structure. Only in this case, a round or square postlike mesa is formed. As in the edge-emitter case, the etch is usually stopped just above the active layer to avoid surface recombination of carriers and reliability problems. Thus, current is confined or apertured to the lateral dimensions of the mesa, but carriers are free to diffuse laterally in the active region. Thus, there is a *lateral leakage current* that becomes important for dimensions less than about 10 μm in diameter. In typical multiple quantum well (MQW) InGaAs 0.98-μm VCSELs, this leakage current has been estimated to account for about half the threshold current for a 6-μm-diameter mesa with a value of about 70–110 μA per quantum well [5].

The etched-mesa structure also provides a lateral index of refraction step over the upper etched portion of the cavity. Because lateral diffraction is small for diameters greater than about 5 μm in typical-length vertical-cavity structures [6], the lateral mode can be approximated as the solution to a *uniform* step-index (vertical) waveguide with an effective index step equal to the fraction of the mode occupying the upper portion times the actual index step in the upper portion [7, 8]. Because this index step tends to be large ($\Delta n \sim 1.5$–2 for nitride or oxide coatings), these devices will

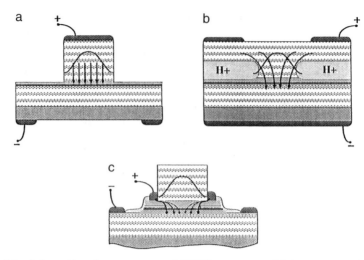

Fig. 6.1 Schematics of some common VCSEL structures. (a) Etched mesa (bottom emitting): the drive current is confined to the mesa width, current is conducted through the mirrors, and the optical index guide is formed by the mesa walls. (b) Proton implanted (top or bottom emitting): the current is confined by a semi-insulating region of the implant, current is conducted through the mirrors, and optical guiding is obtained by gain and thermal profiles. (c) Dielectric apertured with intracavity drive layers (top or bottom emitting): the current is apertured by the dielectric layer, current is conducted laterally along the contact layers, and optical index guiding is accomplished by the lensing action of the dielectric aperture and/or the mesa walls.

support multilateral modes for all but submicron-diameter mesas. For smaller sizes, where finite diffraction does occur, there also tends to be a scattering of energy into higher order modes at the boundary between the upper and lower portions of the axial cavity [8]. However, because the etched surface is not perfect, because the higher order modes tend to suffer higher losses than that of the fundamental mode, and because there is less scattering into higher order modes for larger diameters, single lateral mode operation is typically observed for diameters up to about 8 μm [9, 10]. Of course, this mode filtering action is accompanied by unwanted loss for the fundamental mode, and this increases the threshold currents and decreases the differential quantum efficiency. In fact, for typical MQW 0.98-μm etched-mesa VCSELs, anywhere between 30–50% of the excess current at threshold (about 50–160 μA/well) is due to the increase in loss for a 5-μm diameter compared to a large diameter VCSEL. The amount of excess current is exponentially dependent on the excess loss, which is a structure-dependent parameter and accounts for the wide variation in the contribution to the threshold current. As the device size is reduced the excess loss eventually gets so large that the QWs cannot provide the gain for lasing to occur. At the same time, the external differential efficiency reduction leads to very low output powers for the small VCSELs. It is this excess loss that ultimately limits the smallest achievable etched-mesa VCSEL size. Figure 6.2 gives results from etched-mesa devices, illustrating typical results for this structure [9].

Because of the relatively small area of the etched-mesa top, it is most convenient to construct bottom-emitting etched-mesa structures. These allow the entire top surface to be used for contact formation for relatively low-contact resistance. Also, this bottom-emitting structure is very compatible with flip-chip bonding technology. In this case, the etched-mesa is typically plated over with some metal such as gold, and this is then used to solder bond to a matching pad on the host substrate. Such a configuration provides a lower thermal impedance, and most parasitic capacitance is eliminated, because the bonding pads can be small [10]. Bottom emission implies that the substrate must be transparent to the emitted light. Thus, InGaAs/GaAs structures with emission wavelengths in the 0.9- to 1.0-μm range have been chosen with GaAs substrates. For relatively low data rates (<500 Mb/s), silicon detectors can be used in the 0.9- to 0.95-μm range. In many cases, emission in the 0.85- to 0.93-μm range is desired. For the etched-mesa structure, this implies either using a transparent AlGaAs substrate, removing the GaAs substrate material in the vicinity of the beam, or going to a top-emission structure.

A problem with the simple etched-mesa structure, as well as the other VCSELs that employ conduction through the mirrors, is the series voltage that develops across the mirrors as a result of the potential barriers at the numerous heterointerfaces [2, 3]. This is especially problematic in the p-type mirror because small barriers will inhibit thermionic emission for the heavy holes. To solve this problem, bandgap engineering techniques have been employed [3]. These involve matching a doping gradient to the compositional gradient within the transition between the different mirror layers to flatten the valence band. When these techniques have been used, excess p-mirror voltage drops have been reduced to as low as 0.2 V over 18 mirror periods [11]. The lowest voltages occur in proton-implanted or dielectric

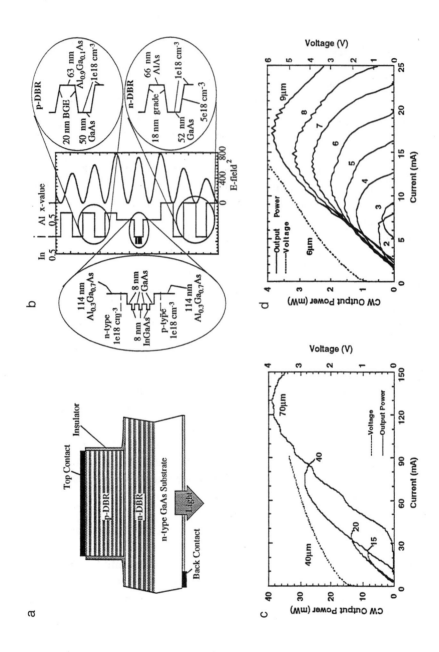

apertured structures, which can have a considerably larger contact area than the current aperture.

An alternative to solving the problem with conduction through the mirrors is to use intracavity contact layers. Figure 6.1c portrays these in combination with dielectric aperturing to avoid shunt current paths outside the cavity [12–14]. However, with the simple etched-mesa structure that employs no dielectric aperture, only a bottom intracavity contact is practical. Figure 6.3a shows an example of such a single-intracavity contact VCSEL [15]. As can be seen, the etch must cut through the active region in this case, but it is possible to use a semi-insulating substrate. Such structures have provided relatively good results, as illustrated in Fig. 6-3b. In this case, conduction through the p-mirror is still included, so bandgap engineering must still be employed to avoid a large series voltage. The use of semi-insulating substrates also reduces electrical parasitics for reduced electrical cross talk and increased modulation bandwidth. Because of the small modal volume of VCSELs, they should have relatively high modulation bandwidths at low current levels [13].

B. PROTON IMPLANTED

Figure 6.1b illustrates the planar, proton-implanted, gain-guided structure [11, 16–20]. This structure is very simple to manufacture, and it has therefore been touted by many as the most important of the VCSEL structures. The proton implant apertures the current to provide a desired current confinement. The planar structure also allows relatively large-area electrodes for low contact resistance. Both top- and bottom-emitting structures are possible, although most work has been with a ring-contacted top-

---

Fig. 6.2   Example results from an etched-mesa VCSEL at 980 nm grown by molecular beam epitaxy (MBE) [3]. (a) Schematic of bottom-emitting structure. (b) Material composition and electric-field-squared standing-wave profiles in the vicinity of the three quantum well InGaAs active region: interfaces in p-DBR (distributed Bragg reflector) are bandgap engineered (BGE) with parabolic compositional gradings and a dipole doping profile; a linear digital alloy grading is used in the n-DBR. (c) Light-out versus drive current for larger devices (numbers give diameters); the voltage-current characteristic is given for a 40-μm-diameter device. (d) Light-current and voltage-current characteristics for smaller devices. CW, continuous wave.

a

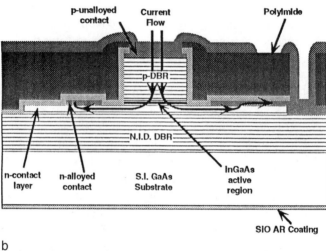

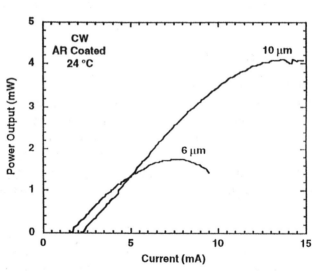

Fig. 6.3 Example results from a single-intracavity-contacted 980-nm VCSEL on a semi-insulating substrate [15]. (a) Device schematic. This structure uses a top p-type mirror of 31 periods with BGE interfaces, a three quantum well InGaAs/GaAs active region, an n-type GaAs contact layer, and an undoped 17.5 bottom period mirror. (b) Power output versus current for two sizes. The 6-μm-diameter device is single mode (>35-dB side mode suppression) over the entire operating range.

emitting configuration to allow 0.85-μm emission with GaAs substrates. The thermal impedance of this planar structure can also be lower than that of the etched-mesa structure, if the mesa is not plated in metal, because heat can now flow laterally from the upper mirror.

The proton-implanted structure shares the problems associated with conduction through the mirrors pointed out previously. Also, it has additional problems in properly aperturing the current as compared with the etched-mesa (or dielectric apertured) case. One can consider two cases: one in which the implant stops above the active region and one in which it penetrates the active region. To ensure that the implant damage does not penetrate the active region, the deep (>2-μm) implant must necessarily stop some distance above it. Thus, *lateral current spreading* tends to occur above the active region, and this adds a significant unproductive shunt current. If the implant proceeds through the active region, the current is effectively apertured, but because the carrier lifetime now goes to zero at the edge of the implant in the active layer, the effective interface recombination velocity is infinite, and this leads to much *larger carrier losses* than in the etched-mesa (or dielectric apertured) case, where the carriers must diffuse away. Reliability problems are also anticipated in such deeply implanted structures.

For many applications, the largest problem with the proton-implanted structure is the lack of any deliberate index guiding. The primary lateral index guiding is due to thermal lensing effects [16]. This can provide a stable mode under continuous-wave (CW) conditions, but under dynamic conditions the lateral mode tends to be dependent upon the data pattern. Thus, data coding schemes that guarantee nearly constant average power have been employed [21].

Modulation experiments have been carried out on these structures [18, 19]. Small signal bandwidths of 14 GHz at 8 mA [18] and 12 GHz at 2.8 mA [19] have been obtained, demonstrating the benefits of the low modal volume in these structures. Bit error rate (BER) measurements with both nominally single-mode and multimode VCSELs have also been investigated with step- and graded-index multimode fibers [21]. The effects of mode selective loss and feedback were studied. Mode selective loss at the transmitter seemed to be most critical, and only high levels of feedback seem to create error floors. Error-free operation at 1 Gb/s with a non-return-to-zero (NRZ) data format has been observed for fiber lengths up to about 500 m. This was obtainable with both multimode and single-mode VCSELs with significant levels of mode selective loss.

Despite the listed problems, the proton-implanted structure has provided attractive terminal characteristics, and it remains a viable candidate for many potential communications applications. Again, its producibility and reproducibility are very attractive in a manufacturing environment. Figure 6.4 gives examples of the characteristics of such structures [20].

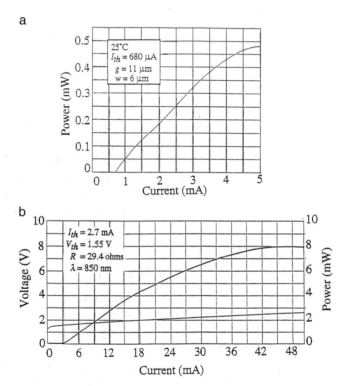

Fig. 6.4 Example results from proton-implanted VCSELs at 850 nm grown by metal-organic chemical vapor deposition (MOCVD) [20]. The device is similar to that in Fig. 6.1b. (a) Low-threshold result using a GaAs three quantum well active region and $Al_{0.16}Ga_{0.84}As/AlAs$ mirrors with 18 top (p-type) and 23 bottom (n-type) periods. Interfaces are graded and heavily doped. The implant and top contact window diameters are $g = 11$ μm and $w = 6$ μm, respectively. (b) Low-voltage results using a similar active region and mirror compositions, but with a special doping and grading profile at the mirror interfaces; $g = 20$ μm and $w = 15$ μm. (Reprinted from Morgan, R. A., et al. 1995. Producible GaAs-based top-surface emitting lasers with record performance. *Electron. Lett.* 31(6):463, with permission of the publisher.)

C. DIELECTRIC APERTURED

Figure 6.1c illustrates a dielectric apertured VCSEL structure with two intracavity contacts [12]. Dielectric aperturing has been found to be desirable even without the intracavity contacts, but this combined structure addresses several of the problem areas introduced previously. Both under-etching [12–14] and oxidation [22–25] of a high aluminum content AlGaAs layer have been used. The first purpose of the dielectric aperture is to block the shunt current that would otherwise flow between the p- and n-regions of the device. This current aperturing is very similar to that in the etched mesa, because the aperture can exist just above the active layer. In either case, the aperture is superior to that resulting from proton implantation, because good aperturing is not accompanied by reduced carrier lifetime in the active region (unless an etch with high ion damage is used). The dielectric aperture also allows for the injection of the current into a smaller region than the optical mode width. In fact, this latter feature was the primary reason that dielectric aperturing was first investigated [26]. More recently, it has become apparent that the *optical lensing* or waveguiding properties of the dielectric aperture also need to be considered [7, 8, 27, 28]. As we discuss subsequently, this index guiding technique provides a lower loss waveguide than other approaches such as with the etched-mesa structures.

Perhaps the most attractive feature of the dielectric apertured VCSEL is that it seems to combine several desirable features into one structure without removing flexibility of design. As illustrated in Fig. 6.1c, it enables both electrical contact layers to be between the mirrors so that conduction through the mirrors is avoided. Of course, either or both may be placed beyond the mirrors if desired. Placing a narrow aperture at a null of the electric-field-squared standing wave allows the current aperture to have little effect on the optical mode [27]. It can thus be moved inside the effective lateral dielectric waveguide that may be formed by another aperture placed away from the standing-wave null. Grading the aluminum content allows both functions to be performed in one layer. Thus, both current and photon confinement can be provided, but with independent control [27, 29]. Of course, lateral carrier diffusion will still occur, unless some lateral potential barrier is provided in the active region.

Figure 6.5 gives experimental results from the device of Fig. 6.1c [12, 13], and Fig. 6.6 gives additional experimental results along with the associated schematics of somewhat simpler versions of the dielectric apertured structure [24, 30, 31]. For comparison, we include results from both under-

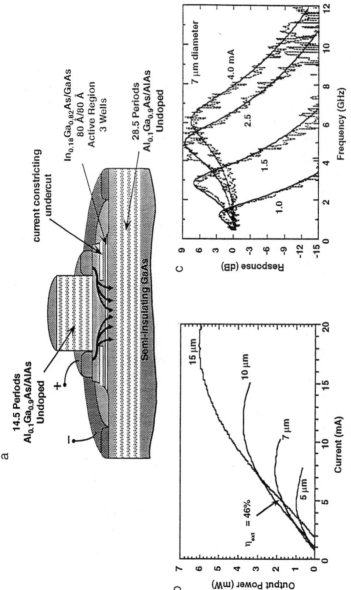

Fig. 6.5 Example results from an intracavity-contacted dielectric aperture 980-nm VCSEL with lateral under-etching [12, 13]. (a) Schematic. (b) Light-out versus current for various indicated diameters. (c) Small-signal frequency response for indicated DC bias currents from a 7-μm-diameter VCSEL. [a, b, Adapted with permission from Scott, J. W., *et al.* 1994. High efficiency submilliamp vertical cavity lasers with intracavity contacts. *IEEE Photon. Tech. Lett.* 6(6):678, 680. © 1994 IEEE.]

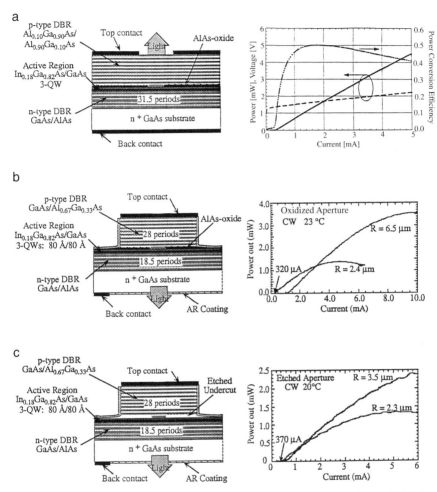

Fig. 6.6 Examples of dielectric apertured VCSELs. (a) Top-emitting VCSEL grown by MOCVD [24] (*left*). (*Right*) Power, voltage, and power conversion efficiency are plotted versus current for a 7-μm-square device. Because of the high-transmission output coupler and low voltages, 50% power conversion efficiency at 1.5 mA is obtained. (b) Bottom-emitting oxide-defined VCSEL grown by MBE [32] (*left*). (*Right*) Power output versus current is shown. Device sizes and threshold currents are reduced from etched pillar devices. (c) Bottom-emitting VCSEL from similar material as in (b), except lateral selective wet etching is used to form the current and optical mode aperture (*left*). (*Right*) The power output curves indicate similar performance as in the oxidized VCSEL. [a, Reprinted from Lear, K. L., *et al.* 1995. Producible GaAs-based top-surface emitting lasers with record performance. *Electron. Lett.* 31(3):208, with permission.]

etched and laterally oxidized versions that use through-the-mirror pumping. Modulation experiments in the under-etched structures, which also combined intracavity contacts on a semi-insulating substrate, have given some of the best examples of high bandwidths at low drive currents. As shown in Fig. 6.5c, the modulation of a 7-μm-diameter device gives bandwidths of more than 7 GHz with only 2.5 mA of total drive current. BER measurements done with *no current prebias* at 2.488 Gb/s using an NRZ data format also indicate no apparent noise floor for all sizes of devices tested [13]. These devices were limited in bandwidth by series voltage drops and size-dependent optical losses. Recent results with oxide-apertured devices that have lower series voltage drops and lower excess losses demonstrate a bandwidth of 15.3 GHz at only 2.1 mA of drive current for 3-μm diameter VCSELs [32].

At this writing, the dielectric apertured devices have provided the lowest threshold currents and highest power converison efficiencies of any VCSEL structure. Figure 6.6a illustrates overall power efficiencies of about 50% at output powers of approximately 1 mW [24]. This performance has never been achieved in edge-emitting lasers. It is a good example of the ease in scaling VCSELs. The back-emitting structures in Figs. 6.6b and 6.6c give comparable thresholds, but because of higher series resistance, they do not have the same efficiency.

The improved performance of the dielectric apertured VCSELs as compared with that of the etched mesas seems to be directly correlated to the obtainable size-dependent optical loss reduction, as indicated in Fig. 6.7. The excess threshold gain as compared with that of a broad-area device varies as approximately $1/r^2$ in all cases. The etched-mesa structures have higher optical losses due to surface roughness of the etched waveguide wall, as compared with the dielectric apertured structures, where the index guiding effect is due to the intracavity lensing action of the aperture [32].

D. VISIBLE

To move to shorter wavelengths, aluminum can be added to the GaAs quantum wells to increase their bandgap. AlGaAs has a direct bandgap for wavelengths down to about 640 nm, but population of the indirect minima, as well as nonradiative recombination associated with the Al, vastly inhibits performance below 780 nm. For viable visible wavelengths (<700 nm), the AlGaInP/GaAs system tends to be superior to AlGaAs/GaAs. This system allows the use of ternary GaInP quantum wells, com-

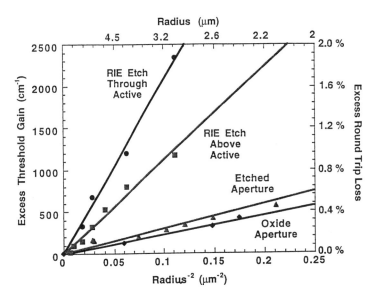

Fig. 6.7 Excess threshold gain (round-trip loss) versus size for etched pillar and dielectrically apertured VCSELs [5, 32]. The plot shows that as the device size is reduced, the round-trip losses increase as a result of optical scattering from the sidewalls or the lenslike dielectric aperture. Because the dielectric apertures reduce the optical scattering loss, the threshold currents are lower, efficiencies are higher, and smaller devices are possible. RIE, reactive ion etching.

pletely avoiding Al in these carrier storage areas. The most successful work in this area at this writing has used the structure indicated in Fig. 6.8 or some close variant of it [34, 35]. Representative data are also given for several wavelengths. As can be seen, a simple dielectric apertured structure has been adopted [35] to reduce the required threshold currents. Prior work used a proton-implanted structure [34], but as seems to be generally the case, the threshold currents were considerably higher, which resulted in additional device heating and reduced output powers.

The primary issue with these red-emitting VCSELs is providing sufficiently high transverse carrier confinement barriers as well as eliminating nonradiative recombination in the active region [34, 35]. Minimizing nonradiative recombination appears to be associated with eliminating or minimizing Al in and near to the quantum wells. Hole confinement is not a significant problem, but the available barrier in the conduction band between GaInP and $Al_x In_y Ga_{1-y} P$ is limited to about 170 meV for $x = 0.67$.

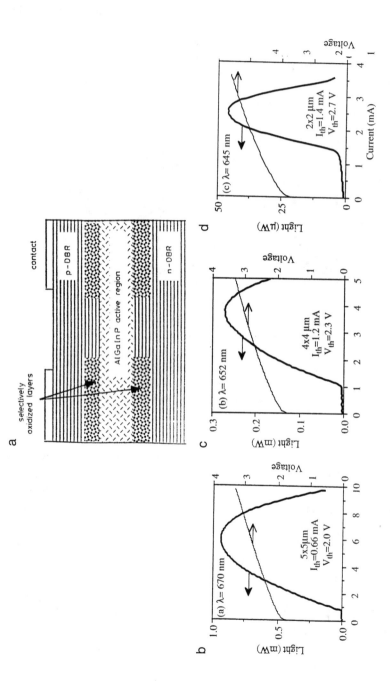

a

selectively oxidized layers

contact

p-DBR

AlGaInP active region

n-DBR

b

(a) λ= 670 nm

5x5μm
I_{th}=0.66 mA
V_{th}=2.0 V

Light (mW)

Voltage

Current (mA)

c

(b) λ= 652 nm

4x4 μm
I_{th}=1.2 mA
V_{th}=2.3 V

Light (mW)

Voltage

d

(c) λ= 645 nm

2x2 μm
I_{th}=1.4 mA
V_{th}=2.7 V

Light (μW)

Voltage

Current (mA)

214

This is not so small, but with quantum-confinement effects used to reduce the wavelength, the lowest electron energy rises, which reduces the net barrier to about 100 meV at 610 nm. With a larger Al content in the barriers, the conduction band offset is actually reduced, because the indirect bandgap is offset in favor of the valence band. To further improve confinement, one possible approach is to use multiple quantum barriers [36], as have been applied with some success in edge-emitters.

To obtain still shorter wavelengths (blue-green), the II-VI ZnSe-related compounds as well as the group III nitrides (e.g., GaN) are being investigated [37–43]. At the time of this writing, intensive work on II-VI edge-emitters has been carried out for a number of years in several laboratories. This has resulted in room-temperature CW operating devices in the blue-green spectral regions [39]. Threshold currents are relatively low, but significant series voltage has led to relatively inefficient operation. GaAs substrates are used for much of this work, so the lattice-mismatched hetero-epitaxy has led to a large number of defects that limit radiative efficiency as well as reliability. CW lifetimes are still limited to hours rather than many years as is desired. VCSELs have been fabricated in the II-VIs, but CW operation has not been achieved, and pulsed outputs are very low [40].

The group III nitride technology is very new at the time of this writing. Work has expanded dramatically since the demonstration of practical, high-efficiency light-emitting diodes (LEDs). High threshold lasers have been demonstrated [40], but lasers with reasonable properties have not been demonstrated. These InGaAlN materials are typically grown on sapphire or SiC, and very high dislocation densities are reported. Surprisingly, this high defect count does not appear to severely limit radiative efficiency or device lifetime in LEDs [42, 43]. Thus, although alternative substrates are

---

Fig. 6.8 Top-emitting red (<680-nm) VCSEL grown by low-pressure MOCVD on (311) GaAs substrate using lateral oxidation [34]. (a) Device schematic showing oxidized layers. The active region uses four compressively strained InGaP quantum wells. (b) Output power and voltage curves for a 5-μm-square 670-nm device. (c) Output power and voltage curves for a 4-μm-square 652-nm device. (d) Output power and voltage curves for a 2-μm-square 645-nm device. (Reprinted from Choquette, K. D., *et al.* 1995. *Electron. Lett.* 31(14):1145, with permission.)

being studied, it is not clear that the defects must be eliminated as is clearly the case in the II-VIs. VCSELs with epitaxial mirrors appear viable because the index difference between GaN and AlN is large. Dielectric mirrors and perhaps oxidized wafer-fused AlAs/GaAlAs may also find use.

E. *InP BASED*

VCSELs that operate in the 1.3- to 1.55-μm wavelength range [44–46] have not evolved as rapidly as the shorter wavelength devices. The reasons are that both highly reflecting mirrors and high-gain active regions are more difficult to obtain. The common InGaAsP/InP and InGaAlAs/InP lattice-matched materials systems both suffer from a relatively small range of achievable optical indices. Thus, very long epitaxial mirror stacks are necessary to theoretically develop the required reflectivity, and because of inherent optical losses, the net reflectivity saturates at about 98% in typical cases [47]. As a result, the epitaxial mirrors have not been used successfully in CW devices. Nevertheless, work continues to try to develop viable epitaxial mirrors in these systems as well as in the AlGaAsSb/InP system [48, 49]. Most effort, however, has turned to either deposited dielectric [44] or wafer-bonded AlGaAs/GaAs mirror stacks [46], which can provide more than 99% of net reflectivity.

The achievable gain in these materials is limited because of significant nonradiative carrier recombination due to Auger processes. These processes tend to vary as the cube of the carrier density and exponentially with temperature [50], which results in runaway nonradiative recombination for carrier densities greater than about 4×10^{18} cm^{-3}, especially above room temperature. Relatively thick ($\sim$1-μm) bulk InGaAsP active layers have been explored in order to obtain reasonable per-pass gain at relatively low carrier densities [44]. However, the high required current densities and the high thermal impedance of the VCSEL structure has made this approach very difficult. From analytical work it has become clear that one must maximize the gain per unit current density for success in these structures [51]. This suggests that strained MQWs, which have been found to provide increased gain at smaller current densities, should be used.

Figure 6.9 gives schematics and results from the first room-temperature CW operating VCSELs in both the 1.3- and 1.55-μm wavelength ranges [43, 45]. As can be seen, the 1.3-μm structure uses deposited dielectric mirrors, and the 1.55-μm structure uses wafer-bonded AlGaAs/GaAs mirrors. The 1.3-μm device uses liquid-phase epitaxy (LPE) regrowth to form

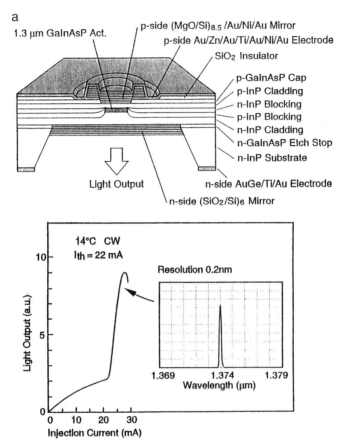

Fig. 6.9 Long-wavelength CW-VCSEL results with InGaAsP/InP active regions.
(a) Schematic (*top*) and results (*bottom*) from a 1.3-μm-wavelength, buried heterostructure VCSEL with dielectric mirrors [43]. The active region is 0.4 μm thick and
5 μm in diameter for the indicated results. (b) Schematic (*top*) and results (*bottom*)
from a 1.55-μm-wavelength, etched-mesa VCSEL with fused AlGaAs/GaAs mirrors
incorporating BGE [45]. The active region consists of seven strain-compensated
quantum wells. Room-temperature light-current results are given for several mesa
diameters as indicated. [a, Reprinted from Baba, T., *et al.* 1993. Near room temperature continuous wave lasing characteristics of GaInAsP/InP surface-emitting laser.
Electron. Lett. 29(10):913, with permission.]

a buried heterostructure active region with lateral current blocking layers.
This, in principle, provides the desired lateral current and carrier confinement as well as the potential for lateral optical waveguiding. The top MgO/

b

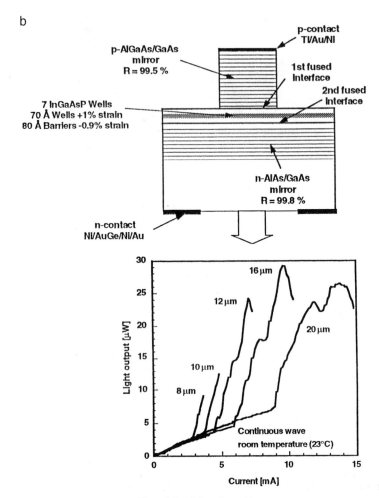

Fig. 6.9 *(Continued)*

Si dielectric mirror is designed to be thermally conductive in this 1.3-μm case. The device runs CW to just below room temperature despite a relatively high threshold current of 22 mA. Besides the use of a thick active region, part of the difficulty in this case may be due to loss in the mirrors associated with the amorphous Si layers.

The double-fused 1.55-μm structure is very analogous to the GaAs-based etched-mesa configuration shown in Fig. 6.1a. In fact, except for the thin InGaAs/InP strained MQW active region, the device is basically the same. It is formed by a two-step wafer-bonding procedure in which epitaxial mirror stacks grown on GaAs are fused to the active region grown on InP. After the epitaxial faces of the InP wafer are fused to the first GaAs wafer (which contains a p-type AlAs/GaAs mirror), the InP substrate is removed and this is fused to a second GaAs wafer with an n-type epitaxial mirror. The first GaAs substrate is then removed and the etched-mesa structure is formed. As in the case of Fig. 6.1a, electrical conduction to the active region is through the GaAs-based mirror stacks. Optical losses have been shown to be even more problematic at 1.3 and 1.55 μm because of increased absorption in p-doped AlGaAs at longer wavelengths [52]. Of course, with this technology any of the other structures introduced previously for GaAs could also be employed. In fact, more recently, double-fused 1.55-μm VCSELs have been constructed with oxide apertures. The result is a room temperature cw threshold current as low as 0.8 mA, cw operation to 64°C, and cw output powers as high as 3 mW in separately optimized devices. Thus, long-wavelength VCSELs are now beginning to look viable [53].

III.   Design Issues

For a viable VCSEL, it is generally understood that one must combine an *active region*, which can supply high levels of gain at reasonable current densities, with a low-loss *optical cavity* created by highly reflecting mirrors, and then provide a scheme for efficient *current injection* by minimizing leakage currents and series resistance. To scale such devices to small sizes, *lateral confinement* of the current, active region carriers, and photons in the desired mode is necessary. For efficient operation, minimizing the thermal impedance is also of prime importance in these small-volume devices.

   In this section, we briefly outline how VCSELs can be designed for a desired level of performance. For the most part, the physics in VCSELs is the same as in other lasers. However, some of the parameters tend to be quite different than those for edge-emitting diode lasers. In fact, some parameters are more similar to gas or solid-state lasers, even though the cavity volume is extremely small by comparison. For example, per-pass gains and losses tend to be about 1%, which are similar to those in He-Ne lasers. Also, there are some new issues that are not dealt with in typical edge-emitters. For example, the net modal gain provided by a quantum well active region can vary from zero to twice as large as expected, depending upon its axial (vertical) location relative to the optical standing wave (square of electric field) [54]. Mirrors are generally formed of multilayer quarter-wave stacks, and their separation must be accurately set to ensure that the single available Fabry–Perot mode is positioned near the center of their reflection bands as well as near the gain peak of the active region [2, 55].

Figures 6.2a and 6.2b show the layer structure and axial electric-field standing-wave profile of a typical all-epitaxial GaAs-based VCSEL that employs strained-layer quantum well gain layers. The device results in Figs. 6.2, 6.3, 6.5, and 6.6 employed similar axial structures. The highly reflective distributed Bragg reflector (DBR) mirrors are formed by quarter-wavelength-thick layers of GaAs and AlAs to provide high reflectivities. As mentioned earlier, dielectrics can also be used for the mirrors, although controlling the index of refraction can be problematic. The cavity spacing between the mirrors must be a multiple of a half wavelength to place the Fabry–Perot resonance in the center of the DBR stop band. In the case illustrated, the spacing is one wavelength thick. The active region is placed at a standing-wave peak in the cavity to provide the desired enhancement in modal gain. (An active region placed at a standing-wave null will provide no gain.) The plot is the output of VCSEL simulation software Vertical by Optical Concepts, Inc. It uses a transmission matrix multiplication approach to determine the standing-wave field as well as the threshold gain level [56].

For the design of any laser, it is typical to use some approximate formulas to arrive at an educated guess at the design, and then apply a more rigorous analysis technique to fine-tune the parameters that can be controlled. With edge-emitters, this latter fine-tuning process is often carried out experimentally with little or no involved analysis. After some experience even the first guess for an edge-emitter design can also be an actual guess. In the VCSEL case, obtaining a good first design guess is very important. The design window is much more narrow, the number of critical parameters is larger, and it is relatively easy to completely miss the parameter space in which the device will work at all. In what follows, we develop some simple analytic design equations.

A.   THRESHOLD CURRENT AND POWER OUT

Figure 6.10 gives a simple schematic of a VCSEL that we use to define our coordinate system and some of the dimensions involved. The lengths l_x, l_y, and l_z refer to the effective optical mode dimensions in the two lateral directions and the axial direction, respectively. The axial length l_z includes the spacing between the DBR mirrors as well as the effective penetration depths into the mirrors. Under the "hard" or "effective" mirror model, l_z

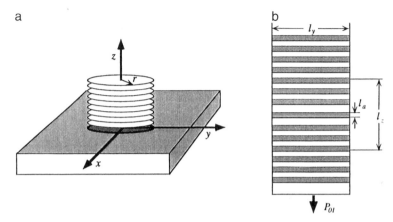

Fig. 6.10 (a) General VCSEL schematic indicating the coordinate system. (b) Cross-sectional schematic indicating active region thickness, l_a, and VCSEL effective cavity length, l_z.

is just the spacing between ideal dispersionless, zero-thickness mirrors that encompass a uniform cavity interrupted only by the active region [47, 57]. The thickness of a single quantum well or the single layer in the active region is l_w, and N_w is the number of wells.

The next thing we need to know is the gain versus current density characteristic of the quantum wells. This can always be measured experimentally with edge-emitting diagnostic lasers, but fortunately it has been found that the experimental results generally can be modeled very well by a two-parameter simple logarithmic expression [58]. That is, for each well,

$$g = g_0 \ln \frac{J}{J_0} \quad \text{or} \quad J = J_0 e^{g/g_0}. \tag{6.1}$$

These experimental results also quantitatively agree with numerical calculations of gain, so we can feel confident in listing the parameters J_0 and g_0 as universal constants for important cases of interest. Table 6.1 gives these values, including Auger, but no surface, recombination. The values of J_0 and g_0 given are at the wavelength of maximum gain. In a VCSEL, because there is only one axial mode near this gain peak, it can easily be misaligned in the manufacturing process or as a result of temperature changes. Such misalignment will result in an effective increase in the transparency current density, J_0, as well as a decrease in g_0, the gain at e times this current [26].

The one-pass threshold gain, G_{th}, is given by the net one-pass losses,

Table 6.1 **Gain Parameters for Common III-Semiconductors**

Active Material[a]	$g = g_0 \ln[J/J_0]^b$	
	J_0	g_0
$J_{sp} + J_{bar} + J_{Aug}$:		
Bulk GaAs	80	700
GaAs/Al$_{0.2}$Ga$_{0.8}$As 80 Å QW	110	1300
In$_{0.2}$Ga$_{0.8}$As/GaAs 80 Å QW	50	1200
J_{sp}:		
Bulk GaAs	75	800
GaAs/Al$_{0.2}$Ga$_{0.8}$As 80 Å QW	105	1500
In$_{0.2}$Ga$_{0.8}$As/GaAs 80 Å QW	50	1440
J_{sp}:		
Bulk In$_{0.53}$Ga$_{0.47}$As	11	500
InGaAs 30 Å QW (+1%)	13	2600
InGaAs 60 Å QW (0%)	17	1200
InGaAs 120 Å QW (−0.37%)	32	1100
InGaAs 150 Å QW (−1%)	35	1500
$J_{sp} + J_{Aug}$:		
In$_{0.53}$Ga$_{0.47}$As/(Q1.25) 70 Å QW	81	583

Source: From Ref. 58.
[a] QW, quantum well.
[b] $[J]$ = A/cm^2, $[g]$ = cm^{-1}.

$$G_{th} = L_i(r) + \ln \frac{1}{R} \approx L_i(r) + T_m, \qquad (6.2)$$

where R is the mean mirror power reflectivity, T_m is the mean mirror transmission, and $L_i(r)$ is the one-pass internal optical loss. The mirror transmission is found from transmission matrix theory, which gives very accurate predictions of the values found experimentally. The net optical loss is given by

$$L_i(r) = \int \psi^*(x, y, z)\alpha_i(x, y, z)\psi(x, y, z) \, dV \approx L_{i0} + L_{i1}/r^2, \quad (6.3)$$

where $\psi(x, y, z)$ is the normalized optical mode field and $\alpha_i(x, y, z)$ is the internal incremental optical loss. For uniform optical losses under the effective mirror model, this reduces simply to $L_i(r) \rightarrow \alpha_i l_z$. The explicit radial dependence in Eq. (6.3) indicates that there is generally some size-dependent optical losses. We have found that these can be approximated by the right-most expression, in which L_{i0} is the one-pass loss for broad-area devices (see Fig. 6.7).

The one-pass gain, G, can also be found from the incremental material gain, $g(x, y, z)$ by integrating over the cavity volume, and if the gain is uniform over the quantum well at g, it can be expressed in terms of a confinement factor as is normally done. That is [59],

$$G = \int \psi^*(x, y, z) g(x, y, z) \psi(x, y, z) \, dV = \Gamma g l_z, \qquad (6.4)$$

where Γ is the volume confinement factor. That is, $\Gamma = N_w \Gamma_{xy} \xi l_w / l_z$, where Γ_{xy} is the lateral confinement factor and ξ is the axial enhancement factor that accounts for position-dependent effects. Equations (6.2) and (6.4) now allow us to solve for g_{th} to get J_{th} from Eq. (6.1).

The optical power out of one mirror, P_{01}, for an input current, I, can be written as [60]

$$P_{01} = F_1 \frac{h\nu}{q} \eta_d (I - I_{th}) = F_1 \frac{h\nu}{q} \left[\eta_i \frac{T_m}{L_i + T_m} \right] (I - I_{th}), \qquad (6.5)$$

where F_1 is the fraction of power coupled out of mirror 1 to the output monitor, $h\nu$ is the photon energy, q is the electronic charge, η_d is the differential quantum efficiency defined as indicated, η_i is the internal quantum efficiency, and I_{th} is the threshold current. The threshold current generally includes some lateral leakage current, I_l, in addition to the current to the active region that can be obtained from Eq. (6.1). Thus, we write the threshold current as

$$\eta_i I_{th} = A_{xy} N_w J_0 e^{g_{th}/g_0} + I_l(n_{th}, r), \qquad (6.6)$$

where A_{xy} is the effective cross-sectional area of the active region, N_w is the number of quantum wells, and $I_l(n_{th}, r) = 2\pi r q N_w l_w n_{th}(r) S_{eff}(n)$. If the leakage is due to lateral diffusion, the effective surface recombination velocity, $S_{eff}(n) = \sqrt{D/\tau(n)}$, where D is the diffusion constant and n is the carrier density at the edge of the pumped region. If we assume primarily radiative recombination, $\tau(n) = \tau_0/n$. Thus, for a uniform carrier density, $I_l \propto r n_{th}^{3/2}(r)$. Now, again for carrier recombination dominated by radiative transition, $n_{th} \propto \sqrt{J_{th}}$, and we find that

$$I_l(n_{th}, r) = N_w K_l r J_{th}'^{3/4}(r), \qquad (6.7)$$

where K_l includes all of the accumulated constants and J_{th}' is the threshold current density with no leakage given by Eq. (6.1).

Now, we are in a position to obtain a completely analytic expression for the output power in terms of the input current for a VCSEL. From Eqs. (6.1) to (6.7), we find

$$P_{01} = F_1 \frac{h\nu}{q} \eta_i \frac{T_m}{L_{i0} + L_{i1}/r^2 + T_m} [I - N_w(I'_{th} + K_l r J_{th}^{'3/4}/\eta i)], \quad (6.8)$$

where $\eta i I'_{th} = A_{xy} J'_{th} = \pi r^2 J_{0} e^{(L_{i0} + L_{i1}/r^2 + T_m)/G_0}$ and $G_0 = g_0 \Gamma l_z$.

Although Eq. (6.8) is a little cumbersome, it does show explicitly the size dependences that are of prime importance in forming VCSELs of small lateral dimensions.

Figure 6.11 plots the one-pass threshold gain (from Eqs. [6.2] and [6.3]), the differential efficiency (from Eq. [6.5]), and the threshold current (from Eqs. [6.6] through [6.8]) versus VCSEL radius for some typical device parameters using three strained InGaAs quantum wells on GaAs. A broad-area per-pass loss of 0.3% was assumed. This high loss level is appropriate for heavily doped p-mirrors, but it may be a little pessimistic if bandgap engineering approaches allow the doping to be kept low. The gain levels required for the smallest devices are probably unrealistic. Thus, results less than about a 1-μm radius should be considered as indicative as trends only. Also, we made several assumptions that may not be entirely true—e.g., that the carrier density is uniform, that the recombination is radiative, and that the mode is aligned with the gain peak. Thus, experimental results may vary somewhat from this simple analytic approach. By following the same steps as outlined previously in a numerical approach, which includes the details of carrier distribution, carrier recombination, gain versus carrier density, mode–gain alignment, and so forth, we can make fairly accurate predictions [5, 26].

Figure 6.12 gives design curves for the current required for both threshold and 1-mW out by inverting Eq. (6.8). These are plotted for a 6-μm-diameter device as a function of mean mirror reflectivity for both unstrained and compressively strained quantum wells on either GaAs or InP. For GaAs, the emission wavelength is either 0.98 or 0.85 μm for strain or no strain, respectively, and for InP the emission wavelength is adjusted to 1.55 μm in both cases. Other various assumed parameters are contained in the caption.

Although the approximate CW active region temperature is indicated in Fig. 6.12, the parameters used for the plots were all room-temperature values assuming alignment of the mode and the gain peak. Thus, the indicated performance is a bit optimistic. This is especially true for the InP

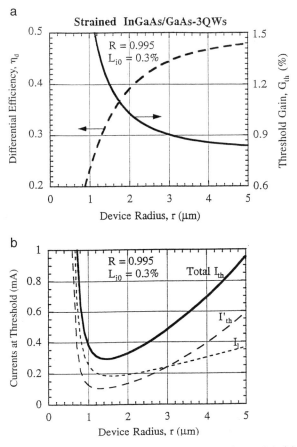

Fig. 6.11 Theoretical predictions from the simple analytic model. (a) Differential efficiency and threshold gain versus VCSEL radius. (b) Current to reach threshold (including lateral leakage current) versus radius. The mean mirror reflectivity and broad area loss parameter are indicated. The size-dependent loss factor is L_{il} = $0.009/\mu m^2$; and the leakage parameter is K_l = 0.95 mA$(\mu m)^{-1}$(mA$/\mu m^2)^{-3/4}$. Gain parameters are given in Table 6.1.

cases where Auger recombination will increase significantly at increased temperatures and higher current density levels. (The current density is 21 kA/cm^2 at 6 mA in this 3-μm-radius case.) Thus, as mentioned previously, the unstrained case may not function well at all. Also, the parameters for the 1% compressively strained case are taken from extrapolated values, and the same Auger coefficient as for the unstrained case was assumed.

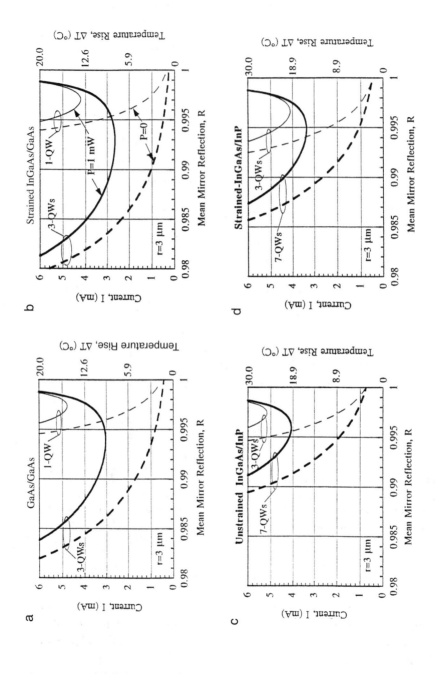

226

Therefore, there is some hope that this case may in practice be better than indicated. In any event, threshold currents of about 1 mA with reasonable powers out (~1 mW) are expected for the long-wavelength VCSELs [53].

Once a desired mirror reflectivity is selected, one must design the number of mirror layers that are needed. This can be determined by inverting a simple expression for the amplitude reflectivity of a dielectric quarter-wave stack, which for mirror i is given by [47]

$$r_i = \frac{1-b}{1+b} \quad \text{where} \quad b = \prod_{j=0}^{N} \left[\frac{n_{Lj}}{n_{Hj}} \right], \quad (6.9)$$

where N is the number of layers. If n_L and n_H remain constant throughout the mirror and $n_{L0}/n_{H0} = 1$, then b reduces to $(n_L/n_H)^{2m}$, where m is the number of mirror periods. Loss can be included by using complex indices. The mean mirror power reflection is just the product of the two laser mirrors ($R = r_1 r_2$). Fortunately, Eq. (6.9) can also be used even when graded composition interfaces are included to improve the electrical conduction properties of the mirrors. In this case, one must raise the index ratio in Eq. 9 to the power of the ratio of the fundamental Fourier coefficients of the graded profile to a square wave profile. For example, if the mirror index profile is sinusoidal, we should raise the index ratio to the power of $\pi/4$. Table 6.2 gives the values for commonly used materials.

B. THERMAL ISSUES

The plots in Fig. 6.12 indicate an internal temperature rise on a right vertical axis. This is approximated by multiplying a calculated power dissipation, P_d, times a thermal impedance, Z_T. That is,

$$\Delta T = P_d Z_T = (IV - P_{01}) Z_T, \quad (6.10)$$

where I and V are the laser terminal current and voltage, respectively. The thermal impedance can be approximated by an analytic formula that is exact for a small disk on an infinite half-space (substrate):

Fig. 6.12 Theoretical current required to reach threshold (*dashed curves*) and 1-mW out (*solid curves*) versus mirror reflectivity for quantum well active regions of various compositions and numbers of wells from the simple analytic model. The device radius is assumed to be 3 μm for all. Other parameters are as indicated in Fig. 6.11 and Table 6.1.

Table 6.2   **Refractive Indices for Common Materials Used in VESELS**

III-V Compounds	n at E_g	n at (λ μm)
GaAs	3.62	3.52 (0.98)
AlGaAs (0.2)	3.64	3.46 (0.87)
		3.39 (0.98)
AlAs[a]	3.2	2.98 (0.87)
		2.95 (0.98)
InGaAs (0.2) compressively strained on GaAs	3.6	
InP	3.41	3.21 (1.3); 3.17 (1.55)
InGaAsP (1.3 μm)	3.52	3.40 (1.55)
InGaAsP (1.55 μm)	3.55	
InGaAs (1.65 μm)	3.56	
InAs	3.52	
GaP[a]	3.5	
AlP[a]	2.97	
AlSb[a]	3.5	
GaSb	3.92	
InSb	3.5	
GaN (hexagonal)	2.67	2.33 (1 eV)
AlN (hexagonal)		2.15 (3 eV)
SiO$_2$ (glass)	1.46 (0.5 μm)	1.45 (1.0 μm)
Si$_3$N$_4$ (noncrystalline)	2.04 (0.5 μm)	2.00 (1.0 μm)
TiO$_2$ (rutile)	2.71 (0.5 μm)	2.49 (1.0 μm)

[a] Indirect gap.

$$Z_T = \frac{1}{4\sigma r_{eff}}, \tag{6.11}$$

where σ is the thermal conductivity of the substrate and r_{eff} is some effective device radius. In the uncovered etched-mesa case, r_{eff} is approximately equal to the mesa radius; in other cases, it tends to be somewhat larger (but usually less than twice the current injection radius) due to heat spreading in either surrounding epitaxial material or deposited heat spreaders [61]. For the case of Fig. 6.12, the conductivity is taken to be that of GaAs or InP, at 0.45 and 0.3 W/cm-°C, respectively. If the device is flip-chip bonded to some heat sink, an additional conductance is added in parallel and the net

result can be found by calculating the parallel sum of the thermal impedances. Heat flow into a directly bonded heat sink may be considered one-
dimensional, so $Z_T(1 - D) = h/(\sigma A)$, where h is the distance to the ideal
heat sink and A is the effective cross-sectional area.

The internal device temperature is important in determining the output
wavelength, the threshold current, and the obtainable maximum output
power. The maximum output power occurs at approximately the same
internal temperature for a variety of device structures, provided they have
the same carrier confinement barriers and the same relative alignment
between the gain peak and the mode wavelength. Two-dimensional finite-
element theoretical work has shown that the primary output limitation in
early VCSELs was due to carrier leakage over confining barriers from the
active region, as well as a misalignment between the gain peak and the
mode wavelength as the device heated [26]. Thus, more recent designs have
used higher barriers and some deliberate gain offset at room temperature,
so that carrier leakage is reduced and the gain moves into alignment with
the mode at higher internal temperatures [62]. Both effects provide higher
output power. This effect is shown both experimentally and analytically in
Fig. 6.13 [55].

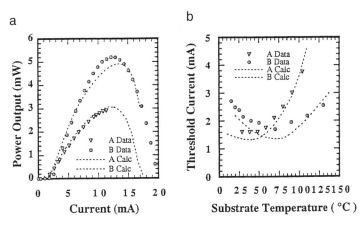

Fig. 6.13 Comparison of experimental and numerically modeled (a) light versus
current and (b) threshold current versus temperature for two VCSELs (A and B).
Temperature-dependent gain and heterobarrier leakage, as well as lateral carrier
diffusion and spatial hole-burning effects were included. Both are etched-mesa
structures as in Fig. 6.1a, but the active region AlGaAs barrier material has 20 and
50% Al for cases A and B, respectively [55].

The deliberate misalignment of the gain from the mode at room temperature also provides a threshold current that decreases slightly as the device is heated. As shown in Fig. 6.13b, it eventually reaches a minimum as the gain peak passes by the mode and then increases with further temperature increases. (The mode wavelength increases at about 0.7 Å/°C, whereas the gain peak wavelength increases at about 3 Å/°C in GaAs.) This technique is now widely used to provide temperature-insensitive devices with threshold currents that remain nearly constant over a 70 to 80°C range [63].

Thermal cross talk is another concern in array configurations. In this case, minimizing the thermal impedance is not necessarily the best thing to do. For example, making use of lateral material to provide enhanced heat spreading for lower thermal impedance will actually increase thermal cross talk between adjacent devices. Thus, as shown by detailed numerical analysis [64], this has been a particular problem in top-emitting proton-implanted structures. The best configuration for low thermal cross talk is probably an etched-mesa design that is flip-chip bonded to some good heat sink, so that lateral heat spreading is short circuited.

C. SERIES VOLTAGE

The net terminal voltage is another key factor in determining the power dissipation and overall power efficiency as well as the device temperature rise. For the right ordinate in Fig. 6.12, we assumed that the voltage could be modeled by a constant offset plus a linear term in current—that is,

$$V = V_s(T) + IR(T). \tag{6.12}$$

Equation (6.12) is useful near and above threshold, but it is less valid below threshold because the diode turn-on characteristic is ignored. The series voltage includes the diode turn-on voltage as well as possible voltage drops across mirror heterobarriers and contacts. The series resistance includes bulk resistance as well as the ohmic contribution from mirror heterobarriers and contacts. An explicit temperature dependence is shown. In VCSELs, this tends to be dominated by thermionic emission processes over various potential barriers. Thus, these voltage drops can increase dramatically at lower temperatures [65]. They can also reduce at higher temperatures to partially compensate for roll-offs in gain, so use of voltage modulation rather than current modulation tends to result in less temperature dependence of the output power [66].

Excess series voltage and resistance in VCSELs have been the subject of intensive investigations because they have been a key limitation on device efficiency. As discussed previously, the two approaches to solving this problem have been to use either (1) bandgap engineered mirrors, in which the composition and doping are carefully graded at the interfaces to flatten the valence band, or (2) intracavity contacts to avoid conduction through the mirrors on one or both sides. Design of bandgap engineered mirrors has been done with self-consistent solutions to Poisson's equation, as indicated in Fig. 6.14. This approach requires a very good calibration of the doping sources during growth. Parabolic compositional grading with two uniform doping segments as well as linear grading with delta doping has been successfully employed [3, 67].

The use of intracavity contacts requires some form of current aperturing to avoid current crowding and leakage at the edges of the active region. For the design of such structures, two-dimensional finite-element approaches using temperature-dependent variables have been employed as already discussed [26, 55]. This analysis has been successful in accurately predicting the characteristics of a variety of VCSEL types. Figure 6.13 gives examples of the comparison between theory and experiment for etched-mesa structures, including both light versus current and the threshold current versus temperature. Figure 6.15 gives the light versus current comparison for double-intracavity-contacted, dielectric apertured structures as in Fig. 6.1c for various diameters [68].

D. MODULATION PROPERTIES

High modulation frequencies are desired for many optoelectronic applications. For low-threshold lasers, the modulation-current efficiency factor (MCEF) is a very useful figure of merit that can be expressed as a function of the internal parameters of the laser [13]. The MCEF is defined as

$$\text{MCEF} \equiv \frac{f_{3dB}}{\sqrt{(I - I_{th})}} = \frac{1.55}{2\pi} \sqrt{\xi \frac{g' v_g \eta_i}{q V_{opt}}}, \tag{6.13}$$

where f_{3dB} is the 3-dB modulation bandwidth, I is the bias current, I_{th} is the threshold current, ξ is the standing-wave enhancement factor, g' is the differential gain (change in gain for a change in carrier density) in square centimeters, v_g is the photon velocity inside the cavity, η_i is the injection efficiency, q is the electronic charge, and V_{opt} is the optical modal volume. The advantages of using VCSELs for high-speed operation at low currents

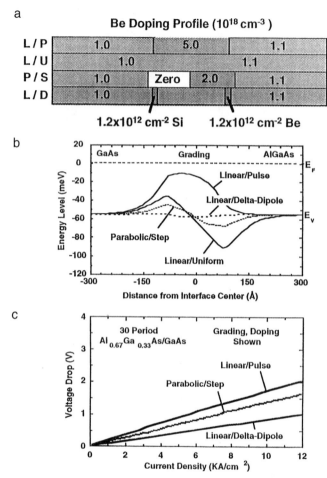

Fig. 6.14 Bandgap engineering [3]. (a) Schematic of interface designs. The numbers indicate doping levels in units of 10^{18} cm^{-3}. L/D, linear grading with delta doping at the interface edges; L/P, linear grading with pulse doping over the entire graded region; L/U, linear grading with uniform doping; P/S, parabolic grading with step doping in the graded region. (b) Results from a self-consistent one-dimensional Poisson equation solver of the interface barriers produced by the doping–grading profiles. (c) Voltage drop over a 30-period DBR using the various techniques simulated. The experiment verifies the simulated results for lowering barrier voltages. [Reprinted with permission from *Appl. Phys. Lett.* 63(25):3411. Copyright 1993 American Institute of Physics.]

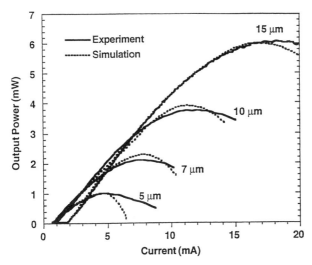

Fig. 6.15 Comparison of the numerical model with the experiment for the intracavity-contacted dielectric apertured VCSEL described in Fig. 6.5. Current crowding effects at aperture edges are important [81].

are clearly defined by Eq. (6.13) and illustrated in Fig. 6.5c. The VCSELs provide a standing-wave enhancement factor that can provide an effective differential gain increase of up to a factor of 2 over in-plane lasers. Also, because of the geometric properties of the device, modal volumes are extremely small for the VCSELs. The differential gain is increased by operating at lower gain levels (low losses) and MQWs. For most VCSELs, however, as the volume is reduced, the optical losses (and threshold gain) increase, which nullifies some of the benefit of the small modal volume. Reducing optical loss is a key factor in improving the dynamic response. Also, temperature rises and gain-peak versus cavity-mode offsets can dramatically affect the performance through a reduced differential gain. Growth control and heat sinking are necessary to realize ultimate performance. Assuming a single quantum well with a differential gain of 5×10^{16} cm^2 (value for a single In$_{0.2}$Ga$_{0.8}$As/GaAs quantum well at a gain level of 700 cm^{-1} with no gain offset [69]), and a radius of 1 μm, an MCEF of 30 GHz/mA$^{1/2}$ can be obtained. Thus, for a threshold current of 0.2 mA and a bias of 1.0 mA, a modulation frequency of 26.8 GHz is predicted. Of course, damping effects may reduce this slightly.

E.   OPTICAL WAVEGUIDING OR LENSING

Lateral photon confinement is desired in all types of lasers to reduce the per-pass optical losses. In edge-emitting semiconductor lasers, this is generally done by incorporating a uniform lateral index step to form a dielectric waveguide. In gas and solid-state lasers, the confinement is usually done by curved mirrors or intracavity lenses. In VCSELs, both kinds of techniques have been employed. The etched-mesa types provide a uniform waveguide over about half the cavity length, the proton-implanted structures rely upon thermal lensing, and the dielectric apertured types provide an intracavity lens. As mentioned previously, because the cavity is so short in VCSELs, even the intracavity lensed structures can be treated as having a uniform index-step lateral waveguide, provided the single-pass diffraction spreading is negligible [7, 8]. That is, the integrated optical path difference over the cavity length is the determining quantity. Thus, for an abrupt aperture of radius a, thickness d, and index n_1 placed within a cavity of averge index n_2 and effective length L, the resulting confined mode can be approximated by the mode of a uniform waveguide of radius a, a core index of n_2, and a uniform cladding index given by [27]

$$n_u = n_2 - \xi d[n_2 - n_1]/L, \tag{6.14}$$

where ξ is the position-dependent standing-wave enhancement factor, which can vary from 0 to 2 for a thin aperture placed between the mirrors as it is moved from the null to the peak of the standing wave. If the aperture is displaced into the DBR mirror its effect decays as the standing wave magnitude. More generally, the net effective lateral index perturbation $\overline{\Delta n}(x)$ caused by an aperture with an index profile given by $\Delta n(x, z)$ is given by [29],

$$\overline{\Delta n}(x) = \frac{\int \Delta n(x, z)|\psi(z)|^2 \, dz}{\int |\psi(z)|^2 \, dz} \tag{6.15}$$

where $\psi(z)$ is the net longitudinal electric field at $x = 0$. For an abrupt aperture, $\overline{\Delta n}(x)$ reduces to $(n_2 - n_u)$, as given by Eq. (6.14).

It has also been shown that Eq. (6.14) can be obtained by calculating the wavelength shift $\Delta \lambda$ for a laterally uniform plane wave with versus without the dielectric layer. Then, if we use $\Delta n/n = \Delta \lambda/\lambda$, the index step Δn can be found. This technique naturally includes the standing-wave effects [7]. The eigenmodes of a cylindrical waveguide with a step index are just the well-known modes of a step-index fiber. Thus, the VCSEL lateral modes are readily obtained.

For small dielectric apertures or long cavities, the uniform waveguide approximation is not useful. Numerical analysis indicates that for $L \geq 0.01\pi n_2 a^2/\lambda$, finite diffraction does occur, and scattering loss occurs at the abrupt aperture. However, if the aperture is tapered somewhat, this scattering is considerably reduced. Figure 6.16 gives results from this numerical analysis and compares it with the uniform waveguide model [29, 70]. For a parabolic-shaped aperture, the exact solution becomes the Hermite–Gaussian modes that are predicted both by the uniform waveguide model and Gaussian resonator theory [71].

IV. Growth and Fabrication Issues

All epitaxially grown GaAs-based VCSELs are now recognized as being truly practical and producible in the 0.8- to 1.0-μm wavelength range. Solid-source molecular beam epitaxy (MBE), gas-source MBE, and organometallic vapor-phase epitaxy (OMVPE) have all been used with good success [3, 72, 73]. However, this has not been a trivial accomplishment. In fact, standard growth technology, as developed for electronic devices or edge-emitting lasers, has not provided sufficient material thickness or uniformity control for VCSELs. This is why many of the major breakthroughs in this field occurred in laboratories where efforts to significantly improve the control of epitaxial growth were taking place. To obtain good wafer yield, some form of *in situ* monitoring is now employed in most facilities.

A. *IN SITU MONITORING*

The most common form of *in situ* monitoring during growth is optical reflectometry [74, 75]. As shown in Fig. 6.17, this can involve a relatively simple apparatus. In this case, a broadband light source is used with multi-mode fiber and a simple optical spectrum analyzer to generate the reflection spectrum from the sample surface. Such monitoring does not necessarily need to be done during the entire growth cycle, although with suitable access ports the optical monitoring can be done continuously with the substrate in the growth position. The most critical issue in VCSEL growth tends to be the placement of the Fabry–Perot mode at the proper wavelength, so that it is approximately in the center of the mirror reflection band and aligned with the desirable offset from the gain peak, as discussed

a

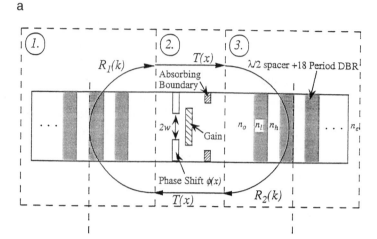

b

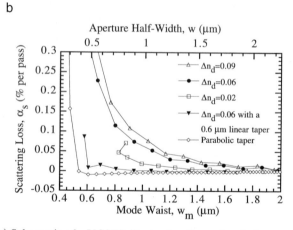

Fig. 6.16 (a) Schematic of a VCSEL illustrating the recirculating numerical calculation that cascades the mirror and cavity transfer functions: $T \times R_2 \times T \times R_1$.... (b) Optical scattering loss versus mode half width for three abrupt apertures as well as linear and graded parabolically tapered apertures. The parameter $\Delta n_d = n_2 - n_u$, and $n_0 = n_2$ in the present discussion. The values $\Delta n_d = 0.06$ and 0.09 correspond to a quarter-wave thick AlAs mirror layer either oxidized or etched [12, 30]. [Reprinted with permission from Hegblom, E. R., *et al.* 1996. Estimation of optical scattering losses in dielectric apertured vertical cavity lasers. *Appl. Phys. Lett.* 68(13):1757. Copyright 1996 American Institute of Physics.]

a

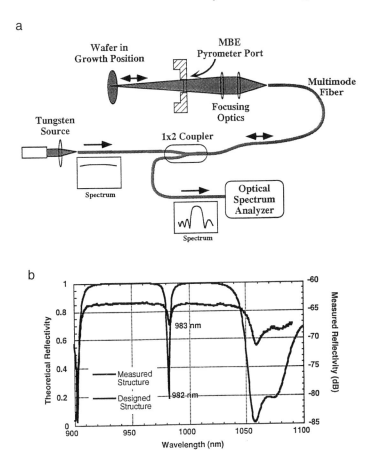

Fig. 6.17 *In situ* reflection spectrum setup for MBE [74]. (a) Schematic illustration of the white-light reflection spectrum setup. At various stages of growth, the reflection spectrum can be checked and modeled, and growth rates can be adjusted. (b) Final spectrum after growth, verifying control of the Fabry–Perot mode in the VCSEL using a reflection setup.

previously. Thus, the apparatus is typically used only after the bottom mirror and active region are grown to calculate what additional trimming layer must be added to center the mode before the upper mirror is grown. This approach can be used with either MBE or OMVPE.

Other monitoring techniques that are used involve either optical pyrometry [76, 77] or molecular beam flux monitoring by optical absorption [78]. The first makes use of the natural blackbody radiation from the sample. Optical pyrometry is commonly used to measure the substrate temperature. However, if a multilayer structure is being grown, this relatively broadband signal is filtered by the transfer function of that structure. Thus, the received optical spectrum can be used to determine the properties of the layer structure by analyzing this spectrum. Commercial instrumentation is available to aid in this task [77]. For a single wavelength channel, several techniques have been employed. In the case of DBR mirrors, one technique involves filtering out the blackbody spectral component that corresponds to the desired Bragg wavelength. Then, the MBE shutters are switched at the maxima and minima of the emission to make the layer structure correspond to the desired standing wave. In another more complex system, the received signal is monitored at several wavelengths simultaneously and a computer program is used to calculate the properties of the films being grown. In the case of the VCSEL, most of the parameters of the grown layers are known, and the technique can easily be applied to fine-tune the growth of the cavity layers to again properly align the mode with the mirror reflection bands as well as the gain peak. The drawback of this Pyritte™ system is that it cannot accurately control thin (<10-nm) layers. This is especially problematic in graded layers.

An *in situ* beam flux monitoring apparatus is illustrated in Fig. 6.18. As can be seen, the system gains sensitivity by using light sources that emit the atomic spectra of the atoms to be monitored. At the receiver end, a relatively crude spectrometer can be used to isolate the lines of interest. The illustrated system can simultaneously monitor the fluxes of Ga, In, and Al because it combines these sources into a single optical path [79]. Using feedback to control the shutter opening time, researchers have used the system to accurately set the bandgap of several quantum wells grown while the source temperatures (and thus fluxes) were being ramped. Because it directly monitors the beam fluxes, it may be the best technique to control the relative composition of multicomponent layers. Of course, integrating the fluxes allows the layer thickness to be controlled also. Unfortunately, the technique cannot be used for OMVPE, but in that case the limitation on layer composition and thickness is usually controlled more by reactor flow and chemistry rather than the source fluxes, which are generally well controlled.

B. PROCESSING FOR LATERAL DEFINITION

Although the initial growth of the VCSEL wafers is undoubtedly the most crucial step in the device's formation, the subsequent processing is also critical to obtain a high-performance, high-yield result. The growth provides the vertical (axial) structure, and the processing defines the lateral structure. Because a number of lateral structures are being considered, a similar number of fabrication procedures are involved.

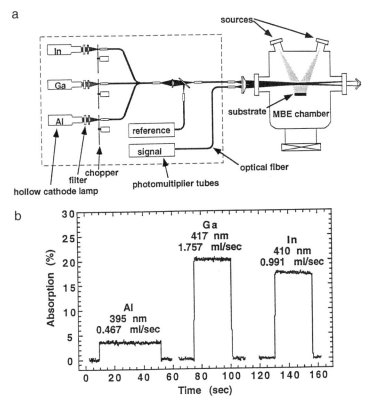

Fig. 6.18 Three-channel optical flux monitoring system [79]. (a) Schematic of the monitoring system. Indium, Ga, and Al hollow cathode lamps provide the wavelengths for absorption. The light passes through the flux twice and is compared with a reference signal to account for light level drifts. (b) A good signal-to-noise ratio is achieved in all three channels. Growth rates are calibrated to absorption levels by reflection electron diffraction (RED) oscillation measurement. [Reprinted with permission from Pinsukanjana, P., *et al.* 1996. Real-time simultaneous optical-based flux monitoring of Al, Ga, and In using atomic absorption for molecular beam epitaxy. *J. Vac. Sci. Technol. B* 14(3):2147. Copyright 1996 American Vacuum Society.]

For the etched-mesa structure, reactive ion etching with Cl_2 has been the principle technique for forming the mesas [80–82]. This provides highly anisotropic etching with vertical and relatively smooth sidewalls. Photoresist, Ni, or dielectric layers can be used for masking. To stop the etch at the desired point, *in situ* optical monitoring is necessary. In this case, a He–Ne or red diode laser is typically reflected off the surface, and the result is monitored with a simple detector. As material is etched away, oscillations in reflectivity from the semiconductor mirror occur, and these are compared with computer simulations to determine the desired stopping point [83, 84]. This monitoring technique is very simple and accurate (~10-nm stopping accuracy after microns of etching). Figure 6.19a provides a scanning electron micrograph of a ridge cross section obtained with this technique. Etchable masks in combination with nonetchable masks have also been used to create accurate double step height structures as required for some of the intracavity-contacted structures [12]. Figure 6.19b gives an example result from this process. In this case, an E-beam evaporated Si layer is used as the etchable mask. After it is patterned to the outer diameter, a smaller Ni dot is centered on it, so that a one-step etching gives the finished result.

As already indicated, the proton-implanted structures are perhaps the easiest to produce because they require only an implant masking, the proton implant, and contact formation. The main issues here are the energy of the implant and the relative diameters of the implant and the iris in the top electrode. A relatively wide variation in device performance has been observed as these parameters are varied [85].

Laterally under-etched or oxidized dielectric apertured VCSELs require accurate control of the etching or oxidation process. In the under-etching case, it is common to include some smaller control structures that will literally fall off when the etching has proceeded a desired amount. This process appears to provide submicron under-etch accuracy even after laterally etching some 10 μm. For AlAs under-etch layers, the best control has

a

b

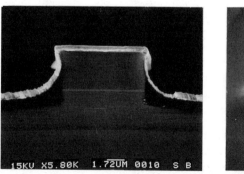

Fig. 6.19 Scanning electron micrographs of (a) a cross section of an etched-mesa VCSEL and (b) a low-angle view of a dielectric apertured VCSEL with step etching for intracavity contacts.

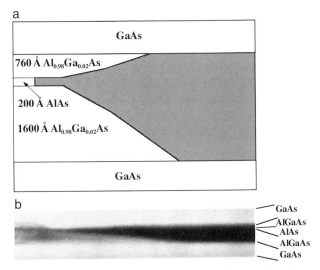

Fig. 6.20 (a) Schematic and (b) scanning electron micrograph of a tapered (lens-like) aperture formed by lateral oxidation [29, 70]. A taper is formed using a thin AlAs layer surrounded by AlGaAs with a slightly lower oxidation rate.

been obtained with a cooled HCl etchant [86]. Following under-etching, the structure is coated with a chemical vapor deposition (CVD) dielectric to provide step coverage and increased mechanical stability. The laterally oxidized structures are formed in flowing wet nitrogen gas at a temperature of about 400°C. The process has been carefully studied [87], and it has been shown to be highly reproducible in AlGaAs layers with small amounts of Ga added. The oxidation rate is strongly dependent upon the amount of Ga [88], and this can be used to provide good oxidation selectivity as well as to form complex lateral structures [29, 70]. Figure 6.20 gives a cross section of an oxidized structure with a tapered oxidation front that is created by a variable Al composition. For both under-etched and oxidized techniques, the quality and size of the larger etched mesa is of less importance because the optical mode is controlled by the aperture.

C. NONEPITAXIAL MIRRORS AND INTEGRATED MICROLENSES

As discussed previously, the all-epitaxial mirror structures are not desired or even practical in some cases. The long-wavelength InP-based VCSELs are important examples. Both conventional dielectric mirrors and wafer-bonding technology have been employed, as shown in Fig. 6.9. The wafer-bonding approach involves atomically fusing two substrates by placing them in contact under pressure at sufficiently high temperatures for a vapor-phase transport to take place. Figure 6.21 shows a transmission electron micrograph of an interface between InP layers fused to GaAs. Although there is a lattice mismatch, the necessary defects are well confined to the interface. Vapor-phase transport of P vapor as well as solid-state lateral diffusion of In appears to do an excellent job of removing high spots and filling in voids. The process is already used in the production of high-efficiency LEDs, where AlGaInP epilayers grown on GaAs are transferred to a transparent GaP substrate [89].

In the GaAs-based system, dielectric mirrors have also been explored. Successful devices have been obtained with TiO_2/SiO_2 and $CaF_2/ZnSe$ [18, 22, 90]. The deposition is generally done by electron-beam evaporation or reactive sputtering. In both cases, reflectivities above 99% can be achieved. Because of the large refractive index difference between CaF_2 and ZnSe, the high reflectivities can be achieved with only four or five mirror periods. Also, if the dielectric mirror is made in the last step, all processing can be done on a planar substrate. VCSEL threshold currents below 100 μA have been achieved with the $CaF_2/ZnSe$ mirrors [22].

Although not a necessary part of VCSEL fabrication, integrated micro-lenses are very attractive embellishments to these devices. They can be added at the wafer level using standard lithographic and etching techniques, and they can provide all the required focusing for a given application. Thus,

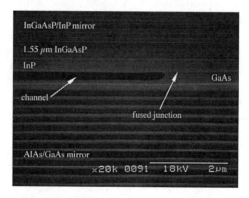

Fig. 6.21 Cross-sectional scanning electron micrograph of the interface between the wafer-fused AlAs/GaAs mirror and the InGaAsP/InP cavity region. Voids are decorated by stain etch [53].

fiber coupling and free-space interconnects can be accomplished with *no additional* external optics. Figure 6.22 illustrates one technique of making such lenses, and some results [91]. This so-called photoresist melting technique involves spinning on a specific thickness of a special photoresist, patterning it to some desired diameter, performing an initial slight etch to define a small confining step at its periphery, heating the resist until it melts and reflows, and finally dry etching with a process to transfer the lens shape into the semiconductor substrate. For the case of Fig. 6.22, the resist was microelectronics SF15, a polymethylglutylimide (PMGI); the dry etch was a pure Cl_2 reactive ion etch (RIE) at 350 V; and the substrate was GaAs. Figure 6.23 shows the effect of such integrated back-side microlenses on the output beams of 0.98-μm etched-mesa VCSELs [89]. A SiO antireflection (AR) coating was also added to prevent unwanted feedback into the laser. As indicated, a nearly collimated beam is possible with a lens focal length equal to the substrate thickness. Far fields indicate that lens aberrations and scattered energy are both small.

V. Integration: Photonic and Optoelectronic

A. *PHOTONIC*

The photonic integration of numerous identical VCSELs into linear and two-dimensional arrays requires no additional growth effort because the required layer structure fills the entire wafer surface. Standard processing technology also produces full wafers of devices naturally arrayed to fill the available space. The primary difference in forming different kinds of VCSEL arrays is in the final electrical contacting steps.

Several contacting schemes have been demonstrated. Arrays of devices have been connected in parallel with a pair of common contacts [92–93]. This has been done to create high-power arrays. Obtaining high power from VCSEL arrays makes sense because of the three-dimensional heat spreading away from small dots. In this sense, they may be viewed as a further extension of the successful philosophy employed in one-dimensional arrays of edge-emitters, which have two-dimensional heat spreading. Attempts to coherently phase-lock arrays of single-mode devices have met with modest success [94], but high-power phase-locked arrays with single-lobed far fields have not been demonstrated.

Separately contacted, individually addressable devices have been studied most widely [95–96]. If grown on a doped substrate, all devices will have

a

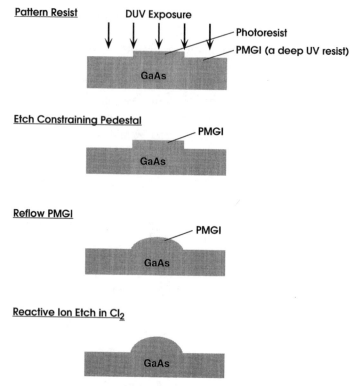

Fig. 6.22 (a) Fabrication process for and (b) scanning electron micrograph of integrated microlenses on GaAs [91]. DUV, deep UV; PMGI, polymethylglutyl-imide.

a common contact (usually the cathode), and the other can be attached to individual bond pads. Two-dimensional arrays up to 32×32 in size have been made, and 8×8 arrays have been offered for sale commercially with the proton-implanted technology.*

Matrix addressing is possible with semi-insulating substrates [96, 97]. Structures as shown in Figs. 6.1c and 6.3 would provide candidate devices. In this case, the p-contacts of devices along a row and the n-contacts of devices down a column are connected. Then, applying a positive voltage to the desired row and an equal negative voltage to the column turns on the VCSEL at the intersection. Other devices along the row and down the

---

* Eight-by-eight arrays have been sold by Vixel Corp., Boulder, CO.

b

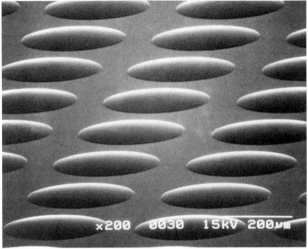

Fig. 6.22   *(Continued)*

column would see only half the voltage of this device. Thus, because it is easy to design devices whose threshold voltage is less than half the desired operating voltage, excellent on–off contrast is possible. Figure 6.24 gives example results [98].

Fabricating arrays of VCSELs with multiple wavelengths does require some additional growth effort to define the necessarily different cavity lengths. Several approaches have been investigated. A thickness gradient can be created across the wafer by stopping rotation during MBE growth. This has been used to make experimental wavelength chirped arrays [99], but the approach is not viable for manufacturing many arrays per wafer because of the nonlinearity of the growth variation across the wafer. To create a wafer with periodic thickness variations during MBE growth, a special grooved wafer holder has been employed to provide periodic surface temperature variations and thus different sticking coefficients for the In and/or Ga species [100, 101]. VCSEL arrays formed across this thickness gradient have different wavelengths. Reproducibility may be a problem with this approach. Perhaps a more reproducible approach uses conventional multiple-step lithography to adjust the reflection phase of one mirror. This has been done on the top surface of a short semiconductor DBR mirror segment prior to completing the mirror either with dielectric–metal layers [102] or with a regrowth of additional semiconductor DBR periods [103].

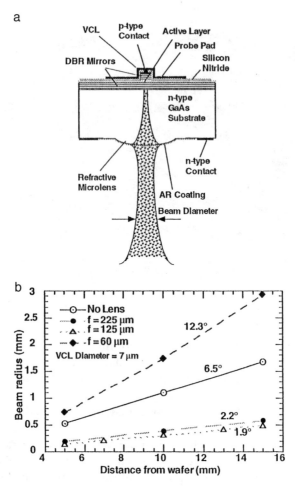

Fig. 6.23 (a) Schematic of a VCSEL (or VCL) with an integrated microlens. (b) Graph of beam radius versus distance from such a structure [91].

One of the key issues with arrays is cross talk. It can derive from thermal, electrical, or optical parasitic paths. It has been shown that thermal cross talk can be minimized with a larger element-to-element spacing or with vertical heat sinking and etched grooves between devices [64]. Electrical cross talk may be capacitive (radiation) or resistive (leakage). Both tend to be aggravated by closely spaced devices or contacting electrodes. In a well-designed array, most of the lateral current confinement techniques

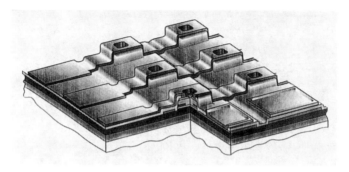

Fig. 6.24 Schematic section of a matrix-addressed VCSEL array on a semi-insulating substrate [98]. [Reprinted from Morgan, R. A., *et al.* 1994. Two-dimensional matrix addressed vertical cavity top-surface emitting laser array display. *IEEE Photon. Tech. Lett.* 6(8):915. © 1994 IEEE.]

should eliminate the resistive paths between devices. Of course, in matrix addressing there is always the half-voltage issue. Radiative or capacitive coupling tends to be the larger problem for high-frequency modulation of elements in an array. Good microwave engineering using thick dielectric spacer layers, thick conductors, and shielding ground planes can address this problem in most cases.

Optical cross talk may be the most difficult problem to solve, especially in free-space interconnected arrays or in cases where the optical coupling loss to the desired medium (e.g., fiber) is large. In both cases, stray light tends to be generated, and unless care is exercised, this may find its way to adjacent receiver elements. It is generally desirable to eliminate large emission angle ray paths from the VCSELs by avoiding significant scattering elements or the generation of very high-order lateral modes. Side lobes generated in the far field by the VCSEL mode or nonideal coupling optics can also create problems. For free-space interconnects, Gaussian modes are desirable. Finally, the entire optical link generally needs to be considered in analyzing optical cross talk. That is, the use of limited numerical aperture fibers or free-space receiver elements may make up for some problems in the emitter. However, it is generally best to avoid generating the unwanted signals in the first place.

Vertical photonic integration is possible by using the back side of the VCSEL wafers, as already shown with the previously discussed microlenses,

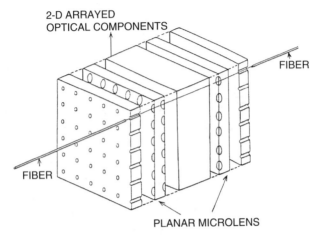

Fig. 6.25 Schematic of stacked planar optics [104]. (Reprinted from Iga, K., Y. Kokubun, and M. Oikawa. 1984. *Fundamentals of microoptics*. New York: Academic Press, 197, with permission of the author.)

or by employing vertically stacked wafers that might be self-aligned with solder bumps, alignment notches, and so forth. As shown in Fig. 6.25, this has been proposed as a means of optically interconnecting a wide array of devices [104]. Perhaps the most attractive feature is the ability to align a vast number of elements by aligning only two, as is typically done in conventional photolithographic mask alignment.

B. OPTOELECTRONIC

Because VCSELs require a relatively small footprint on a semiconductor wafer, they appear to be more compatible with electronic devices than edge emitters are, and this has generated some interest in integrating VCSELs with electronic driver or logic circuits. The growth and processing required for heterojunction bipolar transistors (HBTs) appear to be fairly similar to those of VCSELs. In fact, VCSELs have been directly integrated in the collector circuits of HBTs by simply growing the layers for both sequentially, as shown in Fig. 6.26 [105]. The HBT provides current gain so that very small base currents can drive several milliamperes through the VCSEL. Although the vertical dimension is much larger, the processing is very similar. Direct illumination of the HBT base can also be used to generate the required minority carriers; thus, optical-to-optical gain is possible in a single vertical stack.

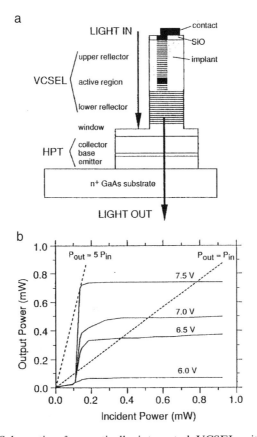

Fig. 6.26 (a) Schematic of a vertically integrated VCSEL with a heterojunction phototransistor (HPT). The structure consists of an emitter-down HPT and a bottom-emitting VCSEL. To avoid optical feedback, the laser emits at a lower energy than that of the detector bandgap. The input and output are codirectional. (b) Light–light characteristics of the structure for 855-nm input light as a function of bias voltage. The laser threshold is at 110 μW. *Dashed lines* represent unity optical power gain and an optical power gain of 5 [105]. (Reprinted with permission from Chan, W. K., *et al.* 1991. Optically controlled surface-emitting lasers. *Appl. Phys. Lett.* 58(21):2343. Copyright 1991 American Institute of Physics.)

Direct vertical integration with *pn-pn* latching diodes has also been explored for logic applications. This so-called V-STEP device is illustrated in Fig. 6.27 [106]. Once a pulse of incident light turns on the *pn-pn* diode, the voltage across it collapses, and the VCSEL turns on. It stays on until the terminal voltage is reduced to reset the device. This is a

a

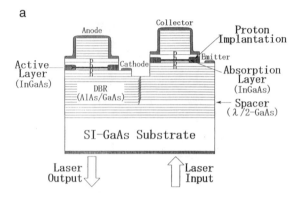

b

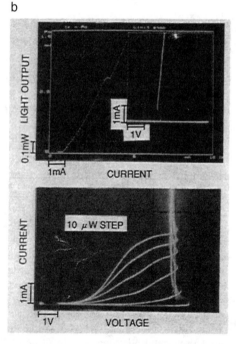

Fig. 6.27 (a) Schematic of the V-STEP laser and photodetector. (b) L-I and I-V for the laser thyristor (*top*) and photocurrent versus voltage for the phototransistor (*bottom*) [106]. (b, Reprinted from Kasaka, H., *et al.* 1993. Pixels consisting of a single vertical-cavity laser thyristor and a double vertical-cavity phototransistor. *IEEE Photon. Tech. Lett.* 5(12):1409–1411. Copyright © 1993 IEEE.)

very simple structure that may find use in optoelectronic logic or image processing.

To enhance flexibility in the processing of integrated optoelectronic circuits composed of VCSELs, transistors, and photodetectors, discrete optimized devices have also been monolithically integrated. VCSEL-MESFET [107] and VCSEL-HBT [108] circuits have been explored. This provides a technology in which the transistors can be interconnected in various ways to form complex logic circuits. Figure 6.28 gives an example of a VCSEL-HBT technology and some experimental results from a 2×2 optical crossbar switching circuit [108].

Hybrid integration using flip-chip bonding and/or multichip module (MCM) technology provides the ultimate in flexibility without sacrificing device and circuit performance. Of course, additional handling and bonding is necessary after the wafers are diced into chips, but the electronic and photonic wafers can be designed and processed without compromise, and chips from both types of wafers can be pretested to reject any defective parts before being integrated. As a result, it has generally been found that unless very large scale production is anticipated, the hybrid form of optoelectronic integration is usually more cost-effective. Figure 6.29 shows a VCSEL array flip-chip bonded to a custom high-speed HBT driver circuit [109]. In this case, the VCSELs are similar to those indicated in Fig. 6.3, and the HBTs are fabricated in a standardized foundry.

VI. Applications

A. DISCRETE DEVICES

Although the advantages of VCSELs for array applications have been widely touted, their potential for low-cost LED-like production coupled to their superiority to LEDs for advanced applications may provide a large LED-upgrade market for discrete VCSELs. VCSELs can be modulated at multigigahertz rates, their output beams are well collimated, and their overall power efficiency can be much higher than that of an analogous LED. Thus, for higher capacity data links, efficient coupling to fibers, low-cross-talk free-space links, or minimizing drive power requirements, VCSELs may offer an attractive evolutionary path to systems designers who want to improve the performance of existing LED-based links.

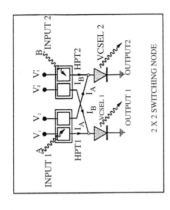

2 X 2 SWITCHING NODE

INPUT 1 A
INPUT 2 B
HPT1
HPT2
V_1 V_2 V_2' V_1'
I_B I_A
I_B I_A
VCSEL 1
VCSEL 2
OUTPUT 1
OUTPUT2

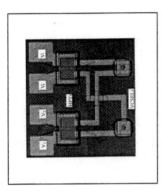

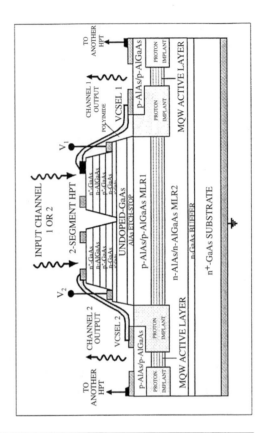

TO ANOTHER HPT
CHANNEL 1 OUTPUT
VCSEL 1
INPUT CHANNEL 1 OR 2
2-SEGMENT HPT
V_1
p-AlAs/p-AlGaAs
PROTON IMPLANT
MQW ACTIVE LAYER
POLYIMIDE
n⁺-GaAs
p-AlGaAs
p⁺-GaAs
n-GaAs
UNDOPED-GaAs
AlAs ETCH-STOP
p-AlAs/p-AlGaAs MLR1
n-AlAs/n-AlGaAs MLR2
n-GaAs BUFFER
n⁺ GaAs SUBSTRATE
CHANNEL 2 OUTPUT
VCSEL 2
V_2
TO ANOTHER HPT
p-AlAs/p-AlGaAs
PROTON IMPLANT
MQW ACTIVE LAYER

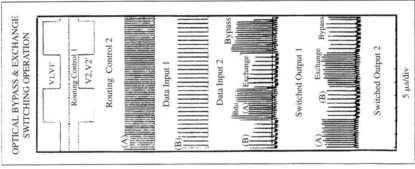

OPTICAL BYPASS & EXCHANGE SWITCHING OPERATION

$V1, V1'$
Routing Control 1
$V2, V2'$
Routing Control 2
(A) Data Input 1
(B) Data Input 2
(A) Exchange / Bypass
Switched Output 1
(B) Exchange / Bypass
Switched Output 2
5 µs/div

The other class of applications involves those in which lasers are required but edge-emitting devices appear too costly. Thus, in this case the focus would be mainly on cost reduction with VCSELs versus edge-emitters. This would derive both from the initial manufacturing of the chips and in their packaging with fibers or other desired media. In fact, their low numerical aperture outputs, either directly or in combination with integrated microlenses, may be an even larger attraction than the wafer-scale fabrication and testing feature of VCSELs. Potential applications in this instance include sources for new low-cost optical sensors, plastic-fiber-based data links, optical disk storage, and fiber to the home (FTTH). For some of these applications, improvements in either short-wavelength (e.g., blue for optical recording) or longer wavelength (e.g., for FTTH) VCSELs will need to be developed. However, it seems clear that VCSELs could have a major impact in both cases.

B. LINEAR ARRAYS

One-dimensional arrays of VCSELs coupled to multimode fiber ribbon cables have been used in prototype parallel data link applications [110–112]. Some illustrations of these results are shown in Fig. 6.30. As indicated, driver and receiver electronics were also developed. These links take advantage of the fact that entire linear arrays of VCSELs can be simultaneously aligned to the elements of a fiber ribbon cable by aligning only the end two lasers and fibers. Thus, these linear arrays significantly leverage the relatively easy alignment of VCSELs to fibers by carrying along the extra elements. Because the uniformity of the arrays proved to be good, it was also decided to monitor only the end VCSELs. This avoided a complex back-side monitor or a beam splitter at the front.

The prototype fiber-based parallel data links of Fig. 6.30 are currently generating some market interest. Predictions suggest that this may reach several tens of millions of dollars by the turn of the 21st century [113].

Fig. 6.28 Two-by-two switching node demonstration by integration of HPTs and VCSELs [108]. A single-epitaxial structure with etch-stop layers is used to form both the HPT and the VCSEL. All light is routed on one side of the substrate but could easily be between many boards by making bottom-emitting VCLs. Packet switching is demonstrated. (Reprinted from Lu, B., *et al.* 1994. Reconfigurable binary optical routing switches with fan-out based on the integration of GaAs/AlGaAs surface-emitting lasers and heterojunction phototransistors. *IEEE Photon. Tech. Lett.* 6(2):222–226, © 1994 IEEE.)

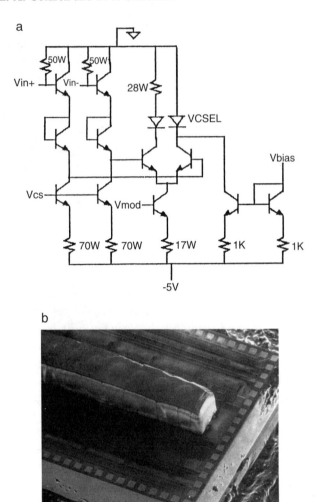

Fig. 6.29 Hybrid integration of VCSELs and an HBT driver circuit by flip-chip bonding. (a) Schematic of the Rockwell driver circuit. (b) Scanning electron micrograph of a VCSEL array with integrated microlenses bonded to the circuit. Cured PMGI is used as the material for structural support.

Currently, efforts are underway to reduce the cost, but it seems that the connectorized fiber ribbon cables and the general packaging are more of a problem than the VCSELs themselves.

Linear arrays are also proposed for parallel board-to-board interconnection using free-space paths [110]. In this case, the devices can be flip-chip mounted, and integrated back-side microlenses will collimate the emission

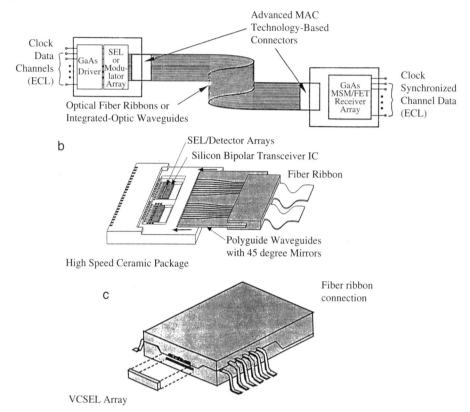

Fig. 6.30 Schematics of parallel guided-wave data links and source modules. (a) Optoelectronics Technology Consortium (OETC) [111] transmit and receive modules with interconnecting fiber ribbon cable. (b) Parallel optical link organization (POLO) [112] transceiver modules. (Courtesy of Hewlett Packard.) (c) Optobus [110] transmit package of Motorola.

to an analogous receiver array on the second board, which also contains integrated back-side microlenses. Because nearly collimated beams are emitted, this system is very tolerant of lateral displacements for movements of up to about half the spacing between array elements.

C. TWO-DIMENSIONAL ARRAYS

Two-dimensional arrays of devices generally are coupled to free-space optical paths. Massively parallel interconnection between processing planes is possible using the same integrated microlenses introduced previously.

These refractive microlenses can be integrated off axis for beam steering, or diffractive optics can also be integrated. For simultaneous fan out and collimation, computer-generated holograms implemented in diffractive optics would be useful [114]. In some cases, simple field lenses external to the chips may be desired, but these will result in much lower tolerances in the placement of the chips to be interconnected. Pairs of aspheric lenses with Risley beam steerers mounted in a barrel assembly have been used in a 4 × 4 VCSEL–MSM detector optical back plane demonstrator with some success [115].

Two important applications for such large numbers of interconnections are in signal processing and switching [116, 117]. For switching, fixed interconnection paths between multiple boards can generally be used to accomplish the desired routing of a signal on one of the inputs to any of the outputs. Thus, all the switching is done with electronically controlled optical switches at each plane. Figure 6.31 illustrates such a multiplane optical

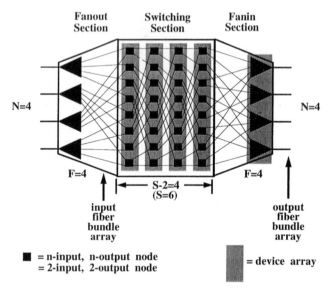

Fig. 6.31 General model for fan-out–switch–fan-in extended generalized shuffle network with four inputs and outputs, a fan out of four, and four concatenated switching planes with two-input/two-output nodes. (Reprinted from Cloonan, T., *et al.* 1993. A complexity analysis of smart pixel switching nodes for photonic extended generalized shuffle switching networks. *IEEE J. Quantum Electron.* 29(2):620. © 1993 IEEE.)

$n \times n$ switch in which the optical paths between nodes are fixed [117]. This architecture would appear to be possible with the current state of the art in VCSELs using flip-chip bonding.

In the signal processing or computing case, a fixed interconnect architecture, as shown in Fig. 6.32, may still be used for some functions. But in other cases, it may also be desirable to alter the interconnection paths between two boards during the processing. A spatial light modulator (SLM) placed in the interconnect path allows selection of one of several possible paths [118]. Alternatively, all the desired interconnection paths from a given node can be set with fixed optics, and the switching can be done in the electronics. However, this greatly increases the number of possible beams that must be spatially separated. A variation is to use wavelength encoding of the source and detector arrays to reduce the spatial cross-talk tolerance. Other proposals involve using tunable sources and diffractive optics to route a particular wavelength to a particular node. Generally, this programmable interconnect architecture will require further research.

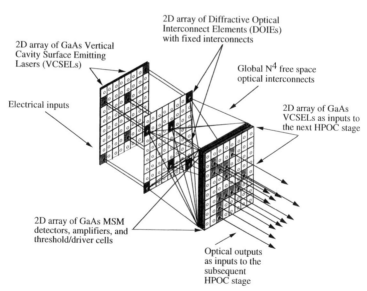

Fig. 6.31 High-performance optical computing (HPOC) stage using two-dimensional arrays of VCSELs, diffractive optical interconnect elements, and detector–amplifier–threshold–driver cells [116]. (Reprinted from Guilfoyle, P. S., *et al.* 1994. "Smart" optical interconnects for high speed photonic computing. *Opt. Comp. 1994 Tech. Ser. Dig.* 7:78–81, with permission of the Optical Society of America.)

Displays are another possible application of two-dimensional arrays of VCSELs. Just as LEDs are currently finding their way into displays [119], the VCSEL could be viewed as an efficient high-power LED for such applications.

References

[1] Jewell, J. L., A. Scherer, S. L. McCall, Y. H. Lee, S. Walker, J. P. Harbison, and L. T. Florez. 1989. Low-threshold electrically pumped vertical-cavity surface-emitting microlasers. *Electron. Lett.* 25:1123–1124.

[2] Geels, R. S., S. W. Corzine, and L. A. Coldren. 1991. InGaAs vertical-cavity surface-emitting lasers. *IEEE J. Quantum Electron.* 27(6):1359–1367.

[3] Peters, M. G., B. J. Thibeault, D. B. Young, J. W. Scott, F. H. Peters, A. C. Gossard, and L. A. Coldren. 1993. Band-gap engineered digital alloy inter-faces for lower resistance vertical-cavity surfce-emitting lasers. *Appl. Phys. Lett.* 63(25):170–171.

[4] Choquette, K. D., G. Hasnain, Y. H. Wang, J. D. Wynn, R. S. Freund, A. Y. Cho, and R. E. Leibenguth. 1991. GaAs vertical-cavity surface emitting lasers fabricated by reactive ion etching. *IEEE Photon. Tech. Lett.* 3(10):859–862.

[5] Thibeault, B. J., T. A. Strand, T. Wipiejewski, M. G. Peters, D. B. Young, S. W. Corzine, L. A. Coldren, and J. W. Scott. 1995. Evaluating the effects of optical and carrier losses in etched post vertical cavity lasers. *J. Appl. Phys.* 78(10):5871–5875.

[6] Babic, D. I., Y. Chung, N. Dagli, and J. E. Bowers. 1993. Modal reflection of quarter-wave mirrors in vertical-cavity lasers. *IEEE J. Quantum Electron.* 29(6):1950–1962.

[7] Hadley, G. R., K. L. Lear, M. E. Warren, K. D. Choquette, J. W. Scott, and S. W. Corzine. 1996. Comprehensive numerical modeling of vertical-cavity surface-emitting lasers. *IEEE J. Quantum Electron.* 32(4):607–616.

[8] Coldren, L. A., J. W. Scott, B. Thibeault, T. Wipiejewski, D. B. Young, M. G. Peters, G. Thompson, E. Strzelecka, and G. Robinson. 1995. Vertical-cavity semiconductor lasers: Advances and opportunities. In *Proceedings of IOOC '95*, 26–27. Washington, DC: Optical Society of America.

[9] Peters, M. G., D. B. Young, F. H. Peters, J. W. Scott, B. J. Thibeault, and L. A. Coldren. 1994. 17.3% Peak wall plug efficiency vertical-cavity surface-emitting lasers using lower barrier mirrors. *IEEE Photon. Tech. Lett.* 6(1):31–33.

[10] Peters, F. H., M. L. Majewski, M. G. Peters, J. W. Scott, B. J. Thibeault, D. B. Young, and L. A. Coldren. 1993. Vertical cavity surface emitting laser technology. *SPIE* 1851:122–127.

[11] Lear, K. L., R. P. Schneider, K. D. Chopquette, S. P. Kilcoyne, J. J. Figiel, and J. C. Zolper. 1994. Vertical cavity surface emitting lasers with 21% efficiency by metalorganic vapor phase epitaxy. *IEEE Photon. Tech. Lett.* 6(9):1053–1055.

[12] Scott, J. W., B. J. Thibeault, D. B. Young, L. A. Coldren, and F. H. Peters. 1994. High efficiency sub-milliamp vertical cavity lasers with intra-cavity contacts. *IEEE Photon. Tech. Lett.* 6(6):678–680.

[13] Scott, J. W., B. J. Thibeault, C. J. Mahon, L. A. Coldren, and F. H. Peters. 1994. High modulation efficiency of intracavity contacted vertical cavity lasers. *Appl. Phys. Lett.* 65(12):1483–1485.

[14] Scott, J. W., B. J. Thibeault, M. G. Peters, D. B. Young, and L. A. Coldren. 1995. Low-power high-speed vertical-cavity lasers for dense array applications. *SPIE Proc.* 2382:280–287.

[15] Thibeault, B. J., J. W. Scott, M. G. Peters, F. H. Peters, D. B. Young, and L. A. Coldren. 1993. Integrable InGaAs/GaAs vertical-cavity surface-emitting lasers. *Electron. Lett.* 29(25):2197–2198.

[16] Tell, B., Y. H. Lee, K. F. Brown-Goebeler, J. L. Jewell, R. E. Leibenguth, M. T. Asom, G. Livescu, L. Luther, and V. D. Mattera. 1990. High-power CW vertical-cavity top surface-emitting GaAs quantum well lasers. *Appl. Phys. Lett.* 57(18):1855–1857.

[17] Morgan, R. A., L. M. F. Chirovsky, M. W. Focht, G. Guth, M. T. Asom, R. E. Leibenguth, K. C. Robinson, Y. H. Lee, and J. L. Jewell. 1991. Progress in planarized vertical cavity surface emitting laser devices and arrays. *SPIE Proc.* 1562:149–159.

[18] Lehman, J. A., R. A. Morgan, M. K. Hibbs-Brenner, D. Carlson. 1995. High-frequency modulation characteristics of hybrid dielectric/AlGaAs mirror single-mode VCSELs. *Electron. Lett.* 31(15):1251–1252.

[19] Shtengel, G., H. Temkin, P. Brusenbach, T. Uchida, M. Kim, C. Parsons, W. E. Quinn, and S. E. Swirhun. 1995. High-speed vertical-cavity surface emitting laser. *IEEE Photon. Tech. Lett.* 5(12):1359–1361.

[20] Morgan, R. A., M. K. Hibbs-Brenner, R. A. Walterson, J. A. Lehman, T. M. Marta, S. Bounnak, E. L. Kalweit, T. Akinwande, and J. C. Nohava. 1995. Producible GaAs-based top-surface emitting lasers with record performance. *Electron. Lett.* 31(6):462–463.

[21] Kupta, D., and C. Mahon. 1994. Mode selective loss penalties in VCSEL optical fiber transmission links. *IEEE Photon. Tech. Lett.* 6(2):288–290.

[22] Deppe, D. G., D. L. Huffaker, J. Shin, and Q. Deng. Very-low-threshold index-confined planar microcavity lasers. *IEEE Photon. Tech. Lett.* 7(9):965–967.

[23] Hayashi, Y., T. Mukaihara, N. Hatori, N. Ohnoki, A. Matsutani, F. Koyama, and K. Iga. 1995. Record low-threshold index-guided InGaAs/GaAlAs vertical-cavity surface-emitting laser with a native oxide confinement structure. *Electron. Lett.* 31(7):560–561.

[24] Lear, K. L., K. D. Choquette, R. P. Schneider, Jr., S. P. Kilcoyne, and K. M. Geib. 1995. Selectively oxidized vertical cavity surface emitting lasers with 50% power conversion efficiency. *Electron. Lett.* 31(3):208–209.

[25] MacDougal, M. H., P. D. Dapkus, V. Pudikov, H. Zhao, and G. M. Yang. 1995. Ultralow threshold current vertical-cavity surface-emitting lasers with AlAs oxide-GaAs distributed Bragg Reflectors. *IEEE Photon. Tech. Lett.* 7(3):229–231.

[26] Scott, J. W., R. S. Geels, S. W. Corzine, and L. A. Coldren. 1993. Modeling temperature effects and spatial hole burning to optimize vertical-cavity surface-emitting laser performance. *IEEE J. Quantum Electron.* 29(5):1295–1308.

[27] Coldren, L. A., B. J. Thibeault, E. R. Hegblom, G. B. Thompson, and J. W. Scott. 1996. Dielectric apertures as intracavity lenses in vertical-cavity lasers. *Appl. Phys. Lett.* 68(3):313–315.

[28] Lear, K. L., K. D. Choquette, R. P. Schneider, Jr., and S. P. Kilcoyne. 1995. Modal analysis of a small surface emitting laser with a selectively oxidized waveguide. *Appl. Phys. Lett.* 66(20):2616–2618.

[29] Hegblom, E. R., D. I. Babic, B. J. Thibeault, J. Ko, R. Naone, and L. A. Coldren. 1996. Estimation of optical scattering losses in dielectric apertured vertical cavity lasers. *Appl. Phys. Lett.* 68(13):1757–1759.

[30] Floyd, P. D., B. J. Thibeault, L. A. Coldren, and J. L. Merz. 1996. Scalable AlAs-oxide vertical cavity lasers. *Electron. Lett.* 32(2):114–116.

[31] Li, G. S., S. F. Lim, W. Yuen, and C. J. Chang-Hasnain. 1995. Polarization and modal behavior of low threshold oxide and air-gap confined vertical cavity lasers. In *Proceedings of IOOC '95,* Paper PD1-6. Washington, DC: Optical Society of America.

[32] Thibeault, B. J., K. Bertillson, E. R. Hegblom, E. Strzelecka, P. D. Floyd, and L. A. Coldren. 1997. High-speed characteristics of low-optical loss oxide-apertured vertical-cavity lasers. *IEEE Photon. Tech. Lett.* 9(1):11–13.

[33] Floyd, P. D., B. J. Thibeault, L. A. Coldren, and J. L. Merz. 1995. Reduced threshold bottom emitting vertical cavity lasers by AlAs oxidation. In *Proceedings of LEOS '95,* 414–415. Paper SCL14.2. Piscataway, NJ: IEEE.

[34] Lott, J. A., R. P. Schneider, Jr., K. J. Malloy, S. P. Kilcoyne, and K. D. Choquette. 1994. Partial top dielectric stack distributed Bragg reflectors for red vertical cavity surface emitting lasers. *IEEE Photon. Tech. Lett.* 6(12):1397–1399.

[35] Choquette, K. D., R. P. Schneider, M. Hagerott Crawford, K. M. Geib, and J. J. Figiel. 1995. Continuous wave operation in 640–660 nm selectively oxidised AlGaInP vertical-cavity lasers. *Electron. Lett.* 31(14):1145–1146.

[36] Takagi, T., and F. Koyama. 1991. Design and photoluminescence study on a multiquantum barrier. *IEEE J. Quantum Electron.* 27(6):1511.

[37] Haase, M. A., J. Qiu, J. M. DePuydt, and H. Cheng. 1991. Blue-green laser diodes. *Appl. Phys. Lett.* 59(11):1272–1274.

[38] Okuyama, H., E. Kato, S. Itoh, N. Nakayama, T. Ohata, and A. Ishibashi. 1995. Operation and dynamics of ZnSe/ZnMgSSe double heterostructure blue laser diode at room temperature. *Appl. Phys. Lett.* 66(6):656–658.

[39] Ishibashi, A., and S. Itoh. 1995. One-hour-long room temperature CW operation of ZnMgSSe-based blue-green laser diodes. In *Proceedings of the 15th International Semiconductor Laser Conference,* Paper PD1.1. Piscataway, NJ: IEEE.

[40] Nakamura, S., M. Senoh, S. Nagahama, N. Iwasa, T. Yamada, T. Matsushita, Y. Sugimoto, and H. Kiyoku. 1996. Ridge-geometry InGaN multi-quantum-well-structure laser diodes. *Appl. Phys. Lett.* 69(10):1477–1479.

[41] Asif Khan, M., S. Krishnankutty, R. A. Skogman, J. N. Kuznia, D. T. Olson, and T. George. 1994. Vertical-cavity stimulated emission from photopumped InGaN/GaN heterojunctions at room temperature. *Appl. Phys. Lett.* 65(5):520–522.

[42] Nakamura, S., 1994. Zn-doped InGaN growth and InGaN/AlGaN double-heterostructure blue-light-emitting diodes. *J. Cryst. Growth* 145:911–917.

[43] Lester, S. D., F. A. Ponce, M. G. Craford, and D. A. Steigerwald. 1995. High dislocation densities in high efficiency GaN-based light-emitting diodes. *Appl. Phys. Lett.* 66(10):1249–1251.

[44] Baba, T., Y. Yogo, K. Suzuki, F. Koyama, and K. Iga. 1993. Near room temperature continuous wave lasing characteristics of GaInAsP/InP surface-emitting laser. *Electron. Lett.* 29(10):913–914.

[45] Dudley, J. J., D. I. Babic, R. Mirin, L. Yang, B. I. Miller, R. J. Ram, T. Reynolds, E. L. Hu, and J. E. Bowers. 1994. Low threshold, wafer fused long wavelength vertical cavity lasers. *Appl. Phys. Lett.* 64(12):1463–1465.

[46] Babic, D. I., K. Streubel, R. P. Mirin, N. M. Margalit, E. L. Hu, J. E. Bowers, D. E. Mars, L. Yang, and K. Carey. 1995. Room-temperature continuous-wave operation of 1.54 μm vertical-cavity lasers. *IEEE Photon. Tech. Lett.* 7(11):1225–1227.

[47] Babic, D. I., and S. W. Corzine. 1992. Analytic expressions for the reflecting delay, penetration depth, and absorptance of quarter-wave dielectric mirrors. *IEEE J. Quantum Electron.* 28(2):514–524.

[48] Blum, O., I. J. Fritz, L. R. Dawson, A. J. Howard, T. J. Headly, J. F. Klem, and T. J. Drummond. 1995. Highly reflective, long wavelength AlAsSb/GaAsSb distributed Bragg reflector grown by molecular beam epitaxy on InP substrates. *Appl. Phys. Lett.* 66(3):329–331.

[49] Lambert, B., Y. Toudic, Y. Rouillard, M. Gauneau, M. Baudet, F. Alard, I. Valiente, and J. C. Simon. 1995. High reflectivity 1.55 μm (Al)GaAsSb/AlAsSb Bragg reflector lattice matched on InP substrates. *Appl. Phys. Lett.* 66(4):442–444.

[50] Agrawal, G. P., and N. K. Dutta. 1993. *Semiconductor lasers.* 2d ed. New York: Van Nostrand-Reinhold, 98–118.

[51] Coldren, L. A., and S. W. Corzine. 1995. *Diode lasers and photonic integrated circuits.* Appendix 17. New York: Wiley.

[52] Babic, D. I. 1995. *Double-fused long-wavelength vertical-cavity lasers.* Santa Barbara: University of California at Santa Barbara. ECE technical report no. 95-20, 95.

[53] Margalit, N. M., D. I. Babic, K. Streubel, R. P. Marin, J. E. Bowers, and E. L. Hu. 1996. Submitting long wavelength vertical-cavity lasers. In *Proceedings of the 15th ISLC.* Paper M3.5. Piscataway, NJ: IEEE.

[54] Corzine, S. W., R. S. Geels, J. W. Scott, R. H. Yan, and L. A. Coldren. 1989. Design of Fabry-Perot surface-emitting lasers with a periodic gain structure. *IEEE J. Quantum Electron.* 25(6):1513–1524.

[55] Scott, J. W., D. B. Young, B. J. Thibeault, M. G. Peters, and L. A. Coldren. 1995. Design of index-guided vertical-cavity lasers for low temperature-sensitivity, sub-milliamp thresholds, and single-mode operation. *IEEE J. Select. Top. Quantum Electron.* 1(2):638–647.

[56] Peters, F. H. 1995. *"Vertical"—Design software for VCSELs and related structures.* Lompoc, CA: Optical Concepts, Inc.

[57] Coldren, L. A., and S. W. Corzine. 1995. *Diode lasers and photonic integrated circuits.* Chapter 3. New York: Wiley.

[58] Coldren, L. A., and S. W. Corzine. 1995. *Diode lasers and photonic integrated circuits.* Chapter 4. New York: Wiley.

[59] Coldren, L. A., and S. W. Corzine. 1995. *Diode lasers and photonic integrated circuits.* Appendix 5. New York: Wiley.

[60] Coldren, L. A., and S. W. Corzine. 1995. *Diode lasers and photonic integrated circuits.* Chapter 2. New York: Wiley.

[61] Wipiejewski, T., D. B. Young, M. G. Peters, B. J. Thibeault, and L. A. Coldren. 1995. Improved performance of vertical-cavity surface-emitting diodes with Au-plated heat spreading layer. *Electron. Lett.* 31(4):279–280.

[62] Young, D. B., J. W. Scott, F. H. Peters, B. J. Thibeault, S. W. Corzine, M. G. Peters, S-L. Lee, and L. A. Coldren. 1993. High-power temperature-insensitive gain-offset InGaAs/GaAs vertical-cavity surface-emitting lasers. *IEEE J. Quantum Electron.* 5(2):129–132.

[63] Morgan, R. A., M. K. Hibbs-Brenner, T. M. Marta, R. A. Walterson, S. Bounnak, E. L. Kalweit, and J. A. Lehman. 1995. 200°C, 96-nm Wavelength range, continuous wave lasing from unbonded GaAs MOVPE grown vertical cavity surface-emitting lasers. *IEEE Photon. Tech. Lett.* 7(5):441–443.

[64] Nakwaski, W., and M. Osinski. 1992. Thermal resistance of top-surface-emitting vertical-cavity semiconductor lasers and monolithic two-dimensional arrays. *Electron. Lett.* 28(6):572–574.

[65] Goobar, E., M. G. Peters, G. Fish, and L. A. Coldren. 1995. Highly efficient vertical-cavity surface-emitting lasers optimized for low-temperature operation. *IEEE Photon. Tech. Lett.* 7(8):851–853.

[66] Wipiejewski, T., D. B. Young, M. G. Peters, B. J. Thibeault, and L. A. Coldren. 1995. Etched-pillar vertical-cavity surface-emitting laser diodes with submilliamp threshold currents and high output power. In *Proceedings of CLEO'95,* Paper CTu/B3. Washington, DC: Optical Society of America.

[67] Schubert, E. F., L. W. Tu, G. J. Zydzik, R. F. Kopf, A. Benvenuti, and M. R. Pinto. 1992. Elimination of heterojunction band discontinuities by modulation doping. *Appl. Phys. Lett.* 60(4):466–468.

[68] Scott, J. W. 1995. Design, fabrication and characterization of high-speed intra-cavity contacted vertical-cavity lasers. Ph.D. diss., University of California at Santa Barbara. ECE technical report no. 95-06, 175.

[69] Corzine, S. W. 1993. Design of vertical-cavity surface-emitting lasers with strained and unstrained quantum well active regions. Ph.D. diss., University of California at Santa Barbara. ECE technical report no. 93-09, 102–109.

[70] Thibeault, B. J., E. R. Hegblom, P. D. Floyd, Y. Akulova, R. L. Naone, and L. A. Coldren. 1995. Reduced optical scattering loss in vertical-cavity lasers with thin or tapered oxide apertures. In *Proceedings of LEOS'95, San Francisco, CA, October 30–November 2,* Paper PD2.1. Piscataway, NJ: IEEE.

[71] Verdeyen, J. 1989. *Laser electronics.* 2d ed. Englewood Cliffs, New Jersey: Prentice Hall, 62–108.

[72] Houng, Y. M., M. R. T. Tan, B. W. Liang, S. Y. Wang, L. Lang, and D. E. Mars. 1994. InGaAs (0.98 μm)/GaAs vertical cavity surface emitting laser grown by gas-source molecular beam epitaxy. *J. Cryst. Growth* 136(1-4):216–220.

[73] Schneider, R. P., Jr., J. A. Lott, K. L. Lear, K. D. Choquette, M. H. Crawford, S. P. Kilcoyne, and J. J. Figiel. 1994. Metalorganic vapor phase epitaxial growth of red and infrared vertical-cavity surface-emitting laser diodes. *J. Cryst. Growth* 145(1-4):838–845.

[74] Chalmers, S. A., and K. P. Killeen. 1994. Method for accurate growth of vertical-cavity surface-emitting lasers. *Appl. Phys. Lett.* 6(6):678–680.

[75] Multilayer-OMS. Pacific Lightwave, San Diego, CA.

[76] Bobel, F. G., H. Moller, A. Wowchak, B. Hertl, J. Van Hove, L. A. Chow, and P. P. Chow. 1994. Pyrometric interferometry for real time molecular beam epitaxy process monitoring. *J. Vacuum Sci. Tech. B* 12(2):1207–1210.

[77] Pyritte. SVT Associates, Inc., Eden Prairie, MN.

[78] Chalmers, S. A., K. P. Killeen, and E. D. Jones. 1994. Accurate multiple-quantum-well growth using real-time optical flux monitoring. *Appl. Phys. Lett.* 65(1):4–6.

[79] Pinsukanjana, P., A. Jackson, J. Tofte, K. Maranowski, S. Campbell, J. English, S. Chalmers, L. Coldren, and A. Gossard. 1995. Real-time simultaneous optical-based flux monitoring of Al, Ga, and In for MBE. *J. Vacuum Sci. Tech. B* 13(6).

[80] Coldren, L. A. 1981. Reactive-ion-etching of III-V compounds. U.S. Patent No. 4,285,763.

[81] Hu, E. L., and R. E. Howard. 1984. Reactive ion etching of GaAs in a chlorine plasma. *J. Vacuum Sci. Tech. B* 2:85.

[82] Scherer, A., J. L. Jewell, Y. H. Lee, J. P. Harbison, and L. T. Florez. 1989. Fabrication of microlasers and microresonator optical switches. *Appl. Phys. Lett.* 55(26):2724–2726.

[83] Scott, J. W. 1995. Design, fabrication and characteristics of high-speed intra-cavity contacted vertical-cavity lasers. Ph.D. diss., University of California at Santa Barbara. ECE technical report no. 95-06, 119–120.

[84] Skidmore, J. A., D. L. Green, D. B. Young, J. A. Olsen, E. L. Hu, L. A. Coldren, and P. M. Petroff. 1991. Investigation of radical-beam etching-induced damage in GaAs/AlGaAs quantum-well structures. *J. Vacuum Sci. Tech. B* 9(6):3516–3520.

[85] Morgan, R. A., G. D. Guth, M. W. Focht, M. T. Asom, K. Kojima, L. E. Rogers, and S. E. Callis. 1993. Transverse mode control of vertical-cavity top-surface-emitting lasers. *IEEE Photon. Tech. Lett.* 4(4):374–376.

[86] Scott, J. W. 1995. Design, fabrication and characterization of high-speed intra-cavity contacted vertical-cavity lasers. Ph.D. diss., University of California at Santa Barbara. ECE technical report no. 95-06, 121.

[87] Dallesasse, J. M., N. Holonyak, Jr., A. R. Snugg, T. A. Richard, and N. El-Zein. 1990. Hydrolization oxidation of $Al_xGa_{1-x}As$-AlAs-GaAs quantum well heterostructures and superlattices. *Appl. Phys. Lett.* 57(26):2844–2846.

[88] Choquette, K. D., R. P. Schneider, Jr., K. L. Lear, and K. M. Geib. 1994. Low threshold voltage vertical-cavity lasers fabricated by selective oxidation. *Electron. Lett.* 30(24):2043–2044.

[89] Kish, F. A., F. M. Steranka, D. C. DeFevere, D. A. Vanderwater, K. G. Park, C. P. Kuo, T. D. Osentowski, M. J. Peanasky, J. G. Yu, R. M. Fletcher, D. A. Steigerwald, M. G. Craford, and V. M. Robbins. 1994. Very high-efficiency semiconductor wafer-bonded transparent-substrate $(Al_xGa_{1-x})_{0.5}In_{0.5}P$/GaP light-emitting diodes. *Appl. Phys. Lett.* 64(21):2839–2841.

[90] Lei, C., T. J. Rogers, D. G. Deppe, and B. G. Streetman. 1991. ZnSe/CaF_2 quarter-wave Bragg reflector for the vertical-cavity surface-emitting laser. *J. Appl. Phys.* 69(11):7430–7434.

[91] Strzelecka, E. M., G. Robinson, M. G. Peters, F. H. Peters, and L. A. Coldren. 1995. Monolithic integration of vertical-cavity laser diodes with refractive GaAs microlenses. *Electron. Lett.* 31(9):724–725.

[92] Morgan, R. A., K. Kojima, L. E. Rogers, G. D. Guth, R. E. Leibenguth, M. W. Focht, M. T. Asom, T. Mullally, and W. A. Gault. 1993. Progress and properties of high-power coherent vertical cavity surface emitting laser arrays. *SPIE, Laser Diode Technology and Applications V* 1850:100–108.

[93] Gourley, P. L., M. E. Warren, G. R. Hadley, G. A. Vawter, T. M. Brennan, and B. E. Hammons. 1991. Coherent beams from high efficiency two-dimensional surface-emitting semiconductor laser arrays. *Appl. Phys. Lett.* 58(9):890–892.

[94] Orenstein, M., and T. Fishman. 1995. Coupling mechanism of two dimensional reflectivity modulated vertical cavity semiconductor laser arrays. In *1995 Proceedings of the International Semiconductor Laser Conference.* 70–71.

[95] Von Lehman, A., C. Chang-Hasnain, J. Wullert, L. Carrion, N. Stoffel, L. Florez, and J. Harbison. 1991. Independently addressable InGaAs/GaAs vertical-cavity surface-emitting laser arrays. *Electron. Lett.* 27(7):583–584.

[96] Moller, B., E. Zeeb, T. Hackbarth, and K. J. Ebeling. 1994. High speed performance of 2-D vertical-cavity laser diode arrays. *IEEE Photon. Tech. Lett.* 6(9):1056–1058.

[97] Orenstein, M., A. C. Von Lehman, C. Chang-Hasnain, N. G. Stoffel, J. P. Harbison, and L. T. Florez. 1991. Matrix addressable vertical cavity surface emitting laser array. *Electron. Lett.* 27(5):437–438.

[98] Morgan, R. A., G. D. Guth, C. Zimmer, R. E. Leibenguth, M. W. Focht, J. M. Freund, K. G. Glogovsky, T. Mullally, F. F. Judd, and M. T. Asom. 1994. Two-dimensional matrix addressed vertical cavity top-surface emitting laser array display. *IEEE Photon. Tech. Lett.* 6(8):913–915.

[99] Chang-Hasnain, C. J., M. W. Maeda, J. P. Harbison, L. T. Florez, and C. Linn. 1991. Monolithic multiple wavelength surface emitting laser arrays. *J. Lightwave Tech.* 9(12):1665–1673.

[100] Goodhue, W., J. Donnelly, and J. Zayhowski. 1989. Technique for monolithically integrating GaAs/AlGaAs lasers of different wavelengths. *J. Vacuum Sci. Tech. B* 7(2):409–411.

[101] Eng, L. E., K. Bacher, Y. Wupen, J. S. Harris, Jr., and C. J. Chang-Hasnain. 1995. Multiple-wavelength vertical cavity laser arrays on patterned substrates. *IEEE J. Select. Top. Quantum Electron.* 1(2):624–628.

[102] Wipiejewski, T., M. G. Peters, E. R. Hegblom, and L. A. Coldren. 1995. Vertical-cavity surface-emitting laser diodes with post-growth wavelength adjustment. *IEEE Photon. Tech. Lett.* 7(7):727–729.

[103] Wipiejewski, T., J. Ko, B. J. Thibeault, D. B. Young, and L. A. Coldren. 1996. Multiple wavelength vertical-cavity laser array employing molecular beam epitaxy regrowth. In *Proceedings of the 46th Electronic Components and Technology Conference.* Piscataway, NJ: IEEE.

[104] Iga, K., Y. Kokubun, and M. Oikawa. 1984. *Fundamentals of microoptics.* New York: Academic Press, 197.

[105] Chan, W. K., J. P. Harbison, A. C. Von Lehman, L. T. Florez, C. K. Nguyen, and S. A. Schwarz. 1991. Optically controlled surface-emitting lasers. *Appl. Phys. Lett.* 58(21):2342–2344.

[106] Kasahara, H., I. Ogura, H. Saito, M. Sugimoto, K. Kurihara, T. Numai, and K. Kasahra. 1993. Pixels consisting of a single vertical-cavity laser thyristor and a double vertical-cavity phototransistor. *IEEE Photon. Tech. Lett.* 5(12):1409–1411.

[107] Yang, Y. J., T. G. Dziura, T. Bardin, S. C. Wang, R. Fernandez, and A. S. H. Liao. 1993. Monolithic integration of a vertical cavity surface emitting laser

and a metal semiconductor field effect transistor. *Appl. Phys. Lett.* 62(6):600–602.

[108] Lu, B., P. Zhou, J. Cheng, R. E. Leibenguth, A. C. Adams, J. L. Zilco, J. C. Zolper, K. L. Lear, S. A. Chalmers, and G. A. Vawter. 1994. Reconfigurable binary optical routing switches with fan-out based on the integration of GaAs/AlGaAs surface-emitting lasers and heterojunction phototransistors. *IEEE Photon. Tech. Lett.* 6(2):222–226.

[109] Thompson, G., B. J. Thibeault, E. Strzelecka, G. Robinson, and L. A. Coldren. VCLs bonded to Rockwell circuits. Unpublished manuscript.

[110] Schwartz, D. B., C. K. Y. Chun, B. M. Foley, D. H. Hartman, M. Lebby, H. C. Lee, C. L. Shieh, S. M. Kuo, S. G. Shook, and B. Webb. 1995. A low cost, high performance optical interconnect. In *Proceedings of the 45th Electronic Components and Technology Conference,* 376–379. Piscataway, NJ: IEEE.

[111] Lewis, D. K., P. J. Anthony, J. D. Crow, and M. Hibbs-Brenner. 1993. The Optoelectronics Technology Consortium (OETC) — Program update. In *LEOS'93 Conference Proceedings,* 7–8.

[112] Hahn, K. H. POLO — Parallel optical links for gigabyte data communications. In *Proceedings of the 45th Electronic Components and Technology Conference,* 368–375. Piscataway, NJ: IEEE.

[113] *Optoelectronic technology roadmap — Conclusions and recommendations.* (OIDA). Washington, DC.

[114] Urquhart, K. S., P. Marchand, Y. Fainman, and S. H. Lee. 1994. Diffractive optics applied to free-space optical interconnects. *Appl. Opt.* 33(17):3670–3682.

[115] Plant, D. V., B. Robertson, H. S. Hinton, M. H. Ayliffe, G. C. Boisset, W. Hsiao, D. Kabal, N. H. Kim, Y. S. Liu, M. R. Otazo, D. Pavlasek, A. Z. Shang, J. Simmons, and W. M. Robertson. 1995. A 4×4 VCSEL/MSM optical backplane demonstrator system. In *LEOS'95 Conference Proceedings,* Paper PD2.4. Piscataway, NJ: IEEE.

[116] Guilfoyle, P. S., F. F. Zeise, and J. M. Hessenbruch. 1994. "Smart" optical interconnects for high speed photonic computing. *Opt. Comput. 1994 Tech. Ser. Dig.* 7:78–81.

[117] Cloonan, T., G. Richards, A. Lentine, F. McCormick, Jr., H. S. Hinton, and S. J. Hinterlong. 1993. A complexity analysis of smart pixel switching nodes for photonic extended generalized shuffle switching networks. *IEEE J. Quantum Electron.* 29(2):619–634.

[118] McCormick, F. B. 1993. Free-space interconnection techniques. In *Photonics in switching,* Vol. II, *Systems,* ed. J. E. Midwinter, San Diego: Academic Press.

[119] Metzger, R. A. 1995. Turning blue to green. *Compound Semiconductor* 1(1):26–28.

Chapter 7 | Optical Fiber Components and Devices

Alice E. White
Stephen G. Grubb*

Lucent Technologies, Bell Laboratories, Murray Hill, New Jersey

Since the mid-1980s, there has been a revolution in the way people think about optical communications. It started with the invention of the optical fiber amplifier and has gathered strength with the discovery and implementation of a number of novel optical fiber components. This chapter begins by summarizing the key developments and properties of optical fiber amplifiers at 1.55 μm (Er^{3+}-doped silica fibers) and 1.31 μm (Pr^{3+}-doped ZBLAN [ZrF_4-BaF_2-LaF_3-AlF_3-NaF] fibers and germanosilicate Raman amplifiers). Fiber dispersion compensators that are necessary to upgrade existing 1.3-μm zero-dispersion fiber routes with Er^{3+}-doped fiber amplifiers (EDFAs) are also discussed.

The emergence of the UV-induced fiber Bragg grating writing process over the past several years has led to numerous novel *in situ* fiber devices. These devices are already demonstrating their vast potential in fiber optic communications systems. Fiber Bragg gratings have been used to construct short-cavity, single-frequency fiber lasers as well as semiconductor laser stabilizers and pump reflectors. Chirped fiber Bragg gratings have been demonstrated as dispersion compensators. Bragg gratings have also been used to construct ultrasharp filters and demultiplexer filters, which are of tremendous importance in dense wavelength-division multiplexing (WDM) systems. Long-period and tilted short-period Bragg gratings have demonstrated their usefulness in the fabrication of complex filter shapes needed to equalize optical amplifier gains. A discussion of the extension of Bragg gratings to planar waveguides is also included in this chapter.

The optical powers in communications systems increased sharply with the introduction of the EDFA. Future systems using higher data rates, WDM, increased passive split architectures, and extended repeater spacings

---

* Present address: SDL Inc., 80 Rose Orchard Way, San Jose, CA.

OPTICAL FIBER TELECOMMUNICATIONS,
VOLUME IIIB

Copyright © 1997 by Lucent Technologies.
All rights of reproduction in any form reserved.
ISBN: 0-12-395171-2

will continue to demand increased optical powers. The technology of high-power Er- and Er/Yb-doped 1.55-μm optical amplifiers is another topic of this chapter.

Finally, cascaded Raman lasers and amplifiers are an alternative method of obtaining high single-mode fiber-coupled powers at wavelengths of interest for communications systems, and they provide a silica-fiber-based alternative for optical amplification at 1.31 μm. Because fiber enables intense pumping over long interaction lengths, it is an ideal medium for up-conversion lasers, where population of the upper lasing or amplifying level requires two or more pump photons. A summary of fiber up-conversion laser and amplifier transitions and performance concludes the chapter.

I. Fiber Amplifiers and Related Components

A. Er-DOPED FIBER AMPLIFIERS

Future communications systems will require higher bit rates, which can be achieved by WDM. In addition, format independence and the possibility of bidirectional transmission would be desirable. The electronic regeneration employed in most of the existing systems is a bottleneck that optical amplification can eliminate. Essentially, an optical amplifier is a one-pass laser. Optical amplification can be achieved through stimulated emission from the excited states of impurity atoms in the glass as long as a population inversion exists. The rare-earth atoms are ideally suited for this application: transitions between electronic levels in the unfilled 4f shell give absorption and emission lines in the infrared,[1] and, because the electrons are shielded from the local environment by the filled 5s and p and 6s shells, the levels are more or less independent of the host glass. For the 1.55-μm communications window, the transition between the $^4I_{13/2}$ level and the ground state of erbium is ideal. The $^4I_{13/2}$ level is relatively long-lived and broad enough in glass so that the atoms can be excited by pumping into the same level with 1480-nm light. Higher excited states with shorter lifetimes are also available for pumping. The Er-doped glass then has a broad gain spectrum that incorporates the 1.55-μm window.

The EDFA is assembled by first incorporating Er into the core of the optical fiber at about the 1000-ppm level during the core deposition step in the fiber fabrication process. Several meters of the Er-doped fiber are then spliced into regular transmission fiber, and the pump light (usually at

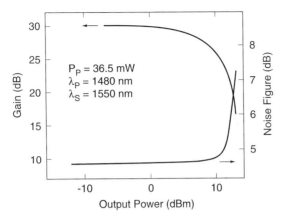

Fig. 7.1   Output power for an Er-doped fiber amplifier (EDFA), where P_P is the pump power, λ_P is the pump wavelength, and λ_S is the signal wavelength. (Reprinted from Ref. 2 with permission.)

1480 or 980 nm) is coupled in. A gain performance curve for a typical EDFA[2] is show in Fig. 7.1. Useful gains are easily achievable by pumping with solid-state laser diodes. Optical amplification with EDFAs is bit-rate independent, modulation-format independent, and, in principle, bidirectional. Amplification of multiple wavelengths over the broad gain spectrum is also possible, which eliminates many of the drawbacks of electronic regenerators. Especially in concatenated systems, noise is an issue because the EDFA does not reshape the signal but rather amplifies noise and signal alike. EDFAs are covered in detail in Chapter 2 in Volume IIIB. Recent advances in high-power EDFAs are discussed later in this chapter.

B.   FIBER DISPERSION COMPENSATORS

One of the barriers to implementation of EDFAs at 1.55 μm is the high dispersion in silica fibers at this wavelength. In fact, the reason that existing optical communications systems were installed at 1.31 μm, where the loss is higher but the dispersion is lower, is that dispersion was thought to be the larger problem. Indeed, as the demand for capacity drives systems to higher and higher bit rates, dispersion is the limiting parameter; however, fiber-based dispersion compensators are now being developed to address this issue. Standard low-delta single-mode fiber (where delta is the normalized core–clad index difference) has a positive dispersion, magnitude

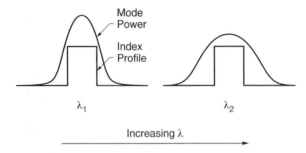

Fig. 7.2 Higher delta fiber has a negative dispersion.

17 ps/nm-km, at 1.55 μm. Higher delta fiber has a negative dispersion. The explanation is illustrated in Fig. 7.2. With increasing wavelength (λ), the mode field broadens, more of the light travels in the cladding rather than in the core of the fiber, and dispersion, which is equal to the time delay, $d\tau/d\lambda$, goes negative. Single-mode dispersion-compensating fibers (DCF) are being designed with more complicated index profiles to tailor the dispersion by, for instance, allowing the guided mode to leak into the lower index cladding. This can give a large negative dispersion, but it also increases the loss. Early DCF designs compensated for the magnitude of the dispersion at a particular wavelength, but their application in multiwavelength systems would be limited. More recent designs[3] can compensate for both the magnitude and the slope of the dispersion at 1.55 μm and have dispersions as high as −200 ps/nm-km. Implementation in an existing system would involve adding approximately a kilometer of DCF for every 10 km of transmission fiber. If this is done in an intermediate stage of a multistage EDFA, the impact of the higher loss can be minimized. DCFs using higher order modes near cutoff have also been demonstrated.[4] In this case, the dispersion can be large, so a relatively small amount of DCF is required; however, there is the additional complication of needing to convert back and forth from the higher order mode. DCF is covered in greater detail in Chapter 7 in Volume IIIA.

C. Pr-DOPED OPTICAL AMPLIFIERS

Upgrading the existing terrestrial communications network, which is optimized for operation at 1.31 μm, could be accomplished by switching to 1.55 μm. This means, however, that in addition to using EDFAs in place of regenerators, all the transmitters and receivers would need to be replaced.

An alternative to this costly upgrade is to develop an optical amplifier at 1.31 μm. Considerable effort has gone into this problem, and some progress has been made. The starting point is the energy-level diagram of the rare-earth atoms. In this situation, we are looking for an excited state that is at least 0.95 eV (1.31 μm) but not more than 750 nm above the ground state (so that it can be pumped with a conventional solid-state laser diode). Three potentially useful transitions exist: promethium (Pm), neodymium (Nd), and praseodymium (Pr). Because Pm is radioactive, it can be eliminated. Neodymium can be pumped at a convenient 820 nm, but it suffers from a poor branching ratio (most of the emission is at 1.06 μm) and excited state absorption (ESA) of the signal, which limits the available gain to longer wavelengths. For these reasons, attention has focused on the $^1G_4$ excited state of Pr. Praseodymium has some advantages: it is a four-level system, which means that it is transparent to the signal wavelength when the pump is turned off; it has a perfect emission spectrum (centered at 1.31 μm); and it has a convenient pump wavelength (1.02 μm). The problem is that the $^1G_4$ level decays nonradiatively to close-lying levels. No 1.31-μm emission is seen in Pr^{3+}-doped silica. To take advantage of the desirable properties of Pr, it is necessary to explore alternative hosts with lower phonon energies. This is because the nonradiative transition probability,

$$W = W_0 \times \left(\frac{e^{\eta\omega/kT}}{e^{\eta\omega/kT} - 1} \right)^{(\Delta E/\eta\omega)} \times \exp\left[-\frac{\Delta E}{\eta\omega} \left(\ln \frac{(\Delta E/\eta\omega)}{\mathbf{g}} - 1 \right) \right], \quad (7.1)$$

decreases exponentially with the number of phonons required to bridge the gap between the excited state and the next-nearest level.[5] In Eq. (7.1), ΔE is the energy spacing between levels, $\eta\omega$ is the highest energy phonon of the host glass, $\mathbf{g}$ is the electron-lattice coupling constant, and the number of phonons required to bridge the gap is $\Delta E/\eta\omega$. For Pr, the gap between the $^1G_4$ and the $^3F_3$ levels is 2700 cm^{-1}. The highest energy phonon in silica is approximately 1000 cm^{-1}. Several candidate glasses exist, but attention has focused on the fluorides, with a maximum phonon energy of about 500 cm^{-1}. The fluorides have been explored for a long time because of their potential for ultralow loss (the same low phonon energy pushes the multiphonon loss edge to longer wavelengths) and are known to be amenable to the manufacture of single-mode fiber.[6]

The low melting and crystallization temperatures of the fluoride glasses mean that they have a tendency to crystallize, and losses in the early fibers were high because of scattering. A composition known as ZBLAN, which, as we noted previously, contains zirconium, barium, lanthanum, aluminum, and sodium fluorides, is the most popular because it is a relatively stable glass composition and it is relatively easy to substitute a rare-earth atom for the La. However, entirely new fiber fabrication techniques had to be developed for ZBLAN because of its low crystallization temperature. Fluorescence experiments showed that although the Pr emission was broad and centered at 1.31 μm in ZBLAN, it was relatively inefficient compared with Er in silica.[7] Nonetheless, in 1991, workers at Nippon Telephone and Telegraph (NTT) and Rutgers University constructed a fiber amplifier.[8] These first Pr-doped fiber amplifier (PDFA) results, shown in Fig. 7.3, were encouraging, but not yet practical. Soon after, another group at NTT[9] was able to achieve tens of decibels of gain with a few hundred milliwatts of pump power at 1.017 μm. Concentration quenching, processes by which adjacent Pr ions lose energy nonradiatively, limits the amplification efficiency that can be achieved in ZBLAN fibers by increasing the number of Pr ions.[10] The NTT results were achieved by going to a small-core, high numerical aperture (NA) fiber, which is technically difficult. Pumping in the first experiments was accomplished with a bench-top Ti:sapphire laser, but, later in 1991, a diode-pumped PDFA was demonstrated.[11] Using two pump units, each consisting of two polarization-multiplexed laser diodes

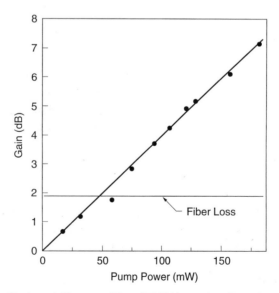

Fig. 7.3 First Pr-doped fiber amplifier (PDFA) results. (Reprinted from Ref. 8 with permission.)

at 1.017 μm, this research group achieved 15.1 dB of gain at 160 mW of pump power. Early system experiments characterizing the PDFAs as a preamplifier, a power amplifier, and a repeater[12] showed no unexpected performance degradations. Because fluoride fibers cannot be fusion spliced to transmission fiber, the PDF was connectorized using index matching fluid between butt-coupled joints secured in connectors. The connector-to-connector gain was 19 dB, and the unpumped loss was 5 dB. In the preamplifier configuration, the PDFA had a gain of 24 dB and a saturated output of 100 mW at 800 mW of pump power. The noise figure of the amplifier was less than 6 dB. Improvements in the host glass loss and reliability may improve the performance of PDFAs to the level where they will be useful for upgrading 1.31-μm communications systems. In 1994, a silica-based 1.31-μm Raman amplifier was invented and, in many ways, has more promise than PDFAs. This amplifier is described in Section III.C.

D. FIBER GRATINGS

An exciting new technology that has tremendous potential for having an impact on fiber optic communications is UV-induced fiber Bragg gratings. The gratings are created directly in the GeO_2-doped core of optical fibers by interfering two UV beams from the side after the coating has been removed (Fig. 7.4). The UV light, usually from a KrF-excimer laser-pumped dye laser operating at 240 nm, is absorbed by the germanium defects in the core, and the resultant periodic index of refraction variation is a Bragg reflector. The Bragg wavelength, λ_B, is given by

$$\lambda_B = 2n_{eff}\Lambda = \lambda_{UV}/2 \sin(\alpha), \qquad (7.2)$$

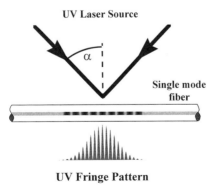

Fig. 7.4   Writing a fiber Bragg grating from the side.

where Λ is the period of the index of refraction variation and α is the angle between the interfering UV beams. Because the fiber geometry is extremely well controlled, the Bragg wavelength is precisely determined (to within a fraction of an angstrom) by α and λ_{UV}. The index change remains after the UV light is removed. The transmission and reflection spectra of a typical grating, shown in Fig. 7.5, reveal why they are so useful. The transmission spectrum (Fig. 7.5a) shows effectively 100% transmission of the light, except at the Bragg wavelength, which is 1557.2 nm in this case. The reflection spectrum (Fig. 7.5b) makes it clear that the light at the Bragg wavelength

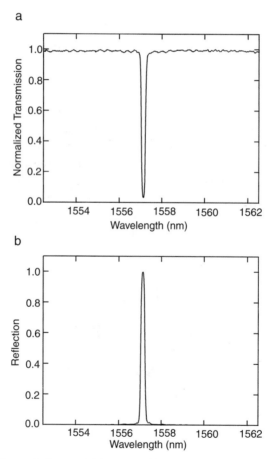

Fig. 7.5 (a) Transmission and (b) reflection spectra of a moderate-strength fiber Bragg grating.

is reflected back down the fiber. These integrated reflectors can replace bulk optic components in transmitters, receivers, filters, and amplifiers.

The first fiber gratings were written using the standing-wave pattern formed by counterpropagating light from an Ar ion laser.[13] The two-photon absorption of the intense visible light caused the index variation with a period set by the laser wavelength. The invention of side writing with UV light in 1989[14] made it possible to conveniently vary the Bragg wavelength throughout the telecommunications windows; however, the photosensitivity of ordinary transmission fiber was too weak to write the strong gratings of interest for applications. The invention of a sensitization process called *hydrogen loading* in 1993[15] made it possible to write useful gratings in standard fiber, enabling a host of practical applications. In this process, the fiber is exposed to high-pressure (20–750 atm) hydrogen or deuterium at moderate temperatures (21–75°C) for up to a week. Hydrogen loading makes any germanium-doped fiber controllably photosensitive. Without H_2 loading, the index changes that are observed are on the order of 10^{-4}. With H_2 loading, index changes as large as 10^{-2} have been achieved. The unreacted H_2 diffuses out during a subsequent anneal.

The mechanism for the index of refraction changes caused by the intense (>10-W/cm^2) UV irradiation is probably a combination of electronic excitations, expansion–compaction due to changes in the defect population of the silica, and stress effects due to heating of the core and not the cladding. The most likely candidate for the electronic transitions are the GeO defects: a $+2$ oxidation state Ge oxygen-deficient center (GODC, $=$Ge:) and a neutral oxygen vacancy (NOV, $=$Ge$-$Ge$=$). The GODC is present in the core of highly germanium-doped fiber (10 mol% Ge, 0.3% Δ) at concentrations of 10^{18}/cm^3, the NOV in concentrations an order of magnitude lower. These defects have a strong absorption at 240 nm and fluoresce strongly at 400 nm. Spectroscopic studies of the core before and after intense UV exposure[16] have been interpreted as pointing to a "color-center" model for photosensitivity, in which the GeO defect band at the exposure wavelength (242 nm) is bleached, and a new absorption, large enough to account for the index change, grows at 195 nm. These changes in the defect population can be reversed by annealing at 900°C, which gives additional support to the color-center model.

Studies under less extreme conditions of heat and UV exposure[17] provide some additional clues about the microscopic mechanisms of photosensitivity. In these experiments, the changes in the glass are monitored with Raman spectroscopy. Greene *et al.*[17] postulated that without H_2 loading, the NOV defects are transformed to Ge E′ centers and the GODCs are not active. Indeed, NOVs were previously identified as the precursor to Ge E′ centers.[18] With H_2 loading, the mechanism for H-induced photosensitivity is a photoinitiated reaction at the GODC centers, giving GeH_2. The signature of the GeH_2 stretch mode is clearly identified in the Raman spectra. This change in the larger defect population (it is estimated that *all* of the Ge in the core reacts with H_2) results in the much larger change in the index of refraction, which explains the enhanced photosensitivity of H_2-loaded fibers. Under heat alone, hydrogen probably adds across a single Ge—O bond, creating Ge—H. This reaction pathway is supported by quantum chemical calculations[19,20] that also show that there are two additional reaction pathways that lead to divalent Ge defects that absorb at 242 nm. This explains the observation that heat treatment can also enhance the photosensitivity of germanium-doped fiber.[21,22]

Section II begins with a description of several of the applications envisioned for fiber gratings, including fiber lasers, demultiplexers, and gain equalizers, and concludes with a discussion of fiber grating reliability and manufacturability, and gratings in planar waveguides.

II. Applications of Fiber Gratings

A. FIBER LASERS

Two fiber gratings matched to the same wavelength written in a fiber with gain can be used to define a laser cavity. This was first demonstrated by Ball, Morey, and Glenn,[23] who used an Er-doped fiber that also contained germanium. Considerable effort was made to match the gratings, which were spaced at 0.5 m in the fiber. Reflectivities of 80 and 72% were chosen for the approximately 1-cm-long gratings in order to minimize the grating bandwidth. When pumped with a Ti:sapphire laser, the Er-doped fiber laser lased at the Bragg wavelength, 1548 nm in this case. A peak output of 5 mW was obtained, and single-mode operation was confirmed. The laser had an extremely narrow linewidth of less than 47 kHz. The primary advantage of a laser made this way is the ability to easily and precisely determine the lasing wavelength using intracore Bragg gratings.

Tuning of the wavelength can be accomplished by the application of temperature (T) or strain.[24] The change in wavelength of the grating[25] is due to the change in period and is a straightforward function of length (L) and T:

$$\Delta\lambda/\lambda = 0.8(\Delta L/L) + (8 \times 10^{-6}/°C)\Delta T. \tag{7.3}$$

At $\lambda = 1.55$ μm, $\Delta\lambda = 1$ Å for $\Delta T = 8°C$ of 7 g or force. Although the temperature dependence is much less than that of a semiconductor diode laser, the strain dependence is considerable; therefore, the gratings must be packaged in a strain-free or strain-controlled package to be useful. Tuning over almost 10 Å was accomplished by stretching the fiber using a piezoelectric translator. Because the change in Bragg wavelength tracks the change in cavity resonance wavelength, no mode hopping was observed.

To ensure the robust single-mode operation of importance for communications systems, the mode spacing of the laser must be comparable to the grating bandwidth. For typical grating bandwidths of more than 0.2 nm, a mode spacing of 10 GHz, corresponding to a cavity length of 1 cm, is necessary. Such short-fiber lasers were first demonstrated by Zyskind *et al.*[26] in 1992. Although the output power is only 57 μW, the pump is hardly absorbed in the laser and can be used to pump a tandem EDFA. Stable single-mode operation in such a tandem short-fiber laser–amplifier was confirmed by Mizrahi and coworkers[27] in 1993 in a bit error rate (BER) measurement. The laser was externally modulated at 5 Gb/s with a pseudo-random bit stream. The transmission experiment resulted in a straight BER line with respect to the received optical power to below 10^{-15} (3 days of error-free operation), and there was no penalty relative to a distributed feedback (DFB) laser. Although DFB lasers are a proven technology, manufacturers are still finding it difficult to hit the target wavelengths with the accuracy required for WDM systems and often resort to binning to meet wavelength specifications. The wavelength accuracy inherent in these grating-based sources may make them an attractive alternative. More recently, output powers of these lasers have been increased by using Er-doped fibers that have been co-doped with ytterbium (Yb) to enhance the pump absorption. Output powers in the milliwatt range have been reported for a single-frequency DFB-like fiber laser[28] and as high as 10 mW for the more conventional two-grating fiber laser.[29]

The hybrid laser shown in Fig. 7.6 combines elements of semiconductor lasers and fiber lasers. Consisting of a semiconductor gain cavity coupled to a passive section of fiber with a grating that serves as the output coupler, this laser can operate continuous wave (CW) or as a pulsed source.[30] In addition to its compact package, this laser has a number of advantages. Directly modulating the gain to the laser permits a mode-locked train of stable, near-transform-limited pulses to be generated that are ideal for soliton systems. The laser can be accurately set to a specific model-locking

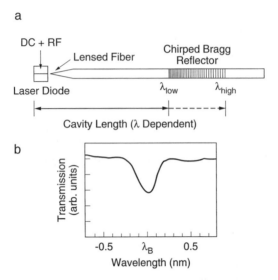

Fig. 7.6   Schematic of (a) the hybrid soliton pulse source and (b) transmission of a chirped Bragg reflector. (Reprinted from Ref. 30 with permission. Copyright © 1993 IEEE.)

frequency, and tuning is achieved automatically through the use of a chirped fiber grating. It can also serve as a compact high-power CW source.

B.   LASER WAVELENGTH STABILIZATION

Even if a grating is not used as part of the gain cavity of a laser, the wavelength accuracy of a grating can be used to advantage to stabilize the wavelengths of semiconductor lasers. Gratings with 1% reflectivity placed directly in the pigtail of 980-nm laser diodes have produced an improvement in the output spectrum by means of coherence collapse.[31] These lasers are used for pumping the 980-nm band of Er, which is only 10–15 nm wide, and without grating stabilization they are notoriously sensitive to back reflections because of their low-reflectivity output facet. The grating reduces the output power of the laser, but optical feedback from the grating reduces the laser threshold in partial compensation. This grating feedback forces the laser to emit at a wavelength between the Bragg wavelength and the nearest mode of the laser chip. Because the modes are spaced at 2 Å, the output changes by only 2 Å over a temperature range of 50°C, which potentially eliminates the need for a thermoelectric cooler for the laser. This grating-stabilized laser was the first commercial application of fiber

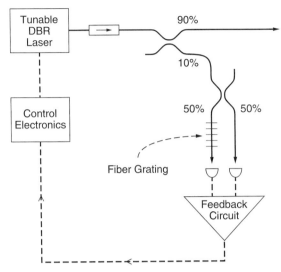

Fig. 7.7 Technique for actively stabilizing the wavelength of a tunable distributed Bragg reflector (DBR) laser with a grating. (Adapted from Ref. 33 with permission. Copyright © 1993 IEEE.)

Bragg gratings. Giles, Erdogan, and Mizrahi[32] showed that a weak grating could also be very effective at locking the pump wavelength of three 980-nm lasers simultaneously. Using a single grating with 64% reflectivity in one arm of a 4 × 4 fused-fiber coupler, they were able to lock the output wavelengths of all three lasers to the Bragg wavelength of the grating, leaving only 6% of the power in the amplified spontaneous emission (ASE) peak.

A scheme for actively stabilizing the wavelength of a tunable distributed Bragg reflector (DBR) laser with a grating[33] is shown in Fig. 7.7. In essence, the grating is used as a null detector. A small part of the signal wavelength is picked off and compared with the Bragg wavelength of a fiber grating. If the wavelength of the laser has drifted away from the target wavelength, light passes through the Bragg reflector and activates a feedback circuit that tunes the laser. The temperature stability of the grating is an asset in this application, but if greater temperature stability is desired, the grating can easily be temperature-stabilized.

C. PUMP REFLECTORS

Hydrogen loading of highly germanium-doped fiber has enabled the fabrication of very broadband gratings.[34] These gratings have index changes as large as 0.02, comparable to the core–clad index difference. In fact, UV light has been used to write the core of a planar waveguide structure.[34] These strong gratings can be used to improve the performance of optically amplified communications systems by reflecting the unused 1480-nm pump power back through the amplifier. The spectral output of 1480-nm pump sources is very broad, 25 nm, so a high bandwidth reflector is required. Using a highly germanium-doped fiber and hydrogen loading, Soccolich *et al.*[35] fabricated a 23-nm full width at half maximum (FWHM) grating. As expected, splicing the pump reflector grating in after the Er^{3+}-doped fiber increases the average inversion over the Er^{3+} fiber and uses the pump light more efficiently, which results in a few decibels of gain and noise-figure improvement. This improvement translates directly into an increased system margin. A similar scheme can be used for 980-nm pumping.

What is not immediately obvious from the reflection spectrum of a strong grating (Fig. 7.8a) is that there are additional features that present a design challenge for more complicated devices. These are apparent, however, in the transmission spectrum, shown for comparison in Fig. 7.8b. Coupling to backward-propagating and cladding modes in the fiber results in a large loss on the short-wavelength side of the grating. The gap between the Bragg wavelength, λ_B, and the longest wavelength at which light can couple to the cladding, λ_L, results from wave-vector matching requirements as the grating starts to grow[36] and is a function of the fiber effective index, n_{eff}, and the index of the cladding, n_{cl}:

$$\lambda_B - \lambda_L = \lambda_B/2(1 - n_{cl}/n_{eff}). \tag{7.4}$$

Use of high-delta fiber can widen the gap and push the loss to even shorter wavelengths. This loss can be further reduced by increasing the grating-mode overlap by either increasing the mode confinement in the core or using a fiber with Ge in the cladding to write a grating across the entire mode.[37]

Another undesirable aspect of the grating transmission spectrum is the features on the short-wavelength edge. These are Fabry–Pérot oscillations: because the Bragg wavelength is slightly shorter in the wings than at the peak of the grating, a resonant cavity is created.[36] The difference in Bragg wavelength is due to a difference in average index that is a consequence of the Gaussian spatial profile laser beam that is used to write the grating. This problem can be solved by using a strongly chirped grating. This is a grating in which the period varies monotonically over the length of the grating. The chirp rate is characterized by the change in period per unit length of the grating. The challenge is in generating the desired chirp. A linearly chirped grating made with different cylindrical lenses in each of

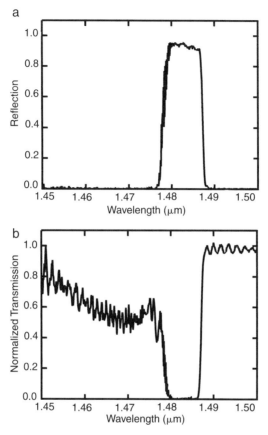

Fig. 7.8 (a) Reflection and (b) transmission spectra of a strong (FWHM = 10 nm) grating.

the arms of the interferometer was used to achieve a very broad, 44-nm-wide, grating.[38]

D.   FILTERS AND DEMULTIPLEXERS

Another application that takes advantage of the wavelength accuracy of fiber Bragg gratings is their use as filters, especially in WDM systems, where channel spacings can be as small as 4 Å. A simple four-channel demultiplexer can be made using a 1 × 4 splitter and fiber gratings.[39] Each arm of the splitter has two gratings that are designed to reflect all channels except the desired signal channel. An isolator is usually already present in the signal path and serves to eliminate the back-reflected wavelengths. A high-delta fiber was used to minimize the deleterious effects of radiation mode loss on the short-wavelength side of the gratings, and the gratings were chirped to eliminate the Fabry–Pérot oscillations. The transmission characteristics of a packaged device (Fig. 7.9) show a very flat passband and superior channel rejection. The insertion loss of this robust, all-fiber device is 7.4 dB, with the only major loss coming from the splitter. Moreover, the device is polarization insensitive.

More complicated filters that take advantage of the clean reflection spectrum of the fiber Bragg grating have also been realized. A narrow band transmission filter can be made by writing identical Bragg gratings in each arm of a 3-dB four-port fused-fiber coupler[40] (Fig. 7.10). Unwanted wavelengths in the broadband input into port 1 would pass through the grating and be lost at ports 3 and 4. Only the desired wavelength is reflected by the grating and appears at the output (port 2). This stop-band characteristic can be converted to a passband if the gratings are written in a Mach–Zehnder interferometer. Another coupler-based device was proposed by

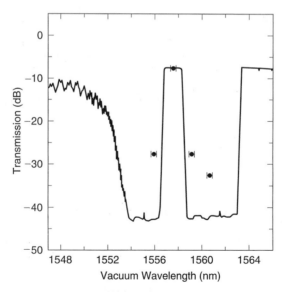

Fig. 7.9 Transmission characteristics of a four-channel fiber grating demultiplexer, showing one channel and including a connector-to-connector insertion loss of 7.4 dB. (Reprinted from Ref. 39 with permission.)

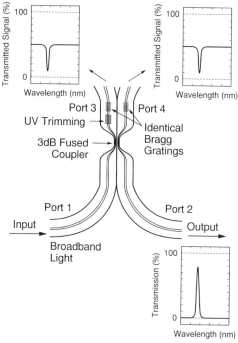

Fig. 7.10 Schematic of a narrow band transmission filter made by writing identical Bragg gratings in each arm of a 3-dB four-port fused-fiber coupler. (Reprinted from Ref. 40 with permission. Copyright © 1994 IEEE.)

Archambault *et al.*[41] A grating in only one arm of the coupling region prevents coupling from one channel to the other over a narrow wavelength range.

E. DISPERSION COMPENSATORS

Another approach to dispersion compensation is to make a resonant device using a chirped fiber Bragg grating. As illustrated in Fig. 7.11, each signal wavelength will be reflected at a different spot in the chirped grating, giving the required differential delay. If the length of the grating is L and the minimum and maximum wavelengths that can be reflected are λ_1 and λ_2, the maximum differential delay time is $\tau = 2nL/c$ and the dispersion is $\tau/(\lambda_1 - \lambda_2)$. The primary limitation of this device is its bandwidth, which is given by $\delta\Lambda L/\tau$, where $\delta\Lambda L$ is known as the chirp parameter, F. Ouellette[42] demonstrated such a device. Using a 10-cm-long grating with $F = 20$,

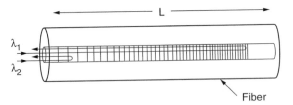

Fig. 7.11 Dispersion compensation using a chirped fiber Bragg grating.

Ouellette was able to flatten the dispersion over about a 16-GHz bandwidth. More recently, Williams *et al.*[43] made a linearly chirped Bragg grating by interfering two UV beams with different wave-front curvatures. They used this to recompress a 2-ps pulse that had been broadened by dispersion and self-phase modulation in a 200-m-long single-mode fiber. Their grating was only 5 mm long, with a bandwidth of 2.5 nm — longer gratings would have given them more bandwidth. Another technique for making the chirped gratings is reported by Hill *et al.*[44] First, the effective index in a short length of fiber is varied linearly by scanning a razor-edge mask along the fiber while irradiating with UV light. Next, a spatially periodic Bragg grating is patterned in the fiber, which, in combination with the first exposure, creates a linearly chirped grating.

F. GAIN EQUALIZATION

With the emergence of WDM systems, flattening of the EDFA gain spectrum to increase the usable bandwidth is a high priority. This can be accomplished using tilted Bragg gratings that couple the guided modes out into the cladding. Originally intended for wavelength-selective taps,[45] a series of such tilted gratings (~6° from normal) has been used to add wavelength-selective loss to an amplifier[46] significantly reducing the variation in the gain. Another approach to this problem is to use long-period fiber gratings.[47] These are gratings with periods of tens of micrometers. Unlike fiber Bragg gratings that reflect light back down the fiber, long-period gratings scatter the light forward into cladding or radiation modes, where it is lost. As a result, they have extremely low back reflections (~−80 dB). In addition, they are easy to fabricate with an amplitude mask or by scanning a slit in front of a UV beam. Because a weak resonance couples light out of the core, these gratings are highly bend sensitive and must be packaged carefully. However, it is possible to tailor the characteristics to closely resemble the inverse of the EDFA gain spectrum.[48]

G. *RELIABILITY AND MANUFACTURABILITY OF FIBER BRAGG GRATINGS*

The reliability requirements for components used in communicatons systems are notoriously demanding. Not surprisingly, then, the most frequently asked questions about fiber Bragg gratings is "How permanent are they?" Long-term stability of the gratings is a major concern. A complete study of the degree of permanence of the gratings was reported by Erdogan *et al.*[49] They monitored the change in the integrated coupling constant ($\eta = \tan h^{-1} [(1 - T_{min})^{1/2}]$, with T_{min} the minimum in the transmission spectrum of the grating) as a function of time for anneals at varying temperatures. Because the integrated coupling constant (ICC) is proportional to the UV-induced index change, η is a measure of the decay of the grating. A typical curve of η versus time shows an initial rapid drop and then levels off. These researchers realized that the thermally induced decay of the grating strength could be fit to a weak power-law-like function, $\eta = 1/[1 + A(t/t_1)^\alpha]$, with $t_1 = 1$ min and $\alpha = T/T_0$ with $T_0 = 5250 \pm 250$ K. A theoretical model in which photoexcited carriers that had been retrapped in non-ground-state sites were being thermally detrapped (Fig. 7.12) is then supported by this empirical observation. It is not necessary to know the exact distribution of energies of the UV-induced defects for the analysis, but the defects discussed previously are certainly consistent with this picture. All that is

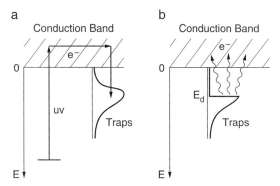

Fig. 7.12 Diagram of the theoretical model in which photoexcited carriers are (a) trapped in a continuous distribution of traps and (b) thermally detrapped. (Reprinted from Ref. 49 with permission.)

necessary is to assume that the detrapping of defects is activated with a rate given by

$$\nu(E) = \nu_0 \exp(-E/kT). \tag{7.5}$$

All traps with activation energy less than $E_d = kT \ln(\nu_0 t)$ are thermally depopulated, and all traps with activation energy greater than E_d are filled. This implies that the population of defects (and hence the index change) depends only on E_d and not on the particular combination of temperature and time that gave E_d. Therefore, if η is plotted versus E_d for a series of anneals, the data should fall on a single curve. That this is indeed the case is shown in Fig. 7.13, a finding that implies that Eq. (7.5) can be rewritten in terms of E_d:

$$\eta = 1/\{1 + \exp[(E_d - \Delta E)/kT_0]\}. \tag{7.6}$$

The importance of Eq. (7.6) is that it is possible to determine the fraction $1 - \eta$ of the UV-induced index change that is eliminated by any combination of temperatures and times. Moreover, it is possible to design a set of conditions under which the aging of the UV-induced index change can be accelerated. This has important practical implications for establishing grating reliability. By eliminating higher activation energy traps with a short anneal at high temperatures, we can render insignificant the decay of the UV-induced index change at lower temperatures over periods of years. For

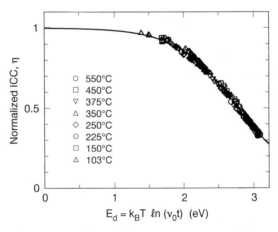

Fig. 7.13 Data for η versus E_d at different temperatures fall on a single curve. The *solid line* is a fit using Eq. (7.6). ICC, integrated coupling constant. (Reprinted from Ref. 49 with permission.)

example, a 1-h anneal at 100°C ensures that the UV-induced index of the grating will decay by only 5% in 100 years at room temperature. Because annealing does change some of the properties of the gratings, careful design of the accelerated aging procedure is required, but the enhanced reliability is worth the effort.

Ensuring reliability is only the first step in determining whether the fiber gratings are manufacturable. A great advance in this area was the development of a phase-mask technique for exposing the gratings.[50-52] A phase mask is a piece of UV-transparent silica with a square-wave surface relief pattern etched in one side. It is illuminated with a single UV beam and generates the interference pattern required to write the fiber Bragg grating. The advantage of this technique is that, to first order, the Bragg wavelength of the grating is insensitive to the wavelength or alignment of the UV writing beam. In practice, once a phase mask has been fabricated, it is much easier to use than an interferometer. However, a separate phase mask must be made for each grating design. Mass production of gratings will clearly be done with phase masks.

Because fiber coatings absorb UV radiation, the coating must be stripped off the fiber before UV exposure of a grating. The coating is vital for maintaining the mechanical strength of the fiber, however, and must be reapplied after the grating is written. One technique that promises to get around this step is to write the grating on the fiber as it is being drawn, before the coating is applied.[53,54] This requires powerful UV pulses, but some success has been reported.[55] Such a manufacturing process would be ideal for grating-based sensor devices in which relatively low optical quality gratings are distributed along many meters of fiber.

H. GRATINGS IN PLANAR WAVEGUIDES

In principle, it is straightforward to extend the work on fiber Bragg gratings to planar waveguide structures. This is true for planar waveguides with germanium-doped cores, where strong Bragg gratings (>4-nm width) have been demonstrated in H_2-loaded guides.[56] Other than strong radiation-mode coupling, which can be at least partly attributed to the low delta of the waveguides, and birefringence, the fabrication and performance of the waveguide gratings are similar to those written in fibers. Because most standard planar waveguide devices have phosphorus-doped (P-glass) cores, considerable effort has gone into developing a technique for fabricating gratings in non-germanium-doped glass.[57] Untreated P glass is insensitive to UV light at 242 nm, but Lemaire *et al.*[21] showed that H_2 loading and heat treatment during UV writing at 242 nm could enhance the photosensitivity of P-doped fibers enough to observe index changes of approximately 10^{-3}. In addition, P-doped waveguides sensitized by heat or H_2 loading can be exposed with 193-nm radiation from an ArF excimer laser.[58] Index changes as great as 3×10^{-3} were reported.

Planar waveguide devices are compact and potentially very low cost. Gratings can enhance the functionality of some standard waveguide structures. For instance, Kashyap, Maxwell, and Ainslie[59] fabricated a four-port band-pass filter using a waveguide Mach–Zehnder interferometer. Identical Bragg gratings were written in the two arms of the interferometer, and the resulting imbalance was corrected with additional UV exposure. A similar device, where the arms of the Mach–Zehnder were close enough together to expose the gratings simultaneously eliminate the need for laser trimming, was reported by Erdogan *et al.*[60] The dropped channel is shown along with the transmitted signal in Fig. 7.14. As designers

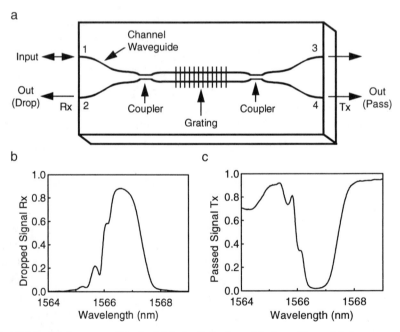

Fig. 7.14   (a) Integrated optical Mach–Zehnder add–drop filter. (b) Spectrum of the dropped channel. (c) Spectrum of the passed signal. (Reprinted from Ref. 60 with permission of the author.)

look to enhance the functionality of integrated optics, gratings will be an obvious tool.

III. High-Power Fiber Lasers and Amplifiers

One result of the erbium amplifier revolution that has taken place since the mid-1980s is the way people view and utilize optical power in communications systems. Prior to the advent of the EDFA, high-quality signal lasers at 1.55 μm were limited to optical powers on the order of 1 mW. Semiconductor optical amplifiers, the favored optical amplifier technology at the time, had low-saturation output powers and were able to boost these transmitters only in the range from 5 to 10 mW. EDFAs, however, are capable of high-output saturation powers and can readily amplify these costly single-frequency signal lasers to powers of tens of milliwatts. It is unlikely that today's EDFA powers will suffice for the systems needs of the near future. There are numerous systems needs and advantages to having more available optical power. The most obvious is in architectures with a high degree of passive splitting. Given a high amount of available optical power, passive splits of 1 $\times$ 16 or higher are possible while maintaining the same amount of power in each arm of the splitter. Such architectures will help the economics and penetration of fiber deployment, especially in the local loop, because power amplifiers can boost the power output of expensive DFB signal lasers by two orders of magnitude and thereby lower the cost per subscriber. High optical powers are also required for analog lightwave transmission systems where 0 dBm of received power is often necessary in order to maintain an adequate carrier-to-noise ratio for 80-channel systems. High-power amplifiers (1.55 μm) and lasers (1.31 μm Nd^{3+} diode-pumped solid state) are already being utilized in supertrunking applications for analog community-antenna television systems. The past several years have also witnessed a dramatic increase in the capacity of lightwave transmission at 1.55 μm, primarily through the increased use of WDM. As WDM becomes more widespread, the powers from EDFAs will be required to continue to steadily increase because a constant optical power is required for each wavelength channel. Finally, there are needs for high optical powers in repeaterless transmission systems. High-power postamplifiers are required at the transmitter end, whereas high-power 1480-nm lasers are often utilized at the receiver end to remotely pump erbium fiber sections in the transmission line and to provide Raman gain for the signal. The same techniques used in repeaterless systems can be used to increase the repeater spacing in amplified systems.

A. CLADDING-PUMPED FIBER LASERS

The output power of EDFAs has, to date, been limited only by the amount of single-mode coupled pump power that one has available for pumping into one of the many absorption bands of erbium-doped fibers. The optical powers that are currently available from single-mode fiber-pigtailed laser diodes are on the order of 100 mW. These powers are limited by intrinsic materials properties of the laser diodes themselves (e.g., facet damage) and are not likely to be significantly improved in the next 10 years. Power scaling with single-stripe diode lasers can be achieved by double pumping (co- and counterpropagating with respect to the signal), but effects such as pump laser diode cross talk have to be considered, and pump laser isolators may be required. Single-stripe laser diodes can be polarization multiplexed so that two pump diodes can be utilized from each single-mode fiber pump port. The insertion losses of the polarization combiners and the need to control each pump polarization tend to limit the usefulness of this approach. Another method of maximizing the pump power from single-stripe laser diodes is to combine both polarization and wavelength multiplexing through WDMs such that as many as four pump lasers can be combined through a single fiber port.[61] Once again, the insertion losses and costs of additional components tend to make this an impractical power scaling approach. The reliability of the overall pump module is of great concern because multiple laser diodes are all being run at their maximum power ratings and there are no apparent methods for design of pump redundancy. Clearly, an approach that is capable of utilizing higher power pump laser diode arrays, is arbitrarily power scalable, and has redundancy and power derating as design parameters is highly desirable.

One method of obtaining scalable single-mode fiber-coupled power has been the diode-pumped solid-state (DPSS) Nd^{3+} laser.[62] A high-power, nondiffraction-limited diode laser array is focused into a Nd:YAG or Nd:YLF crystal, which is surrounded by feedback mirrors that define the laser cavity. The laser cavity defines the spatial mode output of the Nd^{3+} laser, which can be readily made to operate in the fundamental TEM_{00} mode. The diffraction-limited output can then be efficiently ($\sim80\%$) coupled into single-mode fiber. The second method of obtaining high fiber-coupled powers is through the use of double-clad fiber lasers. These cladding-pumped fiber lasers are designed to have two distinct waveguiding regions: a large multimode guiding region for the diode pump light and a rare-earth-doped single-mode core from which the diffraction-limited laser output is extracted. A schematic diagram of a high-power Yb^{3+} cladding-pumped fiber laser is shown in Fig. 7.15. The diode laser pump is contained in a silica ($n = 1.46$) rectangular waveguiding region of dimensions 360×120 μm, usually referred to as the *pump cladding*. The pump cladding region is typically surrounded by a low-index polymer ($n = 1.39$) giving a high NA pump region (NA $= 0.48$) into which to couple diode laser power. The low-index polymer is coated with a second protective polymer. The Yb^{3+}-doped single-mode core is located at the center of the pump cladding. If the background losses of the pump cladding can be neglected, the only loss of mechanism of the pump light is when the rays occasionally cross the rare-earth-doped single-mode core and are absorbed. When feedback elements, either dielectric coating or fiber Bragg gratings, are present, all the laser power can be extracted from the single-mode core. The most important property of the cladding-pumped fiber laser is that a brightness conversion of highly nondiffraction-limited diode laser arrays is obtained. The brightness increase is approximately given by the ratio of areas of the pump cladding to the single-mode core area, a value of 1500 in this example.

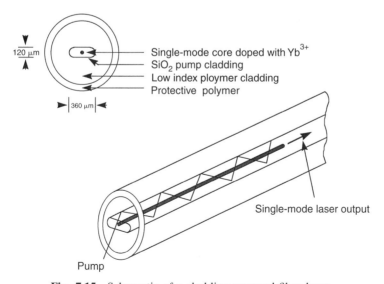

Fig. 7.15 Schematic of a cladding-pumped fiber laser.

The first cladding-pumped fiber laser to be demonstrated used a circular pump cladding.[63] The modes in a fiber of circular cross section are unique in that only the HE_{1m} modes have intensity at the center of a multimode waveguide. In order to dramatically improve the pump absorption, the single-mode rare-earth-doped core was offset to the side of the pump cladding. A second version of a cladding-pumped fiber laser utilized a rectangular pump region in order to break the circular symmetry.[64] The rectangular pump region also better matched the aspect ratio of the broad-area pump laser diode. The cladding-pumped laser was pumped by directly butting the fiber up to the pump laser facet, without the use of any pump coupling optics. Output powers of 5 W at slope efficiencies of 51% have been obtained from diode-pumped Nd^{3+} cladding-pumped fiber lasers.[65] An output power of 9.2 W has recently been obtained from a diode-pumped Nd^{3+} cladding-pumped fiber laser.[66] The slope efficiency was only 25%, a direct result of using a circular geometry for the cladding-pumped structure and the resultant inefficiency of pump light absorption by the single-mode core. Ytterbium-doped cladding-pumped operation has recently been demonstrated with slope efficiencies of greater than 70% and output powers of 6.8 W.[67] The feedback elements were fiber Bragg gratings that were written in the innermost single-mode germanium-doped core. The wavelength of operation of the Yb^{3+} cladding-pumped fiber laser was 1090 nm, where the Yb^{3+} laser behaves primarily as a four-level laser system. In glass, Yb^{3+} ions exhibit such a high degree of Stark splitting that laser operation has been obtained from 975 to 1170 nm from the $^2F_{5/2}$ excited electronic state. An energy-level diagram of Yb^{3+} is shown in Fig. 7.16. Modeling of Yb^{3+}-doped cladding-pumped fiber lasers shows that in the cladding-pumped geometry, laser operation should be readily obtained from 1050 to 1150 nm, where the laser behaves primarily as a four-level or quasi-four-level

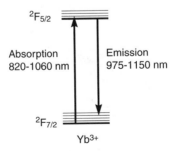

Fig. 7.16   Energy-level diagram of Yb^{3+}.

laser system.[68] Because of the indirect nature of the pumping in the cladding-pumped geometry, it has been generally believed that only four-level laser operation should be possible. A notable exception has been the demonstration of both lasing and amplification by Er^{3+} ions at 1.55 μm in cladding-pumped fibers.[69] Pumping of three-level systems such as Er^{3+} in double-clad fiber structures is difficult because the inherent ground-state absorption must be bleached before gain can be achieved. In the case of Er/Yb double-clad fibers, a higher inversion of Er^{3+} ions was made possibly by co-doping with Yb^{3+} and pumping at 970 nm, where the large Yb^{3+} absorption cross section compensates for the reduction in pump rate. Recently, 980-nm pumped operation of Er^{3+}-doped cladding-pumped amplifiers was demonstrated, although the degree of inversion was not apparent because the noise figure of the amplifier was not reported.[70]

There are numerous advantages to the use of cladding-pumped fiber lasers as a method of obtaining high single-mode fiber-coupled optical powers. The first is that the power is intrinsically single-mode fiber coupled: there is no power lost to this coupling step, nor are there any alignment tolerances associated with single-mode fiber coupling. Because the alignment of the pump diode light is into a highly multimode waveguide, alignment tolerances of tens of microns are typical. The second advantage is the high degree of efficiency with which pump diode light can be converted into single-mode fiber-coupled power: 50% in Nd^{3+} and more than 70% in Yb^{3+}-doped cladding-pumped fibers have been demonstrated. Cladding-pumped fibers also appear to be a preferred method of power scaling for high-power CW lasers. Thermal effects are minimized in the cladding-pumped fiber laser as a result of the high surface-area-to-volume ratio. Given an active gain medium volume of 1 cm^2, a cladding-pumped fiber laser has 40 times more surface area than that of a DPSS laser. DPSS lasers also suffer from thermal problems at CW powers of a few watts, which tends to make operation in the TEM_{00} mode difficult and thereby lowers the fiber-coupling efficiency. Because the output mode quality of the cladding-pumped fiber laser is defined by the single-mode waveguiding core, a diffraction-limited beam is obtained at all power levels of operation. Finally, cladding-pumped fiber lasers offer the possibility of writing integral feedback elements directly in the fiber core, through the use of fiber Bragg gratings.

The power scaling limitations of cladding-pumped fiber lasers have not yet become apparent. A power limit of several tens of watts has been estimated.[71] Nonlinear effects rather than thermal effects will probably be the ultimate limiting mechanism. Stimulated Brillouin gain will probably not be the limiting nonlinearity because of the large number of longitudinal laser modes that are operational. It is much more likely that stimulated Raman scattering will be the ultimate limiting nonlinearity. The cladding-pumped fiber laser power will most likely not be significantly decreased but will be frequency converted by approximately 450 cm^{-1}. The future of cladding-pumped lasers will depend on new methods of efficient coupling of high-power diode laser power into cladding-pumped fibers. The development of higher brightness laser sources will also be important and will increase the efficiency and wavelength range over which cladding-pumped fiber laser operation is possible.

B. Er/Yb AMPLIFIERS AND LASERS

The absorption spectrum of Yb^{3+} in silica fibers consists of an intense, broad peak centered at 975 nm. The absorption spectra of both Er^{3+} and Er^{3+}/Yb^{3+} co-doped fibers are shown in Fig. 7.17. Co-doping with Yb^{3+} provides a much greater spectral region into which to pump these fibers, from approximately 800–1070 nm. A diagram illustrating the Yb → Er energy transfer process is shown in Fig. 7.18. Absorption of a pump photon by Yb^{3+} ions promotes an electron from the $^2F_{7/2}$ ground-state level to the $^2F_{5/2}$ manifold, which is followed by efficient energy transfer from this level to the $^4I_{11/2}$ level of Er^{3+} and nonradiative decay to the $^4I_{13/2}$ amplifying level. This energy transfer can be up to 85% efficient provided that the energy is efficiently funneled from the Yb^{3+} sensitizer network (the Yb^{3+} concentration is usually 10 times the concentration of Er^{3+} ions), and that

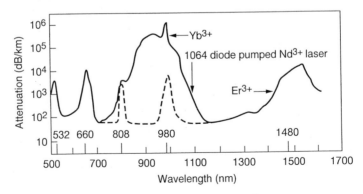

Fig. 7.17 Absorption spectra of Er^{3+} and (Er^{3+}/Yb^{3+}) co-doped silica fibers.

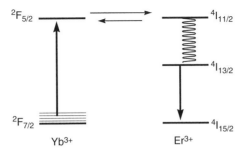

Fig. 7.18   Er/Yb energy-transfer diagram.

the transferred energy remains in the Er^{3+} ion (i.e., there is no significant amount of back transfer of energy from the $^4I_{11/2}$ level of Er^{3+} to the $^2F_{5/2}$ level of Yb^{3+}). The host glass composition has been found to be critically important in controlling the rate of back transfer of energy.[72,73] High phonon energy phosphate glasses and subsequently phosphosilicate fibers have been found to be necessary in order to increase the $^4I_{11/2} \rightarrow {}^4I_{13/2}$ nonradiative relaxation rate compared with the $^4I_{11/2} \rightarrow {}^2F_{5/2}$ back-transfer process.

The sensitization of Er^{3+}-doped silica fiber by Yb^{3+} presents several advantages. Ytterbium sensitization was initially proposed for more efficient pumping of the 800-nm band with AlGaAs diode lasers. Early results with 807-nm pumping of Er^{3+}-doped fibers were disappointing: there was a strong ESA band in this region that degraded the amplifier performance.[74] There is a strong decrease in ESA in the 820- to 830-nm spectral region. Co-doping with Yb^{3+} would allow one to pump in this region with AlGaAs diode lasers without the deleterious effects of ESA. However, the results obtained with pumping of Er/Yb co-doped fibers in the 800-nm band have never appeared attractive enough to compete with 980- or 1480-nm pumping of Er^{3+}-doped fibers.[75] By far, the biggest advantage of co-doped fibers has been the pumping with high-power DPSS or cladding-pumped lasers at wavelengths near 1060 nm. These results, which are summarized next, have allowed almost infinite power scaling of 1.5-μm fiber amplifiers with diode laser-based sources. Another advantage of Er/Yb co-doped fibers has been for 980-nm pumping of short laser and amplifier devices. Because of the large oscillator strength of the Yb^{3+} transition and the high Yb^{3+} concentrations necessary to be in the fast donor diffusion limit, 980-nm pump absorption occurs within 1 cm of these co-doped fibers. This is a tremendous advantage for short-cavity, single-frequency, fiber Bragg grating lasers. An output power of 19 mW was obtained at a slope efficiency of 55% in a 2-cm cavity.[76] This laser had the advantage of a high-output power without using a fiber master oscillator power amplifier (MOPA) structure, which adds significant noise to the source. The short absorption length also has a tremendous advantage in 1.55-μm planar amplifiers, where high gains need to be achieved in short devices. Internal gains in excess of 30 dB have been demonstrated in Er/Yb planar waveguide amplifiers.[77] Furthermore, an analysis shows that for amplifier lengths of 1 m or less, Er/Yb co-doped amplifiers will in all cases exhibit superior performance to Er^{3+}-doped waveguides.[78]

The first long-wavelength pumped (1.06-μm) Er/Yb co-doped fiber laser was reported in 1988 by a group from the University of Southampton.[79] The fiber host was an aluminosilicate glass with a relatively low concentration of Yb^{3+} — 5000 ppm. The optical conversion efficiency from 1.06 → 1.55 μm was reported to be only 4% in this case. The first optical fiber amplifier based on Er/Yb co-doped fibers and a diode-pumped Nd : YAG pump laser at 1064 nm was reported in 1991; output powers of +13 dBm and gains of 35 dB were reported.[80] The fiber was based on a bulk phosphate glass composition and fabricated by the rod-in-tube method. Both the host glass composition effects and the high concentration of Yb^{3+} ions necessary to be in the fast donor diffusion limit were recognized. A major disadvantage of this approach, however, was that the use of the low-melting-point phosphate glass fiber prohibited direct fusion splicing to silica fibers. However, the efficiency of the pure phosphate glass host was soon reproduced in a co-doped phosphosilicate fiber.[81] Output powers of +24 dBm and small-signal gains of 50 dB were reported using a diode-pumped Nd : YLF laser at 1053 nm as the pumping source.[82] The optical conversion efficiency from 1.06 → 1.55 μm was reported to be approaching 40% in these fibers. Further power scaling of Er/Yb co-doped fiber amplifiers came with the use of cladding-pumped fiber lasers as pumping sources. The first report was of a 3-W multistripe AlGaAs diode laser at 805 nm that gave a gain of 45 dB and an output power in excess of +20 dBm.[83] An amplifier based on cladding pumping of Er/Yb fibers with an output power of +17 dBm was demonstrated.[84] A 1-W broad-stripe diode laser at 962 nm was the pump source. Cladding pumping of the Er/Yb core with a pump wavelength of 962 nm gave approximately the same pump absorption per unit length as 1060-nm pumping of the co-doped single-mode core. A Nd^{3+}-doped cladding-pumped fiber laser with an output of 4.2 W was used to demonstrate a 1.5-W (+32.6-dBm) Er/Yb power amplifier.[85] A 1-W Er^{3+} fiber amplifier was demonstrated by pumping with four 980-nm semiconductor MOPA devices.[86] Three cladding-pumped fiber lasers and a three-stage Er/Yb amplifier produced an output power of 4.3 W (+36.2 dBm).[87] The output power of this amplifier was more than 4 W over a 30-nm spectral range from 1535 to 1565 with a 0-dBm input signal.

The noise figure of Er/Yb amplifiers, which is determined by the inversion of Er^{3+} ions at the input end of the amplifier, is highly dependent on the Yb/Er dopant ratio, pump intensity, and wavelength.[88] Er/Yb power amplifiers with noise figures of 4 dB have been reported.[89] Generally, the noise figure of Er/Yb amplifiers is somewhat worse than that of 980-nm pumped Er^{3+} amplifiers but better than that of 1480-nm pumped amplifiers. In the ideal case, the noise figures of Er/Yb amplifiers can approach quantum-limited values within a few tenths of a decibel. Further studies of the host glass compositional effects, pump wavelength dependence, and dopant ratios are needed to fully optimize both the conversion efficiency and the noise figure of Er/Yb co-doped fiber amplifiers.

C. FIBER RAMAN LASERS AND AMPLIFIERS

Prior to the advent of Er^{3+}-doped fiber optical amplifiers, the two main technologies directed toward optical amplifiers were semiconductor doped optical amplifiers and fiber Raman amplifiers. There was a significant amount of work in the mid- to late-1980s on the use of Raman amplification in long lengths of germanosilicate fibers.[90] Although these amplifiers possessed many attractive features, such as low noise, polarization insensitive gain, and the ability to achieve amplification in ordinary germanosilicate transmission fiber, it was primarily the unavailability of high-power diode laser pump sources that prevented their acceptance.

In stimulated Raman scattering, light is scattered by optical vibrational modes (optical phonons) of the material, which results in frequency down-shifted Stokes light. In optical fibers doped with the index of refraction modifying element GeO_2, this shift occurs at approximately 450 cm^{-1} (or 13.2 THz),[91] as shown in Fig. 7.19. Although the nonlinear cross section for this process is relatively weak in germanosilicate fibers, the long lengths and low loss of optical fibers more than compensate for the weak cross section. The potential of both fiber amplifiers and lasers based on Raman scattering was first demonstrated in the 1980s by Stolen and Lin,[92] who constructed Raman lasers operating between 0.3 and 2.0 μm. However, it was not clear from the early work that a Raman fiber laser could be pumped

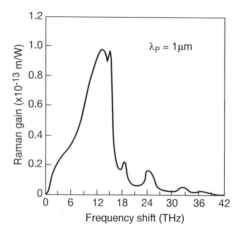

Fig. 7.19 Frequency dependence of the Raman scattering cross section in german-osilicate optical fibers.

by a practical laser source (i.e., semiconductor laser based) or that an efficient CW pumped Raman fiber laser would be possible. The recent availability of high single-mode fiber-coupled output powers from cladding-pumped fiber lasers and the ability to construct ultra-low-loss fiber cavities through the use of fiber Bragg gratings have dramatically changed this situation.

The emergence of fiber Bragg grating technology has made it possible to fabricate highly reflecting elements directly in the core of germanosilicate fibers with reflection widths of several nanometers and out-of-band insertion losses of a few hundredths of a decibel. This technology, coupled with that of cladding-pumped fiber lasers, has made a whole new class of fiber lasers based on intracavity pumping possible. Intracavity-pumped fiber lasers based on multiple rare-earth fiber laser cavities or nonlinear effects in germanosilicate fibers become possible, as shown in Fig. 7.20. In the case of stimulated Raman conversion, pump light is introduced through one set of highly reflecting fiber Bragg gratings. The cavity consists of several

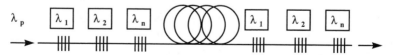

Fig. 7.20 Schematic diagram of an intracavity-pumped fiber laser.

hundred meters to a kilometer of germanosilicate fiber. An output set of fiber Bragg gratings consists of a set of high reflecting gratings through Raman order $n - 1$. The output wavelength of Raman order n is coupled out by means of a partially reflecting ($R \sim 20\%$) fiber grating. The intermediate Raman–Stokes orders are contained by sets of highly reflecting fiber Bragg gratings, and this power is circulated until it is nearly entirely converted to the next successive Raman–Stokes order. These resonant laser cavities have been termed *cascaded Raman lasers*. Modeling of these cascaded Raman resonators has highlighted the high CW conversion efficiencies that can be achieved in low-loss fiber cavities.[93]

It has been shown that it is possible to efficiently convert the output of a Yb^{3+} cladding-pumped fiber laser at 1117 nm by five Raman–Stokes orders to 1480 nm with a cascaded Raman laser resonator, as shown schematically in Fig. 7.21. Diode laser pumped, single-mode fiber output powers of 1.7 W at a slope conversion efficiency of 46% have been obtained.[94] The spectral output of the Raman fiber laser is between 1 and 2 nm wide and is controlled by the widths of the fiber Bragg gratings. The suppression ratio between the desired final-output Raman order and the intermediate Raman orders is typically 20 dB.

High single-mode fiber-coupled powers at around 1480 nm are desired for pumping of high-power erbium-doped postamplifiers as well as remote pumping of in-line erbium-doped fibers.[95] In remote pumping, where the goal is to maximize the distance between active repeaters, one desires high-power 1480-nm pump sources at the terminal ends. The evolution of repeaterless transmission experiments is shown in Fig. 7.22. A high-power postamplifier, usually an Er/Yb co-doped fiber amplifier, is used as a power amplifier at the transmitter end. Up to +26 dBm at 1558 nm has been transmitted, through the use of a stimulated Brillouin scattering (SBS)

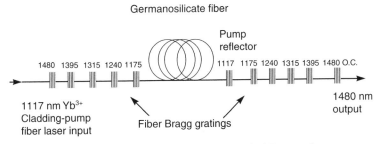

Fig. 7.21   Diagram of a 1480-nm cascaded Raman laser.

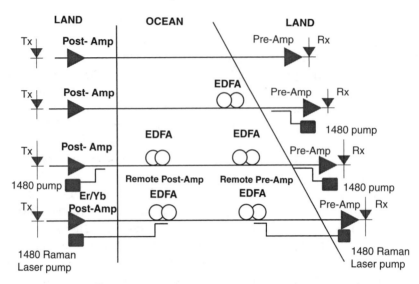

Fig. 7.22 Evolution of repeaterless systems.

suppression technique that applied a series of tones on a phase modulator. High-power 1480-nm lasers can be used at both the transmitter and receiver terminals in order to pump remote Er^{3+} postamplifiers and preamplifiers, respectively. Using a high-power Er/Yb postamplifier and three high-power Raman lasers at 1480 nm, as shown in Fig. 7.23, Hansen et al.[96] achieved a repeaterless transmission distance of 529 km at 2.5 Gb/s.

Amplification can also be achieved by use of the cascaded Raman resonator approach. In particular, an amplifier at 1.31 μm has been demonstrated.[97] This was the first silica-fiber-based optical fiber amplifier to be demonstrated at 1.31 μm. Gains of 40 dB and output powers of +24 dBm were obtained in an intracavity-pumped Raman amplifier. As shown schematically in Fig. 7.24, a high-power cladding-pumped laser at 1.06 μm is injected into a long length of germanosilicate fiber. At each end of the germanosilicate fiber are three highly reflecting fiber Bragg gratings, at the first three Stokes frequencies from 1060 nm. The pump light at 1060 nm is therefore efficiently converted to pump light at 1240 nm. A 1.3-μm signal injected through this structure will experience amplification because it is at the next Stokes–Raman shift from the 1240-nm pump light. The efficiency of the Raman amplification process is controlled both by the amount of germanium dopant in the fiber core and by the cross-sectional area of the fiber core. Gains of

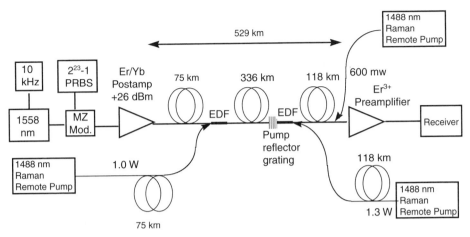

Fig. 7.23 Diagram of a 529-km repeaterless transmission experiment at 2.5 Gb/s. MZ, Mach–Zehnder; PRBS, pseudo-random bit sequence.

25 dB for only 350 mW of pump power have been obtained in highly germanium-doped, small-core fibers.[98] A novel ring geometry that did not utilize fiber Bragg gratings was also used to generate third Stokes light at 1240 nm and amplification at 1.31 μm.[99]

The theoretical noise-figure contribution from signal–spontaneous beating for Raman amplifiers has been shown to be 3 dB.[100] However, systems tests of Raman amplifiers have uncovered other sources of noise that generally are not important in Er^{3+}-doped fiber amplifiers. The first source is the coupling of intensity fluctuations from the pump light to the signal. The fundamental cause of this noise is the lack of a long upper-state lifetime to buffer the Raman gain from fluctuations in the pump intensity. It has been shown that when a counterpropagating amplifier geometry is used,

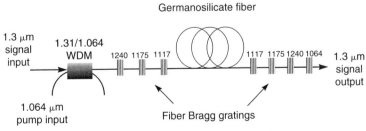

Fig. 7.24 Schematic diagram of a 1.3-μm cascaded Raman amplifier.

the transit time of the amplifier can be used to average gain fluctuations due to the pump.[101] Last, double-Raleigh and SBS can also give significant contributions to the noise figure of Raman amplifiers because of the long lengths of fiber used. However, the noise figure of the amplifier can be controlled by limiting the fiber lengths used and constructing multistage amplifiers, as has been shown in a 2.5-Gb/s systems test of a 1.3-μm Raman amplifier with a gain of 30 dB and an output power of +15 dBm.[102] Analog grade performance has been demonstrated in a +23-dBm Raman power amplifier at 1.31 μm.[103] Raman amplifiers have also been proposed for applications in WDM systems at both 1.31 and 1.55 μm because of their potential for achieving distributed gain, large bandwidth, and low noise. However, cross talk between the channels in a Raman amplifier has always been a concern. Cross talk in Raman amplifiers is mediated by the pump. Each wavelength channel causes a patterned pump depletion that is super-imposed on the other channels in the amplification process. It has been shown that cross talk depends on the modulation frequency of the channels and the pump. Cross talk in the forward and backward configurations strongly differ because of the walk off between the signal and the pump.[104] Backward pumping has been shown to result in a dramatic reduction of the cross-talk bandwidth such that Raman amplifiers operated in this con-figuration should exhibit adequate performance in high-capacity WDM systems. The amount of cross talk in a counterpropagating Raman amplifier WDM experiment has been shown to be neglible.[105] In this experiment, a single-channel 10-Gb/s system was upgraded to a 4 $\times$ 10 Gb/s system purely by Raman amplification in an existing fiber span.

Because one is no longer constrained to particular transitions of rare-earth ions in the cascaded Raman approach, lasing or amplification should be possible from 1.1 to 2.0 μm. Because the bandwidth of the Raman process is broad and the pump wavelength obtained from a Nd^{3+} or Yb^{3+}-doped cladding-pumped laser can be varied by nearly 100 nm, one can efficiently down-convert to virtually any arbitrary wavelength. Numerous additional applications for these lasers and amplifiers are likely to emerge.

IV. Up-Conversion Fiber Lasers and Amplifiers

Few compact, efficient, CW diode-based sources of visible and UV coherent radiation are currently available. Such sources are desired for optical memo-ries, reprographics, and displays, and as sources for short-haul polymer fiber communications. Current approaches directed toward a practical source generally fall into one of three categories: (1) frequency doubling of semiconductor lasers or semiconductor laser pumped solid-state lasers typically through intracavity or resonant doubling techniques, (2) the development of short-wavelength semiconductor diode lasers fabricated from II-VI and III-V materials such as ZnSe and GaN, and (3) up-conversion pumped lasers. *Up-conversion* is a term that is associated with a variety of processes whereby the gain medium, a trivalent rare-earth ion, in a crystal or glass host absorbs two or more photons to populate high-lying electronic states. Commercial products based on frequency doubling of diodes or diode-pumped solid-state lasers have been available for several years but are generally expensive and hence the applications are currently limited. This limitation is due to the relative complexity and large number of components in frequency-doubled lasers and the sensitivities of the intracavity or external cavity frequency-doubling process. Since diode lasers based on II-VI ternary compounds were demonstrated in 1991, device lifetimes have steadily increased; however, they are currently limited at about 1 h at room temperature.[106] InGaN light-emitting diodes (LEDs) with up to 3-mW output power at 450 nm are commercially available, but laser diodes have yet to be demonstrated.[107] Up-conversion lasers appear to be attractive candidates for compact, efficient visible laser sources because of the relative simplicity (the gain and frequency conversion material are one and the same) of these devices.[108] Up-conversion lasers can also be useful in obtaining infrared laser and/or amplifier operation at wavelengths of interest for telecommunications applications. Multiphoton-pumped up-conversion laser operation is sometimes required even when the desired emission is of lower energy than that of the excitation photons. This occurs when the infrared emission is from a high, excited, rare-earth level, as occurs in the 1.48-μm Tm^{3+} up-conversion laser, which is pumped by sequential absorption of two 1.06-μm photons.

The first up-conversion laser was reported in 1971, when stimulated emission at 670 and 551 nm was observed in flashlamp-pumped BaY_2F_8 crystals that were co-doped with Er/Yb and Ho/Yb, respectively.[109] The first near-infrared up-conversion laser that would potentially be diode laser pumpable was not demonstrated until 1987, when CW lasing at 550 nm was demonstrated in Er^{3+}-doped $YAlO_3$ crystals by a two-color pumping scheme.[110] Several other crystalline lasers have been demonstrated since that time. Nevertheless, the performance of most up-conversion lasers in rare-earth crystalline hosts remains limited, and these lasers appear to suffer from several disadvantages: (1) operation of up-conversion lasers in crystalline hosts generally occurs at cryogenic temperatures ($<90°K$) (two notable exceptions, the $Tm:BaY_2F_8$ and $Er:YLiF$ lasers, have recently been operated at room temperature), (2) the limited pump-gain medium interaction lengths due to pump focusing considerations and crystal size limit the amount of gain that can be achieved, and (3) narrow absorption linewidths associated with the electronic transitions of rare-earth ions in crystals limit the simultaneous matching of ground-state absorption and ESA necessary to achieve efficient up-conversion laser operation.

The advantages of rare-earth-doped optical fibers as the gain medium for up-conversion lasers started to become apparent in 1990 with the demonstration of several CW, room-temperature up-conversion lasers in ZBLAN single-mode fibers. In addition to their simplicity and compactness, up-conversion fiber lasers are noted for their efficiency and tunability. Slope efficiencies (pump power to output power conversion) of up to 50% have been obtained in the two-photon pumped 550-nm Er^{3+} ZBLAN fiber laser and 32% in the three-photon pumped Tm^{3+} ZBLAN fluoride laser. The most notable characteristic of up-conversion fiber lasers is that unlike the majority of their crystalline host counterparts, all operate CW and at room temperature. A summary of up-conversion fiber lasers that have been demonstrated in rare-earth-doped fluorozirconate glass is shown in Table 7.1. The key to up-conversion laser operation in single-mode optical fibers has been the use of the low phonon energy ($h\omega < 600$ cm^{-1}) fluorozirconate glasses as hosts for the rare-earth ions. No up-conversion lasers have been demonstrated to date in rare-earth-doped silica fibers where the lowest energy phonon is 1300 cm^{-1}. Because operation of an up-conversion laser requires the efficient excitation of highly excited states of trivalent rare-earth ions ($>20,000$ cm^{-1}) with near-infrared diode laser photons in the range of 10,000 cm^{-1}, the sequential absorption of two or more photons is required, and it is essential that the energy in the intermediate energy levels is not dissipated by nonradiative decay, which is dominant in rare-earth-doped silica fibers.

The first advantage of the single-mode optical fibers as a choice for the gain medium in up-conversion lasers is that high excitation intensities are possible and, hence, a high degree of inversion of the rare-earth ions. Even when single-stripe diode lasers with output powers of tens of milliwatts are used, excitation densities of up to 10^6 W-cm^{-1} are possible because the cross-sectional area of single-mode fibers is on the order of 25 μm^2. Perhaps the largest advantage of the fiber geometry for up-conversion laser opera-

Table 7.1   **Summary of Rare-Earth-Doped Fluorozirconate (ZBLAN) Up-Conversion Fiber Lasers**[a]

Rare Earth(s)	Pump Wavelength(s) (nm)	Laser Wavelength(s) (nm)	Temperature (°K)	Slope Efficiency (%)
Er	801	544, 546	300	15
Er	970	544, 546	300	>40
Tm	647, 676	455, 480	77	—
Tm	1064, 645	455	300	1.5
Tm	1112, 1116, 1123	480, 650	300	32 (480 nm)
Tm	1114–1137	480	300	13
Ho	643–652	547.6–549.4	300	36
Nd	582–596	381, 412	300	0.5 (412 nm)
Pr	1010, 835	491, 520, 605, 635	300	12 (491 nm)
Pr/Yb	780–885	491, 520, 605, 635	300	3 (491 nm) to 52 (635 nm)

[a] From Ref. 108.

tion is that the pump spot size is decoupled from the device length. As a consequence, the high excitation density present in the single-mode core is maintained over the entire device length, which is typically several meters or even tens of meters. This leads to a high degree of flexibility in rare-earth ion concentration that is not possible in bulk up-conversion devices. This flexibility in concentration can be critical in obtaining efficient up-conversion laser operation when there are unfavorable ion–ion interactions that lead to a decreased pumping efficiency or an increase in the upper level deactivation rate. The length of the pump-gain medium interaction in bulk crystalline up-conversion lasers is limited both by crystalline growth techniques, which limit the size of the gain medium, and by the confocal beam distance over which a Gaussian pump beam can be focused. The flexibility with device length is also extremely important in balancing the excitation rates associated with ground-state absorption and ESA, which is critical to obtaining efficient population of the upper lasing level. In bulk crystalline lasers, the pump wavelength is primarily chosen so as to obtain sufficient ground-state absorption to effectively absorb all the pump radiation within the gain medium interaction length, typically a few centimeters. It is extremely unlikely that the ESA of the rare-earth transition is optimized at that same wavelength. A goal in obtaining a practical laser diode pumped up-conversion laser is that a single excitation wavelength be utilized, unless two excitation wavelengths can be derived from a single excitation laser diode, perhaps through the use of an intermediate fiber laser pump. In the fiber laser geometry, the excitation wavelength can often be different from the optimal ground-state absorption wavelength such that a balance is obtained between the ground-state absorption and one or more ESA steps. The length of the resonator is merely increased by using a longer fiber resonator, with no penalty in excitation pumping intensity. Furthermore, in the fiber laser geometry, pump bands of rare-earth ions with extremely weak absorption cross sections can be utilized.

The fiber laser geometry also has an advantage in heat-removal efficiency over bulk lasers. For a given volume of gain medium, the fiber laser has greater than two orders of magnitude more surface area over which to dissipate heat than bulk crystalline lasers do. This allows for both efficient operation and the ability to scale to high operating powers for up-conversion fiber lasers. The ability to control the core size in fiber lasers also allows one to utilize a fiber that is single mode at the wavelength of laser operation. In this way, a diffraction-limited output, the LP_{01} mode, is automatically obtained.

Because the fluorozirconate host is a disordered medium, rare-earth-doped fibers fabricated from this glass exhibit absorption and emission profiles that are broad compared with those characteristic of a crystalline host. The broad emission profile has a negative impact on the stimulated emission cross section but is more than compensated for by the advantages of the fiber laser geometry, the high pump intensities, and the maintaining of this intensity over the entire device length. The broad emission linewidths have allowed several of the up-conversion lasers to be tuned continuously over 10 nm. The broad absorption features coupled with the flexibility in device length both make single-wavelength pumping with diode laser sources practical in nearly all cases.

Perhaps the most dramatic of all the up-conversion fiber lasers is the Pr^{3+}-doped ZBLAN fiber laser.[111] When pumped simultaneously with 1010- and 835-nm pump light, the fiber exhibits an intense white glow and with the appropriate feedback mirrors can be made to lase in the blue, green, and red spectral regions. CW lasing at room temperature has been obtained at 491, 520, 605, and 695 nm. The energy levels of Pr^{3+} and the up-conversion laser transitions are shown in Fig. 7.25. Two methods of extending this work to use a single-wavelength pump have been reported. The first uses co-doping with Yb^{3+} to sensitize the second up-conversion step.[112] With

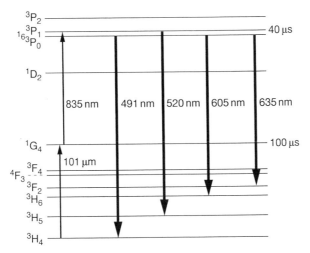

Fig. 7.25 Energy-level diagram of Pr^{3+} and associated up-conversion laser transitions.

the use of a co-doped Yb/Pr fiber, single-wavelength pump operation has been demonstrated from 780 to 885 nm. The second method involves 830-nm pumping of a short section of Yb^{3+} fiber.[113] The Yb^{3+} lasing at 1020 nm together with excess pump at 830 nm pump the Pr^{3+} up-conversion fiber laser. Thulium-doped fluoride fiber lasers have also produced a variety of wavelengths by up-conversion laser operation, as shown in Fig. 7.26. Perhaps the most surprising example is that of the three-photon pumped Tm^{3+} up-conversion fiber laser at 480 and 650 nm.[114] This remains the sole example of a three-photon pumped up-conversion laser. Slope efficiencies of up to 32% and output powers in excess of 100 mW have been obtained at 480 nm. Operation of a Tm^{3+} up-conversion fiber laser at 455 nm has also been reported, although two-wavelength pumping, 647 and 676 nm, and low-temperature operation (77°K) are required.[115] Efficient operation of Er^{3+} up-conversion fiber lasers at 545 nm has been reported by single-wavelength pumping at either 801 or 970 nm.[116,117] Diode-pumped operation with an output power of 18 mW at a slope efficiency of 25% has been reported. Holmium-doped ZBLAN fiber lasers have presented another route to a green up-conversion laser. Output powers of 50 mW at a slope efficiency of 36% have been obtained with 650-nm pumping.[118] Recently, up-conversion lasing has been reported at 381 and 412 nm from the $^4D_{3/2} \rightarrow {}^4I_{11/2}$ and $^2P_{3/2} \rightarrow {}^4I_{11/2}$ transitions, respectively, in Nd^{3+}-doped

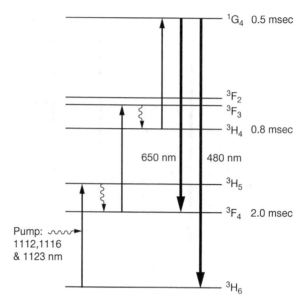

Fig. 7.26   Energy-level diagram of Tm^{3+} and associated up-conversion laser transitions.

ZBLAN fibers.[119] This is the first report of an ultraviolet fiber laser. The pump laser wavelength was 580 nm, and the slope efficiency for the 412-nm laser was 0.5% at 300°K.

There are several examples of up-conversion fiber devices that are directly applicable to wavelengths of interest for telecommunications. The first is the Tm^{3+}-doped up-conversion laser and amplifier, which has been demonstrated at 1.48 μm.[120–122] This up-conversion system has been pumped by high-power diode-pumped Nd^{3+} lasers at 1.06 μm. The first 1.06-μm pump photon excites a Tm^{3+} ion from the ground state to the $^3H_5$ level, from which it relaxes to the $^3H_4$ level. The second 1.06-μm pump photon provides excitation from the $^3H_4$ level to the $^3F_3$ level, from which relaxation to the $^3F_4$ upper laser level takes place. The 1.48-μm transition takes place from the $^3F_4$ to the $^3H_4$ level. Normally, because the lower $^3H_4$ has a longer lifetime than the upper $^3F_4$ level, this is a self-terminating laser transition. However, because there is a large ESA cross section at 1.06 μm from the $^3H_4$ level, there is efficient depletion of the lower laser level and repopulation of the upper laser level. Ideally, the first photon in the up-conversion pumping scheme is deposited only once and the ions are recycled by the second pump photon through repeated stimulated emission and

ESA. When operated as a laser, output powers of 1W at a slope efficiency of 50% have been obtained at 1.48 μm.[123] As an amplifier, gains of more than 10 dB have been obtained from 1.44 to 1.51 μm, with a peak gain of 23 dB at 1.47 μm. Noise figures as low as 3.5 dB have been obtained. Although the bandwidth of the EDFA is not likely to pose any limitation on lightwave transmission capacity in the near future, the Tm^{3+} fiber amplifier provides gain over a broader and complementary spectral region in the low-loss 1.5-μm region. Operation of an up-conversion amplifier in the first telecommunications window at 850 nm has also been reported in Er^{3+}-doped ZBLAN fibers.[124] The 850-nm emission originated from the $^4S_{3/2}$ level of Er^{3+} and required the sequential absorption of two 800-nm photons. Gains of 23 dB at 850 nm were reported. Although efficient amplification in the 800-nm band has also been reported in Tm^{3+}-doped fibers, up-conversion pumping has not yet been demonstrated.

The requirement of a low phonon energy fluoride glass host (ZBLAN) for room-temperature up-conversion fiber laser operation has been a severe barrier to the serious acceptance of up-conversion fiber devices. Fluoride glass fibers are at a tremendous disadvantage relative to silica fibers with regard to strength, environmental stability, background loss values, and the ability to fusion splice sections of fiber together. In addition, some of these rare-earth-doped ZBLAN fibers have exhibited a peculiar photodarkening effect at wavelengths less than 500 nm.[125] Although this effect appears to be absent in the lower doped Tm^{3+} ZBLAN fibers (~1000 ppm) used in the original blue up-conversion laser demonstration, a severe photodarkening appears at higher Tm^{3+} concentrations (3000–10,000 ppm). Remarkably, this loss can be removed by irradiation at the same pump wavelength at lower powers. These observations suggest the formation of color centers formed from highly excited states populated by cross relaxation of Tm^{3+} ions. An understanding of these processes as well as a dramatic improvement of fluoride glass fiber properties will be needed in order for up-conversion lasers and amplifiers to succeed as commercial devices. Perhaps an alternative host material, such as the intermediate phonon energy germanate glasses will make up-conversion devices practical.

Acknowledgments

We would like to thank Turan Erdogan, Tom Strasser, Andrew Stentz, Glenn Kohnke, and Don Monroe for their knowledgeable advice in preparing this chapter.

References

1. Dicke, G. H. 1968. Spectra and energy levels of rare earth ions in crystals. New York: Wiley Interscience.
2. DiGiovanni, D. J. 1992. Materials aspects of optical amplifiers. In *Materials Research Society Symposium Proceedings*, vol. 244, M. M. Broer, G. H. Sigel, Jr., R. Th. Kersten, and H. Kawazoe, eds., 137. Pittsburgh: Materials Research Society.
3. Vengsarkar, A. M., A. E. Miller, M. Haner, A. H. Gnauck, W. A. Reed, and K. L. Walker. 1994. Fundamental-mode dispersion-compensating fibers: Design considerations and experiments. In *Optical Fiber Conference (OFC'94)*, 225. Paper ThK2. Technical Digest. Washington, DC: Optical Society of America.
4. Poole, C. D., J. M. Wiesenfeld, D. J. DiGiovanni, and A. M. Vengsarkar. 1994. Optical fiber-based dispersion compensation using higher order modes near cutoff. *J. Lightwave Tech.* 12:1746–1758.
5. Layne, C. B., W. H. Lowdermilk, and M. J. Weber. 1977. Multiphonon relaxation of rare-earth ions in oxide glasses. *Phys. Rev. B* 16:10.
6. France, P. W., M. G. Drexhage, J. M. Parker, M. W. Moore, S. F. Carter, and J. V. Wright. 1990. Fluoride glass optical fibers. Glasgow and London: Blackie.
7. Carter, S. F., D. Szebesta, S. Davey, R. Wyatt, M. C. Brierly, and P. W. France. 1991. Amplification at 1.3 μm in a Pr^{3+}-doped single-mode fluorozirconate fibre. *Electron. Lett.* 27:628–629.
8. Ohishi, Y., T. Kanamori, T. Kitagawa, S. Takahashi, E. Snitzer, and G. H. Sigel, Jr. 1991. Pr^{3+}-doped fluoride fiber amplifier operating at 1.3μm. *Opt. Lett.* 16:1747–1749.
9. Miyajima, Y., T. Sugawa, and Y. Fukasaku. 1991. 38.2 dB Amplification at 1.31 μm and possibility of 0.98 μm pumping in Pr^{3+}-doped fluoride fiber. *Electron. Lett.* 27:1706–1707.
10. Ohishi, Y., T. Kanamori, T. Nishi, and S. Takahashi. 1991. A high gain, high output saturation power Pr^{3+}-doped fluoride fiber amplifier operating at 1.3μm. *IEEE Photon. Tech. Lett.* 3:715–717.
11. Ohishi, Y., T. Kanamori, J. Temmyo, M. Wada, M. Yamada, M. Shimizu, K. Yoshino, H. Hanafusa, M. Horiguchi, and S. Takahashi. 1991. Laser diode pumped Pr^{3+}-doped and Pr^{3+}-Yb^{3+}-codoped fluoride fiber amplifiers operating at 1.3μm. *Electron. Lett.* 27:1995–1996.
12. Lobbett, R., R. Wyatt, P. Eardley, T. J. Whitley, P. Smyth, D. Szebesta, S. F. Carter, S. T. Davey, C. A. Millar, and M. C. Brierly. 1991. System characterization of high gain and high saturated output power Pr^{3+}-doped fluorozirconate fiber amplifier at 1.3μm. *Electron. Lett.* 27:1472–1474.
13. Hill, K. O., Y. Fujii, D. C. Johnson, and B. S. Kawasaki. 1978. Photosensitivity in optical fiber waveguides: Application to reflection filter fabrication. *Appl. Phys. Lett.* 32:647.

14. Meltz, G., W. W. Morey, and W. H. Glenn. 1989. Formation of Bragg gratings in optical fibers by a transverse holographic method. *Opt. Lett.* 14:823–825.
15. Lemaire, P. J., R. M. Atkins, V. Mizrahi, and W. A. Reed. 1993. High pressure H_2 loading as a technique for achieving ultrahigh UV photosensitivity and thermal sensitivity in GeO_2 doped optical fibers. *Electron. Lett.* 29:1191–1193.
16. Atkins, R. M., V. Mizrahi, and T. Erdogan. 1993. 248 nm Induced vacuum UV spectral changes in optical fibre preform cores: Support for a colour centre model of photosensitivity. *Electron. Lett.* 29:385–387.
17. Greene, B. I., D. M. Krol, S. G. Kosinski, P. J. Lemaire, and P. N. Saeta. 1993. Thermal and photo-initiated reactions of H_2 with germanosilicate optical fibers. *J. Non-Cryst. Solids* 168:195–199.
18. Hosono, H., Y. Abe, D. L. Kinser, R. A. Weeks, K. Muta, and H. Kawazoe. 1992. Nature and origin of the 5-eV band in $SiO_2 : GeO_2$ glasses. *Phys. Rev. B* 46:445–451.
19. Raghavachari, K., and B. L. Zhang. 1994. First principles study of the thermal reactions of H_2 with germanosilicate optical fibers. *J. Non-Cryst. Solids* 180:80.
20. Zhang, B. L., and K. Raghavachari. 1995. Microscopic reaction mechanisms in hydrogen-loaded germanosilicate fibers: Formation of divalent Ge defects. *Phys. Rev. B* 51:7946–7949.
21. Lemaire, P. J, A. M. Vengsarkar, W. A. Reed, and D. J. DiGiovanni. 1995. Thermally enhanced ultraviolet photosensitivity in GeO_2 and P_2O_5 doped optical fibers. *Appl. Phys. Lett.* 66:2034–2036.
22. Atkins, R. M., P. J. Lemaire, T. Erdogan, and V. Mizrahi. 1993. Mechanisms of enhanced UV photosensitivity via hydrogen loading in germanosilicate glass. *Electron. Lett.* 29:1234–1235.
23. Ball, G. A., W. W. Morey, and W. H. Glenn. 1991. Standing-wave monomode erbium fiber laser. *IEEE Photon. Tech. Lett.* 3:613–615.
24. Ball, G. A., and W. W. Morey. 1992. Continuously tunable single-mode erbium fiber laser. *Opt. Lett.* 17:420–422.
25. Meltz, G., and W. W. Morey. 1991. Bragg grating formation and germanosilicate fiber photosensitivity. *SPIE, International Workshop on Photoinduced Self-Organization Effects in Optical Fiber*, 1516:185–199.
26. Zyskind, J. L., V. Mizrahi, D. J. DiGiovanni, and J. W. Sulhoff. 1992. Short single frequency erbium doped fibre laser. *Electron. Lett.* 28:1385–1387.
27. Mizrahi, V., D. J. DiGiovanni, R. M. Atkins, S. G. Grubb, Y-K. Park, and J-M. Delavaux. 1993. Stable single-mode erbium fiber-grating laser for digital communication. *J. Lightwave Tech.* 11:2021–2025.
28. Kringlebotn, J. T., J-L. Archambault, L. Reekie, and D. N. Payne. 1994. $Er^{3+} : Yb^{3+}$-codoped fiber distributed-feedback laser. *Opt. Lett.* 19:2101–2103.
29. Kringlebotn, J. T., J. L. Archambault, L. Reekie, J. E. Townsend, G. G. Vienne, and D. N. Payne. 1994. Highly-efficient, low-noise grating-feedback $Er^{3+} : Yb^{3+}$ codoped fibre laser. *Electron. Lett.* 30(12):972–973.

30. Morton, P. A., V. Mizrahi, P. A. Andrekson, T. Tanbun-Ek, R. A. Logan, P. Lemaire, D. L. Coblentz, A. M. Sergent, K. W. Wecht, and P. F. Sciortino, Jr. 1993. Mode-locked hybrid soliton pulse source with extremely wide operating frequency range. *IEEE Photon. Tech. Lett.* 5:28–31.

31. Ventrudo, B. F., G. A. Rogers, G. S. Lick, D. Hargreaves, and T. N. Demayo. 1994. Wavelength and intensity stabilization of 980 nm diode lasers coupled to fibre Bragg gratings. *Electron. Lett.* 30:2147–2149.

32. Giles, C. R., T. Erdogan, and V. Mizrahi. 1994. Simultaneous wavelength-stabilization of 980-nm pump lasers. *IEEE Photon. Tech. Lett.* 6:907–909.

33. Woodward, S. L., V. Mizrahi, T. L. Koch, U. Koren, and P. J. Lemaire. 1993. Wavelength stabilization of a DBR laser using an in-fiber Bragg filter. *IEEE Photon. Tech. Lett.* 5:628–630.

34. Mizrahi, V., P. J. Lemaire, T. Erdogan, W. A. Reed, D. J. DiGiovanni, and R. M. Atkins. 1993. Ultraviolet laser fabrication of ultrastrong optical fiber gratings and of germania-doped channel waveguides. *Appl. Phys. Lett.* 63:1727–1729.

35. Soccolich, C. E., V. Mizrahi, T. Erdogan, P. J. Lemaire, and P. Wysocki. 1994. In *Conference on Optical Fiber Communication*, 277–278. Paper FA7. Technical Digest. Washington, DC: Optical Society of America.

36. Mizrahi, V., and J. E. Sipe. 1993. Optical properties of photosensitive fiber phase gratings. *J. Lightwave Tech.* 11:1513–1517.

37. Delvaque, E., S. Boj, J. F. Bayon, H. Poignant, J. LeMellot, M. Monerie, P. Niay, and P. Bernage. 1995. Optical fiber design for strong gratings photoim-printing with radiation mode suppression. In *Conference on Optical Fiber Communication*, PD5-1–PD5-5. Paper PD5. Technical Digest. Washington, DC: Optical Society of America.

38. Farries, M. C., K. Sugden, D. C. J. Reid, I. Bennion, A. Molony, and M. J. Goodwin. 1994. Very broad reflection bandwidth (44nm) chirped fibre gratings and narrow bandpass filters produced by the use of an amplitude mask. *Electron. Lett.* 30:891–892.

39. Mizrahi, V., T. Erdogan, D. J. DiGiovanni, P. J. Lemaire, W. M. MacDonald, S. G. Kosinski, S. Cabot, and J. E. Sipe. 1994. Four channel fibre grating demultiplexer. *Electron. Lett.* 30:780–781.

40. Bilodeau, F., K. O. Hill, B. Malo, D. C. Johnson, and J. Albert. 1994. High-return-loss narrowband all-fiber bandpass Bragg transmission filter. *IEEE Photon. Tech. Lett.* 6:80–82.

41. Archambault, J-L., P. St. J. Russell, S. Barcelos, P. Hua, and L. Reekie. 1994. Grating-frustrated coupler: A novel channel-dropping filter in single-mode optical fiber. *Opt. Lett.* 19:180–182.

42. Ouellette, F. 1987. Dispersion cancellation using linearly chirped Bragg filters in optical waveguides. *Opt. Lett.* 12:847–850.

43. Williams, J. A. R., I. Bennion, K. Sugden, and N. J. Doran. 1994. Fibre dispersion compensation using a chirped in-fibre Bragg grating. *Electron. Lett.* 30:985–987.

44. Hill, K. O., F. Bilodeau, B. Malo, T. Kitagawa, S. Theriault, D. C. Johnson, and J. Albert. 1994. Chirped in-fiber Bragg gratings for compensation of optical-fiber dispersion. *Opt. Lett.* 19:1314–1316.

45. Meltz, G., W. W. Morey, and W. H. Glenn. 1990. In-fiber Bragg grating tap. In *Optical Fiber Conference 1990 (OFC'90)*, 24. Washington, DC: Optical Society of America.

46. Kashyap, R., R. Wyatt, and P. F. McKee. 1993. Wavelength flattened saturated erbium amplifier using multiple side-tap Bragg gratings. *Electron. Lett.* 29:1025–1026.

47. Hill, K. O., B. Malo, K. A. Vineberg, F. Bilodeau, D. C. Johnson, and I. Skinner. 1990. Efficient mode conversion in telecommunication fibre using externally written gratings. *Electron. Lett.* 26:1270–1272.

48. Vengsarkar, A. M., J. R. Pedrazzani, J. B. Judkins, P. J. Lemaire, N. S. Bergano, and C. R. Davidson. 1996. Long-period fiber-grating-based gain equalizers. *Opt. Lett.* 21:336–338.

49. Erdogan, T., V. Mizrahi, P. J. Lemaire, and D. Monroe. 1994. Decay of ultraviolet-induced fiber Bragg gratings. *J. Appl. Phys.* 76:73–80.

50. Hill, K. O., B. Malo, F. Bilodeau, D. C. Johnson, and J. Albert. 1993. Bragg gratings fabricated in monomode photosensitive optical fiber by UV exposure through a phase mask. *Appl. Phys. Lett.* 62:1035–1037.

51. Anderson, D. Z., V. Mizrahi, T. Erdogan, and A. E. White. 1993. Production of in-fibre gratings using a diffractive optical element. *Electron. Lett.* 29:566–568.

52. Prohaska, J.D., E. Snitzer, S. Rishton, and V. Boegli. 1993. Magnification of mask fabricated fibre Bragg gratings. *Electron. Lett.* 29:1614–1616.

53. Archambault, J-L., L. Reekie, and P. St. J. Russell. 1993. 100% Reflectivity Bragg reflectors produced in optical fibres by single excimer laser pulses. *Electron. Lett.* 29:453–455.

54. Askins, G. G., T-E. Tsai, G. M. Williams, M. A. Putnam, M. Bashkansky, and E. J. Friebele. 1992. Fiber Bragg reflectors prepared by a single excimer pulse. *Opt. Lett.* 17:833–835.

55. Dong, L., J-L. Archambault, L. Reekie, P. St. J. Russell, and D. N. Payne. 1993. Single pulse Bragg gratings written during fibre drawing. *Electron. Lett.* 29:1577–1578.

56. Strasser, T. A., T. Erdogan, A. E. White, V. Mizrahi, and P. J. Lemaire. 1994. Ultraviolet laser fabrication of strong, nearly polarization-independent Bragg reflectors in germanium-doped silica waveguides on silica substrates. *Appl. Phys. Lett.* 65:3308–3310.

57. Strasser, T. A. 1996. Photosensitivity in phosphorus doped fibers. In *Conference on Optical Fiber Communication*, 81–82. Paper TuO1. Technical Digest. Washington, DC: Optical Society of America.

58. Malo, B., J. Albert, F. Bilodeau, T. Kitagawa, D. C. Johnson, K. O. Hill, K. Hattori, Y. Hibino, and S. Gujrathi. 1994. Photosensitivity in phosphorus-doped silica glass and optical waveguides. *Appl. Phys. Lett.* 65:394–396.

59. Kashyap, R., G. D. Maxwell, and B. J. Ainslie. 1993. Laser-trimmed four-port bandpass filter fabricated in single-mode photosensitive Ge-doped planar waveguide. *IEEE Photon. Tech. Lett.* 5:191–194.

60. Erdogan, T., T. A. Strasser, M. A. Milbrodt, E. J. Laskowski, C. H. Henry, and G. E. Kohnke. 1996. Integrated-optical Mach–Zehnder add–drop filter fabricated by a single uv-induced grating exposure. (to be published).

61. Bousselet, P., R. Meilleur, A. Coquelin, P. Garabedian, and J. L. Beylat. 1995. +25.2 dBm Output power from an Er-doped fiber amplifier with 1.48 μm SMQW laser diode modules. In *Conference on Optical Fiber Communication.* Paper TuJ2. Technical Digest. Washington, DC: Optical Society of America.

62. Fan, T. Y., and R. L. Byer. 1988. Diode laser-pumped solid-state lasers. *IEEE J. Quantum. Electron.* QE-24(6):895–912.

63. Snitzer, E., H. Po, F. Hakimi, R. Tumminelli, and B. C. McCollum. 1988. Double clad offset core Nd fiber laser. In *Conference on Optical Fiber Communication.* Paper PD5. Technical Digest. Washington, DC: Optical Society of America.

64. Po, H., E. Snitzer, R. Tumminelli, F. Hakimi, N. M. Chu, and T. Haw. 1989. Doubly clad high brightness Nd fiber laser pumped by GaAlAs phased array. In *Conference on Optical Fiber Communication.* Paper PD5. Technical Digest. Washington, DC: Optical Society of America.

65. Po, H., J. D. Cao, B. M. Laliberte, R. A. Minns, R. F. Robinson, B. H. Rockney, R. R. Tricca, and Y. H. Zhang. 1993. High power neodymium-doped single transverse mode fibre laser. *Electron. Lett.* 29(17):1500–1501.

66. Zellmer, H., A. Willamowski, A. Tunnermann, H. Welling, S. Unger, V. Reichel, H. Muller, J. Kirchhof, and P. Albers. 1995. High-power cw neodymium-doped fiber laser operating at 9.2 W with high beam quality. *Opt. Lett.* 20(6):578–580.

67. Grubb, S. G. 1995. High-power diode-pumped fiber lasers and amplifiers. In *Conference on Optical Fiber Communication.* Paper TuJ1. Technical Digest. Washington, DC: Optical Society of America.

68. Pask, H. M., R. J. Carman, D. C. Hanna, A. C. Tropper, C. J. Mackechnie, P. R. Barber, and J. Dawes. 1995. Ytterbium-doped silica fiber laser: Versatile sources for the 1–1.2 μm region. *IEEE J. Select. Top. Quantum Electron.* 1(1):2–12.

69. Minelly, J. D., W. L. Barnes, R. I. Laming, P. R. Morkel, J. E. Townsend, S. G. Grubb, and D. N. Payne. 1993. Diode-array pumping of Er^{3+}/Yb^{3+} co-doped fiber lasers and amplifiers. *IEEE Photon. Tech. Lett.* 5(3):301–303.

70. Minelly, J. D., Z. J. Chen, R. I. Laming, and J. E. Caplen. 1995. Efficient cladding pumping of an Er^{3+} fibre. In *Proceedings of the 21st European Conference on Optical Communications (ECOC'95),* Florence, Italy, 917–920. Paper Th.L.1.2.

71. Zentano, L. A. 1993. High-power double-clad fiber lasers. *J. Lightwave Tech.* 11:1435–1446.

72. Gapontsev, V. P. 1982. Erbium glass lasers and their applications. *Opt. Laser Tech.* 189–196.

73. Townsend, J. E., W. L. Barnes, K. P. Jedrzejewski, and S. G. Grubb. 1991. Yb^{3+} sensitized Er^{3+} doped silica optical fibre with ultrahigh transfer efficiency and gain. *Electron. Lett.* 27(21):1958–1959.

74. Laming, R. I., S. B. Poole, and E. J. Tarbox. 1988. Pump excited state absorption in erbium-doped fibers. *Opt. Lett.* 13(12):1084–1088.

75. Barnes, W. L., S. B. Poole, J. E. Townsend, L. Reekie, D. J. Taylor, and D. N. Payne. 1989. Er^{3+}-Yb^{3+} and Er^{3+} doped fiber lasers. *IEEE J. Lightwave Tech.* LT-7(10):1461–1465.

76. Kringlebotn, J. T., J. L. Archambault, L. Reekie, J. E. Townsend, G. G. Vienne, and D. N. Payne. 1994. Highly-efficient, low-noise grating-feedback Er^{3+} : Yb^{3+} codoped fibre laser. *Electron. Lett.* 30(12):972–973.

77. Barbier, D., J. M. Delavaux, A. Kevorkian, P. Gastaldo, and J. M. Jouanno. 1995. Yb/Er integrated optics amplifiers on phosphate glass in single and double pass configurations. In *Conference on Optical Fiber Communication.* Paper PD3. Technical Digest. Washington DC: Optical Society of America.

78. Lester, C., A. Bjarklev, T. Rasmussen, and P. Dinesen. 1995. Modeling of Yb^{3+} sensitized Er^{3+} doped silica waveguide amplifiers. *J. Lightwave Tech.* 13(5):740–743.

79. Fermann, M. E., D. C. Hanna, D. P. Shepherd, P. J. Sini, and J. E. Townsend. 1988. Efficient operation of an Yb-sensitized Er fibre laser at 1.56 μm. *Electron. Lett.* 24(18):1135–1136.

80. Grubb, S. G., R. S. Cannon, T. W. Windhorn, S. W. Vendetta, P. A. Leilabady, D. W. Anthon, K. L. Sweeney, W. L. Barnes, E. R. Taylor, and J. E. Townsend. 1991. High power sensitized erbium doped fiber amplifier. In *Conference on Optical Fiber Communication.* Paper PD7. Technical Digest. Washington, DC: Optical Society of America.

81. Townsend, J. E., W. L. Barnes, K. P. Jedrzejewski, and S. G. Grubb. 1991. Yb^{3+} sensitized Er^{3+} doped silica optical fibre with ultrahigh transfer efficiency and gain. *Electron. Lett.* 27(21):1958–1959.

82. Grubb, S. G., W. F. Humer, R. S. Cannon, S. W. Vendetta, K. L. Sweeney, P. A. Leilabady, M. R. Kuer, J. G. Kwasegroch, T. C. Munks, and D. W. Anthon. 1992. +24.6 dBm Output power Er/Yb codoped optical amplifier pumped by a diode-pumped Nd : YLF laser. *Electron. Lett.* 28(13):1275–1276.

83. Minelly, J. D., R. I. Laming, J. E. Townsend, W. L. Barnes, E. R. Taylor, K. P. Jedrzejewski, and D. N. Payne. 1992. High gain power amplifier tandem pumped by a 3 W multistripe diode. In *Conference on Optical Fiber Communication.* Paper TuG2. Technical Digest. Washington, DC: Optical Society of America.

84. Minelly, J. D., W. L. Barnes, R. I. Laming, P. R. Morkel, J. E. Townsend, S. G. Grubb, and D. N. Payne. 1993. Diode-array pumping of Er^{3+}/Yb^{3+} co-doped fiber lasers and amplifiers. *IEEE Photon. Tech. Lett.* 5(3):301–303.

85. Grubb, S. G. 1995. High-power diode-pumped fiber lasers and amplifiers. In *Conference on Optical Fiber Communication.* Paper TuJ1. Technical Digest. Washington, DC: Optical Society of America.

86. Livas, J. C., S. R. Chinn, E. S. Kintzer, J. N. Walpole, C. A. Wang, and L. J. Missaggia. 1994. High-power erbium-doped fiber amplifier with 975 nm tapered-gain-region laser pumps. *Electron. Lett.* 30:1054–1055.

87. Grubb, S. G., D. J. DiGiovanni, J. R. Simpson, W. Y. Cheung, S. Sanders, D. F. Welch, and B. Rockney. 1996. Ultrahigh power diode-pumped 1.5 μm fiber amplifiers. In *Conference on Optical Fiber Communication.* Paper TuG4. Technical Digest. Washington, DC: Optical Society of America.

88. Wysocki, P. F., G. Nykolak, D. S. Shenk, and K. Eason. 1996. Noise figure limitations in ytterbium-codoped erbium-doped fiber amplifiers pumped at 1064 nm. In *Conference on Optical Fiber Communication.* Paper TuG6. Technical Digest. Washington, DC: Optical Society of America.

89. Grubb, S. G., P. A. Leilabady, and D. E. Frymyer. 1993. Solid-state laser pumping of 1.5 μm optical amplifiers and sources for lightwave video transmission. *J. Lightwave Tech.* 11(1):27–32.

90. Aoki, Y. 1988. Properties of fiber Raman amplifiers and their applicability to digital optical communication systems. *J. Lightwave Tech.* 6(7):1225.

91. Aggrawal, G. P. 1995. *Nonlinear fiber optics.* 2d ed. New York: Academic Press.

92. Stolen, R. H., and C. Lin. 1991. Fiber Raman lasers. In *CRC handbook of laser science and technology.* Suppl. 1, *Lasers.* Boca Raton, FL: CRC Press.

93. Reed, W. A., W. M. Coughran, and S. G. Grubb. 1995. Modeling of cascaded Raman resonators. In *Conference on Optical Fiber Communication.* Paper WD1. Technical Digest. Washington, DC: Optical Society of America.

94. Grubb, S. G. 1995. High-power diode-pumped fiber lasers and amplifiers. In *Conference on Optical Fiber Communication.* Paper TuJ1. Technical Digest. Washington, DC: Optical Society of America.

95. Stafford, E. K., J. Mariano, and M. M. Sanders. 1995. Undersea non-repeatered technologies, challenges, and products. *AT&T Tech. J.* 74(1):47–59.

96. Hansen, P. B., L. Eskilden, S. G. Grubb, A. M. Vengsarkar, S. K. Korotky, T. A. Strasser, J. E. J. Alphonsus, J. J. Veselka, D. J. DiGiovanni, D. W. Peckham, E. C. Beck, D. Truxal, W. Y. Cheung, S. G. Kosinski, D. Gasper, P. F. Wysocki, V. L. da Silva, and J. R. Simpson. 1995. 529 km Unrepeatered transmission at 2.488 Gbit/s using dispersion compensation, forward error correction, and remote post- and pre-amplifiers pumped by diode-pumped Raman lasers. *Electron. Lett.* 31(17):1460–1461.

97. Grubb, S. G., T. Erdogan, V. Mizrahi, T. Strasser, W. Y. Cheung, W. A. Reed, P. J. Lemaire, A. E. Miller, S. G. Kosinski, G. Nykolak, and P. C. Becker. 1993. 1.3 μm Cascaded Raman amplifier in germanosilicate fibers. In *Optical amplifiers and their applications.* Paper PD3. Technical Digest. Washington, DC: Optical Society of America.

98. Dianov, E. M., A. A. Abramov, M. M. Bubnov, A. V. Shipulin, A. M. Prokhorov, S. L. Semjonov, and A. G. Schebunjaev. 1995. Demonstration of 1.3 μm Raman fiber amplifier gain of 25 dB at a pumping power of 300 mW. *Opt. Fiber Tech.* 1:236–238.

99. Chernikov, S. V., Y. Zhu, R. Kashyap, and J. R. Taylor. 1995. High-gain, monolithic, cascaded fibre Raman amplifier operating at 1.3 μm. *Electron. Lett.* 31(6):472–473.

100. Desurvire, E. 1994. *Erbium-doped fiber amplifiers.* New York: Wiley.

101. Dykaar, D. R., S. G. Grubb, J. R. Simpson, T. A. Strasser, A. M. Vengsarkar, J. M. Borick, W. Y. Cheung, and S. B. Darack. 1995. In *Conference on Optical Fiber Communication.* Paper PD1. Technical Digest. Washington, DC: Optical Society of America.

102. Stentz, A. J., S. G. Grubb, C. E. Headley III, J. R. Simpson, T. A. Strasser, and N. Park. 1996. Raman amplifier with improved system performance. In *Conference on Optical Fiber Communication.* Paper TuD3. Technical Digest. Washington, DC: Optical Society of America.

103. Stentz, A. J., T. Nielsen, S. G. Grubb, T. A. Strasser, and J. R. Pedrazzani. 1996. Raman ring amplifier at 1.3 μm with analog grade noise performance and an output power of +23 dBm. In *Conference on Optical Fiber Communication.* Paper PD16. Technical Digest. Washington, DC: Optical Society of America.

104. Forghieri, F., R. W. Tkach, and A. R. Chraplyvy. 1994. Bandwidth of crosstalk in Raman amplifiers. In *Conference on Optical Fiber Communication.* Paper FC6. Technical Digest. Washington, DC: Optical Society of America.

105. Hansen, P. B., L. Eskildson, S. G. Grubb, A. J. Stentz, T. A. Strasser, J. Judkins, J. J. DeMarco, J. R. Pedrazzani, and D. J. DiGiovanni. 1996. Capacity upgrade of transmission systems by Raman amplification. In *Optical amplifiers and their applications.* Paper ThB4. Technical Digest. Washington, DC: Optical Society of America.

106. Haase, M. A., J. Qiu, J. M. DePuydt, and H. Cheng. 1991. Blue-green laser diodes. *Appl. Phys. Lett.* 59:1272–1274.

107. Nakamura, S., T. Mukai, and M. Senoh. 1994. Candela-class high brightness InGaN/AlGaN double heterostructure blue light emitting diodes. *Appl. Phys. Lett.* 64:1687–1689.

108. Funk, D. S., and J. G. Eden. 1995. Glass-fiber lasers in the ultraviolet and visible. *IEEE J. Select. Top. Quantum Electron.* 1(3):784–791.

109. Johnson, L. F., and H. J. Guggenheim. 1971. Infrared pumped visible laser. *Appl. Phys. Lett.* 19:44–47.

110. Silversmith, A. J., W. Lenth, and R. M. McFarlane. 1987. Green infrared-pumped erbium upconversion laser. *Appl. Phys. Lett.* 51:1977–1979.

111. Smart, R. G., D. C. Hanna, A. C. Tropper, S. T. Davey, S. F. Carter, and D. Szebesta. 1991. CW room temperature upconversion lasing at blue, green, and red wavelengths in infrared-pumped praeseodymium-doped fluoride fibre. *Electron. Lett.* 27:1307–1309.

112. Piehler, D., D. Craven, N. K. Kwong, and H. Zarem. 1993. Laser diode pumped red and green upconversion fibre lasers. *Electron. Lett.* 29:1857–1858.

113. Pask, H. M., A. C. Tropper, D. C. Hanna, B. N. Samson, R. D. T. Lauder, P. R. Barber, L. Reekie, J-L. Archambault, S. T. Davey, and D. Szebesta. 1994. *Adv. Solid State Lasers* 20:352–354.

114. Grubb, S. G., K. W. Bennett, R. S. Cannon, and W. F. Humer. 1992. CW room-temperature blue upconversion fibre laser. *Electron. Lett.* 28:1243–1244.

115. Allain, J. Y., M. Monerie, and H. Poignant. 1990. Blue upconversion fibre laser. *Electron. Lett.* 26:166–168.

116. Whitley, T. J., A. A. Millar, R. Wyatt, M. C. Brierley, and D. Szebesta. 1991. Upconversion pumped green lasing in erbium-doped fluorozirconate fibre. *Electron. Lett.* 27:1785–1786.

117. Allain, J. Y., M. Monerie, and H. Poignant. 1992. Tunable green upconversion erbium fibre laser. *Electron. Lett.* 28:111–113.

118. Funk, D. S., S. B. Stevens, and J. G. Eden. 1993. Excitation spectra of the green Ho:fluorozirconate glass fiber laser. *IEEE Photon. Tech. Lett.* 5:154–157.

119. Funk, D. S., J. W. Carlson, and J. G. Eden. 1994. Ultraviolet (381 nm), room temperature laser in neodymium-doped fluorozirconate fibre. *Electron. Lett.* 30:1859–1860.

120. Komukai, T., T. Yamamoto, T. Sugawa, and Y. Miyajima. 1993. 1.47 μm Band Tm^{3+} doped fluoride fibre amplifier using a 1.064 μm upconversion pumping scheme. *Electron. Lett.* 29:110–112.

121. Percival, R. M., D. Szebesta, and J. R. Williams. 1994. Highly efficient 1.064 μm upconversion pumped 1.47 μm thulium doped fluoride fibre laser. *Electron. Lett.* 30:1057–1058.

122. Komukai, T., T. Yamamoto, T. Sugawa, and Y. Miyajima. 1995. Upconversion pumped thulium-doped fluoride fiber amplifier and laser operating at 1.47 μm. *IEEE J. Quantum. Electron.* 31:1880–1889.

123 Miyajima, Y., T. Komukai, and T. Sugawa. 1993. 1-W cw Tm-doped fluoride fiber laser at 1.47 μm. *Electron. Lett.* 29:660–661.

124. Whitley, T. J., C. A. Millar, M. C. Brierley, and S. F. Carter. 1991. 23 dB Gain upconversion pumped erbium doped fibre amplifier operating at 850 nm. *Electron. Lett.* 27:184–185.

125. Barber, P. R., R. Paschotta, A. C. Tropper, and D. C. Hanna. 1995. Infrared-induced photodarkening in Tm-doped fluoride fibers. 20:2195–2197.

Chapter 8 | Silicon Optical Bench Waveguide Technology

Yuan P. Li

Lucent Technologies, Bell Laboratories, Norcross, Georgia

Charles H. Henry

Lucent Technologies, Bell Laboratories, Murray Hill, New Jersey

I. Introduction

The needs of communications encompass a spectrum of optical technologies (Agrawal, 1992). At one end of the spectrum are compound semiconductor devices, lasers and detectors, which are evolving into photonic integrated circuits. At the other end are optical fibers, which are also used as discrete passive components such as 3-dB couplers and Bragg reflection filters. In between, there are other technologies such as optical isolators and thin film filters. In this chapter, we describe another technology that is beginning to supply components for optical communications: passive optical waveguide circuits made from doped silica films deposited on planar substrates (silicon in particular). We refer to this technology as *silicon optical bench (SiOB) optical integrated circuits (OICs)*. The SiOB OIC technology is being pursued by many laboratories around the world, including the Nippon Telephone and Telegraph (NTT) Optoelectronics Laboratory, Hitachi Cable, Furukawa Cable, Laboratoire d'Electronique et de Technologie de L'Informatique (LETI) in Grenoble, British Telecom Research Laboratories, and Lucent Technologies, Bell Laboratories.

A. ROLE OF PASSIVE CIRCUITS IN OPTICAL COMMUNICATIONS

As optical communications advances, more and more passive optical components are needed. For example, broadband multiplexers are needed for delivering voice and video to the home, for combining pump and signals

319

OPTICAL FIBER TELECOMMUNICATIONS,
VOLUME IIIB

Copyright © 1997 by Lucent Technologies.
All rights of reproduction in any form reserved.
ISBN: 0-12-395171-2

in an optical amplifier, and for adding a monitoring signal to the traffic on optical fibers. Dense wavelength-division multiplexing (WDM) systems need multiplexers to combine and separate channels of different wavelengths and need add–drop filters to partially alter the traffic. Splitters and star couplers are used in broadcast applications. Low-speed optical switches are needed for sparing applications and network reconfiguration.

B. FROM OPTICAL FIBERS TO PLANAR WAVEGUIDES: ADVANTAGES OF INTEGRATION

One might think that there is no need for passive OICs because optical fiber components can perform many optical functions with unsurpassed low loss. However, planar integration brings a number of unique new advantages. *Compactness* becomes important when we take the technology from the laboratory to applications, where all functions are squeezed into small modular containers inserted into relay racks — e.g., making 1×16 splitters on a single chip instead of fusing discrete optical fiber 1×2 splitters. *Reduced cost* is achieved when complex functions are integrated, which saves manual interconnections, and when many chips are made in the processing of a single wafer. This advantage will become very important as optical communications begins to serve individual subscribers. *Increased complexity* occurs because photolithography allows the design of components with numerous optical paths — e.g., single-mode star couplers. The great precision of photolithography enables the relative delay and coupling between different paths to be *accurately controlled*. This makes new device configurations possible — e.g., waveguide grating routers (WGRs) and Fourier filters. Passive components are often used in conjunction with lasers and detectors. In this case, it is attractive for the waveguide chip to become a platform onto which the other components are attached (Henry, Blonder, and Kazarinov 1989). The concept of the waveguide chip as a platform led to the name *silicon optical bench*.

C. SURVEY OF CURRENT PLANAR WAVEGUIDE TECHNOLOGIES

Waveguides are made with many other materials besides silica on silicon. There are extensive reviews on integrated optics that treat the earlier development of the planar waveguide technology (Hunsperger 1991; Marz 1994; Kashima 1995). Planar waveguides are part of semiconductor lasers and are used in semiconductor photonic integrated circuits (Leonberger and Donnelly 1990; Koren 1994). Optical waveguides are made in lithium niobate by titanium diffusion, where they are used to build modulators, switching arrays, and polarization controllers (Korotky and Alferness 1987). Passive waveguides have been made by ion exchange in alkali–silicate glass (Findakly 1985; Ramaswamy and Srivastava 1988). These waveguides have been used commercially to manufacture splitters and other components. Recently, there has been much interest in waveguides made from polymer materials (Neyer, Knoche, and Muller 1993).

D. ADVANTAGES OF SILICA WAVEGUIDES ON A SILICON SUBSTRATE

In view of these alternatives, why single out silica on silicon for the fabrication of passive OICs? We believe that SiOB OICs have a number of advantages. OICs require low-cost substrates that are flat, extremely smooth, and large in area. The commercially available silicon substrates are ideal for this application. In our SiOB program at Bell Laboratories, we currently use 5-in.-diameter wafers, and larger diameters are available. The area cost of silicon is nearly two orders of magnitude less than that of indium phosphide.

Silica (SiO_2) is extensively used in the silicon integrated circuit industry and for the manufacture of optical fibers. It has a stable, well-controlled refractive index and is highly transparent. Thus, SiOB waveguides can be entirely processed with the commercially available equipment of this industry. In many cases, the processes are chemically compatible with silicon technology, which makes it possible to share a facility that is used for conventional silicon integrated circuit processing. In view of the great cost of processing equipment, this is a practical advantage.

Optical fibers are also made from silica, so that use of an index-matching oil or elastomer between the waveguide and the optical fiber results in a nonreflecting interface. As abundantly demonstrated by optical fiber performance, silica is an inherently very low-loss optical material. This may not be the case for polymer materials, which can have weak vibronic absorption bands at optical communications wavelengths. The waveguide and optical fiber are also matched in thermal expansion coefficient so that, if necessary, they can be fused. The thermal expansion coefficient of silicon is greater than that of doped silica, a feature that causes the film to be compressively strained after annealing. A film in compression is more stable against cracking. This stability is in contrast with the case of films in tension occurring for many lower temperature glasses deposited on silicon and most doped silica films deposited on fused silica substrates.

Silicon also has good thermal conductivity and can be used as a laser submount. The silicon substrate, acting as a heat sink, makes possible hybrid laser–waveguide applications and stable thermooptic switches. Silicon has etch stop planes that enable the chemical etching of precision V grooves that are used for aligning optical fiber arrays to IOCs. The V grooves can be made in the substrate of the optical waveguide. Although it requires more complex processing, this approach should result in a more precise alignment and more secure attachment of optical fibers to chips than when the optical fibers are held by a separate part.

Silica waveguides can also be formed on fused silica substrates. In some cases, this can result in reduced polarization-dependent splittings of narrow band filters (Suzuki, Inoue, and Ohmori 1994). This substrate, which has low thermal conductivity, is also convenient for fusing optical fibers to waveguides (Imoto 1994).

Our chapter reviews SiOB waveguide technology. We attempt to cover all aspects of this subject. However, because our backgrounds are in OIC design and characterization, we emphasize this aspect of optical waveguide technology. We try to explain the concepts underlying various components, especially if this information is not readily available. Our examples of this technology are taken mainly from the work at Bell Laboratories, with which we are most familiar. A briefer version of this review was published by the authors (Li and Henry 1996).

II. Materials and Fabrication

A. WAVEGUIDE CROSS SECTION

In this section, we describe the standard waveguide made at Bell Laboratories as an example. A representative waveguide cross section is shown in Fig. 8.1. The waveguide is formed from three layers: base, core, and cladding. The base layer isolates the fundamental mode from the silicon substrate and thereby prevents leakage through the silica–substrate interface, which, unlike other waveguide interfaces, is not totally reflecting.

The refractive index of the cladding layer is chosen to be nearly equal to that of the base layer. To achieve optical confinement, the core layer refractive index is increased by a small amount. This is normally described by delta (Δ), the percentage increase in refractive index of the core, relative

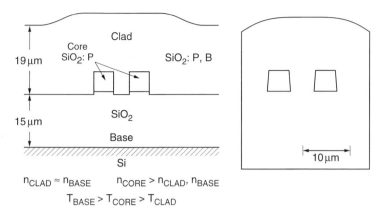

Fig. 8.1 Cross-sectional layout of coupled optical waveguides and etch-stained cross section. The base, core, and cladding layers are shown. The core thickness is 5 μm and can range from 3.5 to 7 μm in different applications.

to the cladding. With a P-doped core, delta is approximately 0.60–0.70%. For Ge-doped waveguides, pioneered by NTT (Kominato *et al.* 1990), higher values of delta are possible. Increasing delta produces a smaller optical mode, a smaller bend radius, and more compact OICs, but usually with increased loss (see Section IV.B).

The base layer is made of undoped silica. This is the most rigid layer, and it keeps the core, which is adhered to it, from moving after it is patterned. The other glasses are made of doped silica and are flowed during annealing, a process that helps to form homogeneous low-loss material. To promote filling in between closely spaced cores, the cladding should flow readily, while the core and the base layer remain rigid. The cladding should also match the base layer in refractive index. These demanding requirements are met by doping the cladding layer with B and P. The addition of B lowers both the flow temperature and the refractive index, compensating for the refractive index increase of P.

B. THICK FILM FORMATION

The processing of SiOB OICs involves steps similar to those encountered in silicon processing, except the films are much thicker. The cross section of two coupled waveguides shown in Fig. 8.1 is representative of waveguides currently made at Bell Laboratories. The total thickness of the films composing these waveguides is about 38 μm. There are currently two main ways of forming these thick silica films for optical waveguides: chemical vapor deposition (CVD) and flame hydrolysis deposition (FHD).

Low-pressure CVD is the method of film deposition primarily used in silicon technology. Deposition rates are about 1 μm per hour. Deposition is done on both sides of the wafer, which minimizes wafer bowing. At Bell Laboratories, uniform depositions can be made on as many as 50 5-in. wafers. Base layers 16 μm thick and core layers 6 μm thick can be deposited in a single step (M. R. Serbin, personal communication, 1996). Deposition temperatures range from 400 to 800°C, depending on the reactants. After deposition, the films are annealed to form a dense homogeneous film. The annealing is done at temperatures of about 900–1100°C for several hours. If necessary, annealing can be done on hundreds of wafers at a time.

FHD has its origin in the vertical axial deposition process, a method used in the optical fiber industry to make fiber preforms (Kawachi, Yasu, and Kobayashi 1983; Kawachi 1990). Fine doped silica particles (of order 0.1 μm in size) are formed in the flame of a torch and are driven onto a cooler silicon substrate. The deposition rate is about 1 μm per minute per wafer. The flame is rastered over the wafer to provide uniformity. FHD suffers from the disadvantage of one-sided deposition, which causes wafer bowing. It is necessary to sinter the particles into a homogeneous film, which requires very high temperatures, about 1100–1300°C. At NTT, which developed this process, a rotating turntable 1 m in diameter is used; this enables 50 wafers to be deposited at a time (Yasu, Kawachi, and Kobayashi 1985). When large numbers of wafers are used, the deposition times are similar for both CVD and FHD, because the time for FHD is proportional to the number of wafers, but FHD is much faster for depositions on small numbers of wafers.

Another important method is electron beam deposition. The throughput of this process is more limited than that of CVD because not as many wafers can be deposited on at one time. Deposition is one sided. The method gives very good control of both thickness and uniformity. Hitachi Cable prefers to use this method for fabrication of the titanium-doped core of its wafers (K. Imoto, personal communication, 1992).

At Bell Laboratories, the base layer is formed either by CVD deposition or by high-pressure steam oxidation of the silicon substrate. The latter process is attractive because as many as 200 wafers can be done at one time with extremely good control of layer thickness. This is a slow process because oxidation is limited by diffusion and the layer thickness increases as the square root of time. A 15-μm-thick base layer takes 53 h of oxidation.

As this discussion indicates, there is not yet a preferred method of thick film formation. One reason for this is that different approaches can be made to work and problems are not openly discussed. An advantage of CVD is that many commercial reactors are available and the silicon industry is continuing to develop this technology (Singer 1995).

C. PROCESSING STEPS

Processing of an OIC utilizes some of the steps of silicon technology. At Bell Laboratories, all processes are carried out in a silicon facility with conventional silicon processing equipment. A set of 12 wafers can be processed simultaneously. In silicon production, all processing steps are usually done in cassettes of 25 wafers.

First, a base layer is formed by low-pressure CVD or high-pressure steam oxidation followed by an anneal that densifies the glass, stabilizes the refractive index, and reduces strain. Next, a core layer is deposited and annealed to flow the glass. The waveguide pattern is then formed by photolithography. Photolithography is done with 1:1 projection covering the whole wafer with 1.5- to 2.0-μm resolution. The higher resolution steppers used in the silicon industry have much smaller fields. Because of the thickness of the core layer, a hard mask must be deposited on top of the core layer and patterned before reactive ion etching through the core. Silicon, chrome, or a composition of hard-baked photoresist and silica layers are used for this process. Next, the cladding layer is deposited in several steps with anneals to flow the glass. The first cladding deposition is a thin layer that helps to fill in between closely spaced waveguides. Finally, the wafer is sawed into chips.

D. FIBER PIGTAIL ATTACHMENT

For practical applications, the light is usually coupled in and out a waveguide circuit by fiber pigtails attached to the waveguide chip. The fiber attachment must be suitable for low-cost mass production, have low insertion loss and low back reflection, and be durable and reliable. There are two major technical challenges in attaching fibers to waveguides: high-precision alignment and durable bonding. These problems have been addressed by a number of authors (Presby and Edwards 1992; Presby *et al.* 1992; Yamada *et al.* 1992; Kato *et al.* 1993; Sugita *et al.*, 1993; Hibino *et al.* 1995).

In order to attain low insertion loss, the core of the fiber must be aligned to that of the waveguide to within a small fraction of the mode size in the transverse direction (e.g., within ± 0.2 μm) and to within 10 μm in the longitudinal direction (see Section IV.A). Active alignment is currently used in most fabrication facilities. The fiber pigtail at one end of the chip is excited with a monitor light signal, and the fiber pigtail at the other end is connected to a detector. The positions of both fibers are adjusted until the transmitted light is maximum or the scattered light is minimum. High-precision motor and piezoelectric translation stages with automatic feed-back control are commonly used for the fiber positioning. To increase the efficiency, ribbons of 8–16 fibers instead of single fibers are often aligned simultaneously. The fibers are placed in V grooves etched in Si or machined in ceramic chips for high-precision lateral spacings that match the spacings of the waveguides. To reduce back reflection, both the waveguide chip and the fiber ribbon chips are polished at an angle of 8–12°. The transverse positioning of the ribbon involves translation along the x and y axis and rotation around the z axis.

Although active alignment is capable of high accuracy and low insertion loss, it is relatively slow and requires numerous piece parts. Passive align-ment, a less costly approach that does not require the monitor light signal and feedback controlled moving stages, is therefore under active research and development. In a popular method, V grooves are etched in the OIC substrate (Day et al. 1992). Whereas the in-plane alignment accuracy is determined by the photolithographic mask and is not difficult to control, the vertical accuracy depends on the thickness of the waveguide materials and is more difficult to control.

After the fiber alignment, the fibers must be bonded to the waveguide chip without affecting the alignment. Ideally, the bonding should be suffi-ciently durable that the packaged device can be used in the outdoor environ-ment for tens of years. If the bonding adhesive fills the light path between the fiber and the waveguide, it should have a refractive index close to that of silica (1.45) and should be transparent. UV cure or thermal cure epoxies are commonly used as the bonding adhesive. The fibers can also be fused to the waveguide chip by using high-power lasers or electric arcs (Imoto 1994). The fused bonding is believed to be highly reliable, but this technique is currently limited to silica waveguides on silica substrates.

E. OPTICAL CHARACTERIZATION

The transmission loss, the transmission spectrum, and their polarization dependence are the basic parameters used to characterize the optical perfor-mance of waveguide devices. These are measured both on diced wafers and on devices with fiber pigtails attached. To measure a wafer, fibers are butt coupled to the waveguides using computer-controlled translation stages, similar to the active fiber alignment described earlier. An index-matching oil is used to fill the gap between the fiber and the waveguide to reduce back reflection. The transmission loss at a given wavelength can be measured using a laser as the light source and a power meter as the detector. The transmission spectrum can be measured using an optical spectrum analyzer and one or more light-emitting diodes (LEDs) as a high-power broadband light source. Alternatively, a tunable light source (e.g., an external cavity laser) and a power meter can be used. To measure the polarization dependence of a device on an unpackaged wafer, the source light is usually launched into the waveguide with a predefined polarization TE (E field parallel substrate) or TM (E field normal substrate), using, for example, a polarization splitter and a polarization-maintaining fiber. For a packaged device, even if the light launched into the pigtail fiber has a known polarization, the polarization at the waveguide device depends on the arrangement of the pigtail fiber. Therefore, a computer-controlled polarization rotator is often used to measure the transmission of many polarization states. The results are then used to derive the polarization dependence of the waveguide device transmission.

Besides measurements of components, the far-field pattern is sometimes used as a characterization of the waveguide. It can be measured by scanning a detector or using a detector array far away from the waveguide end. The dimensions and refractive indices of waveguide material layers are often measured to describe the fabrication process.

F. COST ISSUES IN MASS PRODUCTION

The main contributions to the cost of a packaged waveguide circuit are mask design, wafer fabrication, chip dicing, fiber attachment, packaging, testing, and qualification. (Development of waveguide devices is not included here.) For large-scale mass production, the cost of one-time mask design is relatively small. If the chip size is small so that many devices can be reproduced on a single wafer, the wafer fabrication cost is also a relatively small contribution. The major cost in this case is in fiber attachment and packaging, thus inexpensive (most likely passive) fiber alignment becomes a key issue in reducing the total cost. Testing all devices from each wafer can also be time-consuming and expensive, depending upon the type of parameters to be measured. However, the testing cost can be significantly reduced if the wafer uniformity and reproducibility are good, because then only one device on each wafer need be measured.

III. Design

A. OPTICAL WAVEGUIDE MODES

The theory of dielectric waveguides is well described in the literature (Snyder and Love 1983; Marcuse 1982, 1991; Tamir 1990; Marz 1994). In this section, we summarize several topics important to the design and simulation of planar waveguide devices, and emphasize the physical principles.

An optical waveguide directs the propagation of an electromagnetic field through a nearly lossless doped silica medium. We refer to the propagating field as the *lightwave*. The field obeys Maxwell's equations. Because the changes in refractive index $n(\mathbf{x})$ are small, Maxwell's equations reduce to a scalar wave equation for the transverse electric field. At a single wavelength λ and polarization, the scalar wave equation is

$$\nabla^2 E(\mathbf{x}) + \frac{4\pi^2}{\lambda^2} n(\mathbf{x})^2 E(\mathbf{x}) = 0. \tag{8.1}$$

For propagation along a waveguide in the z direction, the refractive index depends only on x and y. The field can be written as the sum of propagating transverse modes. Each mode field has the form $E(\mathbf{x}) = \Phi(x,y)e^{ik_z z}$, where the propagation constant k_z can be expressed in terms of an effective refractive index $\bar{n}$, $k_z = 2\pi\bar{n}/\lambda$. The wave equation for the transverse mode reduces to

$$\frac{\lambda^2}{4\pi^2}\left(\frac{\partial^2}{\partial x^2} + \frac{\partial^2}{\partial y^2}\right)\Phi(x,y) + n(x,y)^2\Phi(x,y) = \bar{n}^2\Phi(x,y). \tag{8.2}$$

Equation (8.2) is equivalent to a two-dimensional steady-state Schrödinger equation with the potential $V(x,y)$ proportional to $-n(x,y)^2$, the mass m proportional to λ^{-2}, and the energy eigenvalue E proportional to $-\bar{n}^2$:

$$\frac{\hbar^2}{2m}\left(\frac{\partial^2}{\partial x^2} + \frac{\partial^2}{\partial y^2}\right)\Psi(x,y) - V(x,y)\Psi(x,y) = -E\Psi(x,y). \tag{8.3}$$

This equivalence helps us understand the behavior of optical waveguide modes, which is illustrated in Fig. 8.2. A local increase in refractive index acts as an attractive potential well that confines the light into a bound mode in the same way that such a potential well confines an electron into a bound state. This analogy is illustrated in Fig. 8.2a. Bound states occur with only $\bar{n}$ between n_{core} and n_{clad}. For smaller values of $\bar{n}$, there is a continuum of unbound or radiation modes.

A waveguide coupler is formed by two identical waveguides running side by side, as shown in Fig. 8.2b. The interaction between the two guides results in a splitting of the mode eigenvalues, associated with even and odd combinations of the waveguide modes. Side-by-side waveguide couplers are discussed in Section V.A.

Leakage into a high refractive index medium (such as the Si substrate) is illustrated in Fig. 8.2c. It is a tunneling process and is exponentially reduced by the width of the cladding layer and depth of the bound mode

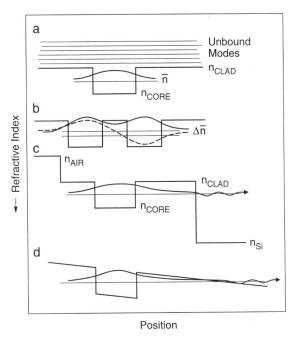

Fig. 8.2   The refractive index medium acts like a potential well to confine the field into a bound mode. (a) Refractive index profile of an optical waveguide with bound and continuum modes. (b) Coupler with two identical waveguides and even and odd supermodes. (c) Waveguide with leakage into the high refractive index silicon substrate. (d) Waveguide with bend loss versus radial position.

in the potential well. Examples are leakage into the silicon substrate and leakage into a coating of epoxy of refractive index higher than that of silica. Calculation of leakage requires introduction of an absorbing or transparent boundary into the mode equation that prevents the outward propagating lost energy from being reflected.

Bend loss can be found by calculating the mode of a circular waveguide of radius R described by $n(r,y)$. This is done by writing the wave equation in a cylindrical coordinate system with $\nabla^2 = \partial^2/\partial r^2 + (1/r)(\partial/\partial r) + (1/r^2)(\partial^2/\partial \theta^2) + \partial^2/\partial y^2$. The change in phase of the mode during propagation depends only on the angular coordinate $E(\mathbf{x}) = \Phi(r,y)\exp(ik_0 R\theta)$. The equation for the mode is

$$\frac{\lambda^2}{4\pi^2}\left(\frac{r^2}{R^2}\right)\left[\frac{\partial^2}{\partial r^2} + \frac{1}{r}\left(\frac{\partial}{\partial r}\right) + \frac{\partial^2}{\partial y^2}\right]\Phi(r,y) + \frac{r^2}{R^2}n(r,y)^2\Phi(r,y) = \bar{n}^2\Phi(r,y). \qquad (8.4)$$

In the vicinity of the waveguide, the effect of the bend is to effectively increase the refractive index by an amount that increases linearly with radius, as shown in Fig. 8.2d. Other changes of the kinetic energy terms of the mode equation are negligible for conventional waveguides (with Δ of order 1% and R of a few millimeters). If we neglect these changes, the waveguide bend is equivalent to a bound electron in a linear potential. This causes the mode to shift outward and to tunnel out in a manner similar to that occurring in leakage, resulting in loss.

B. ADIABATIC CHANGE

Let us consider the propagation of light in a complex waveguide that is gradually changing in the direction of propagation, such as the couplers shown in Figs. 8.3a and 8.3b. We refer to such modes of coupled waveguides as *supermodes*. The waveguides support two supermodes having different $\bar{n}$ — i.e., the two modes are propagating with different wavelengths in the medium. Changes in the dielectric function along z couple the modes, but only changes with Fourier components equal to the difference in the wavelengths alter the supermode occupation. If the change is sufficiently gradual, these Fourier components are negligible and the occupation of each supermode remains constant. This is known as an *adiabatic change*. It has the same meaning as adiabatic change in quantum mechanics or thermodynamics, where the level occupations remain constant. Adiabatic change permits waveguides to change in cross section without loss (Shani

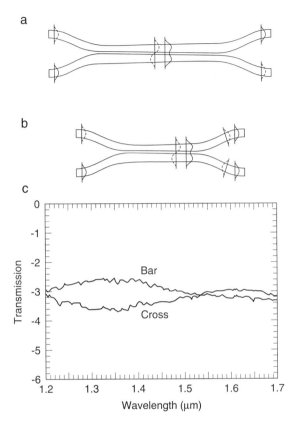

Fig. 8.3   Layout of (a) an adiabatic full coupler and (b) an adiabatic 3-dB coupler. The components are 12 and 9 mm long, respectively, and 30 rm μm wide. (c) Transmission of the adiabatic 3-dB coupler in the bar and cross states relative to that of a straight waveguide. (After Adar, Henry, Kazarinov, *et al.* 1992.)

et al. 1992). A bend can be thought of as an adiabatic change, and it must be gradual to prevent loss.

C.   NUMERICAL SIMULATION

Mode computation can be done analytically for only very simple geometries such as a slab waveguide. Numerical calculation is often necessary for more complex waveguides using, for example, Galerkin's method (Chiang 1994). The field is expanded in a complete set of functions converting the wave equation for the mode, Eq. (8.2), into a matrix eigenvalue equation that may be solved with standard subroutine packages such as EISPACK (Smith *et al.* 1976). The matrix order is equal to the number of basis functions used to describe the mode.

At Bell Laboratories, we use two expansions. The first is essentially a two-dimensional Fourier expansion in which the field is expressed as a sum of products of sine waves of x and y (Henry and Verbeek 1989). This method is useful if the mode can be described by a relatively small number of waves — e.g., by 10 x-waves and 20 y-waves for a waveguide with symmetry in the x direction. The matrix is of order 200, and the computation takes about 1 s using a modern workstation.

For more complex problems, we use the finite element method. The space enclosing the waveguide is divided into triangles and the field is represented by linear functions of x and y in each triangle that are joined continuously. The triangular mesh needs to be finely divided only where the details are needed, such as near the core. This method results in a sparse matrix to solve. The computation time for this matrix increases slightly superlinearly with matrix order, whereas the computation time for the dense matrix sine wave method increases as the cube of matrix order. Smith (1995) has extended the finite element method to rigorously solve Maxwell's equations for waveguide modes. The power of the finite element method is illustrated in Fig. 8.4 (R. K. Smith, personal communication, 1995), in which the scalar wave equation is solved for a waveguide resembling the AT&T logo. The structure has low symmetry and nonplanar layers. The lowest six modes were found in about 30 s of computation.

It is often essential to simulate the propagation of light through complex waveguide structures such as the adiabatic couplers shown in Figs. 8.3a and 8.3b. Propagation is used to determine the required length of the device and the expected values of insertion loss and cross talk. Many algorithms are available to simulate field propagation (Hadley 1992; Yevick 1994; Nolting and Marz 1995). A method that we have used extensively is a direct extension of the sine wave method of mode computation (Henry and Shani 1991). In simple geometries, a three-dimensional computation is possible. For complex structures of large lateral size, calculations are restricted to two dimensions using the effective index approximation to obtain reasonable computation times. The waveguide cross sections are simulated by vertical slabs having an effective delta that approximates the lateral behavior of the actual waveguides — e.g., that correctly simulates the mutual coupling.

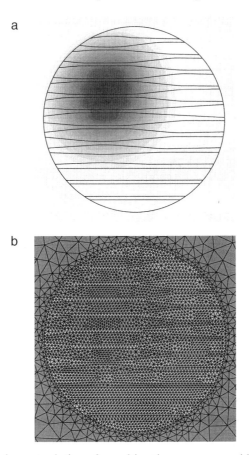

Fig. 8.4 Finite element solution of a multisection core waveguide in the shape of the AT&T logo. (a) Waveguide structure and fundamental mode. (b) Finite element mesh. (After R. K. Smith, personal communication, 1995.)

Mode calculations and propagation are used to determine the detailed behavior of specific components — e.g., the wavelength dependence of a side-by-side waveguide coupler. It is also important to simulate the transmission spectrum of larger structures such as a Fourier filter composed of a series of waveguide couplers separated by Mach–Zehnder (MZ) delaying arms (see Section VI.B). For larger structures, the spectral performance is found by making an approximate analytical description of the single-mode transmission amplitudes between the ports of each component and then calculating the combined transmission of all elements.

D. MASK LAYOUT

As discussed in Section II.C, the geometry of waveguide cores is patterned by the photo-lithographic mask through the photolithography and chemical etching processes. Mask layout connects the optical design to the device fabrication by generating the geometric patterns required by the waveguide device in a format suitable for mask fabrication. Some tools and methods for mask layout are adapted from the electronic integrated circuit industry. However, because of the different propagation characteristics of lightwave and electric current, waveguide mask layout has several unique aspects not present in electronic integrated circuit designs.

Unlike the electric current in a conductor, the lightwave in a waveguide cannot make a sharp bend unless a mirror is used. This requires that along a waveguide, the width, angle, and curvature should be continuous (i.e., smooth connection), and the bend radius R larger than a minimum value. The minimum bend radius is limited by bend loss and mode conversion (for multimode waveguides) and is determined by the delta and dimension of the waveguide. For example, $R \gtrsim 10$ mm can be used for $\Delta = 0.6\%$.

Arcs, raised cosine curves $y = h \sin^2(\pi x/2l)$, and power-law curves $y = h(x/l)^p$ are often used for smooth bending of a waveguide. A Bezier curve (Farin 1988), which is a parametric polynomial curve of the form $x = \sum_{i=0}^{5} a_i t^i$ and $y = \sum_{i=0}^{5} b_i t^i$, can be used to make a smooth connection between two waveguide ends. The waveguide ends are specified by their coordinates, tangent angle, and curvature. The fifth-order Bezier connection is continuous up to the second derivative. The continuity of the curvature is often relaxed, and the resultant insertion loss can be partially compensated for by a relative transverse shift of the waveguides at the connection.

Many optical devices, such as the Fourier filter and the WGR, to be discussed later, work by multipath interference of the lightwave. The differences among the lengths of the waveguides forming these paths determine the interference. In mask layout, the waveguides (which can be ~2 cm long) must be rendered with great precision so that these path differences are accurate to a fraction of the optical wavelength ($\lambda/20 \sim 0.05 \ \mu$m). To achieve this, the waveguides are usually described by mathematical curves with adjustable parameters that are calculated iteratively on a computer.

Although the waveguide patterns can be described with great accuracy both analytically and numerically, mask fabrication instruments have finite accuracy. Several factors determine the minimum accuracy required. One factor is the minimum feature size of the waveguide patterns, such as waveguide width and gap size in a coupler. A more subtle aspect is the effect of the width irregularities in the form of uniform change or fluctuations. These irregularities result in waveguide loss, split-ratio error in a Y branch, coupling ratio error in a coupler, and phase error in a multipath interference device.

IV. Transmission Loss

A. COUPLING LOSS

The fiber–waveguide coupling efficiency is given approximately by the overlap of the normalized field distribution function of the fiber (Φ_f) and that of the waveguide (Φ_w):

$$T = |\iint dx \, dy \, \Phi_f(x,y)\Phi_w(x,y)|^2. \tag{8.5}$$

This follows from the continuity of the field at the interface, mode orthogonality, and the approximation that the reflected field can be neglected. In engineering, the term *coupling loss* refers to the value of T in decibels — i.e., -10 (dB) $\log_{10}T$.

The coupling loss results from mode mismatch between the waveguide and the fiber and from misalignment of the fiber relative to the waveguide. The former is due to the differences in the delta and size between the waveguide and the fiber cores. The most commonly used single-mode fibers for 1.3- and 1.55-μm wavelength communications, such as the AT&T/Lucent 5D, have a delta of about 0.4% and a core diameter of about 8 μm. However, the delta of planar waveguides is preferably higher — e.g., 0.6% for a P-doped silica waveguide — because there are limitations in practical fabrication with a lower delta, and a higher delta permits a higher integration density. As a result of the higher delta, the core size and the size of the guided optical mode in a typical single-mode waveguide are also smaller than those in a commonly used fiber.

For weakly guiding single-mode waveguides and fibers, the optical fields can be approximated by Gaussian beams. The coupling loss in this case can be calculated in simple closed forms, and is representative for the general dependence on mismatch and misalignment parameters.

For a waveguide and a fiber with (intensity $1/e^2$) mode radii a_w and a_f, respectively, the mode mismatch loss is given by

$$T_m = \left(\frac{2a_f a_w}{a_f^2 + a_w^2} \right)^2. \tag{8.6}$$

For transverse offset d,

$$T_t = T_m e^{-(d/a_0)^2}, \tag{8.7}$$

where $a_0 = \sqrt{(a_w^2 + a_f^2)/2}$. The dependence is Gaussian and the characteristic length for transverse misalignment is comparable to the average mode radius. For angular tilt θ,

$$T_a = T_m e^{-(\theta/\theta_0)^2}, \tag{8.8}$$

where the characteristic angle is $\theta_0 = a_0 \lambda/\pi n_0 a_w a_f \approx \lambda/\pi n_0 a_0$, which is comparable to the half far-field angle, and n_0 is a refractive index of the material between the fiber and the waveguide. The dependence of coupling loss on tilt is also Gaussian. For a longitudinal gap $z \gg a_0$,

$$T_g \approx \frac{1}{1 + (z/z_0)^2}, \tag{8.9}$$

where $z_0 = 2\pi n_0 a_f a_w/\lambda \approx 2\pi n_0 a_0^2/\lambda = 2a_0/\theta_0$. The dependence is $1/z^2$ and usually $z_0 \gg a_0$. Therefore, longitudinal misalignment is more tolerable than transverse or angular misalignment. Finally, note that a large mode size reduces the coupling loss due to transverse and longitudinal misalignment, but increases the loss due to angular misalignment.

Most practical waveguides have a rectangular cross section, and the coupling efficiency is usually calculated numerically for more accurate modeling. As an illustration, Fig. 8.5 is a contour plot of the calculated fiber–waveguide–fiber (two interfaces) mode field mismatch loss of a P-doped silica waveguide ($\Delta \approx 0.6\%$) and an AT&T/Lucent 5D single-mode fiber at $\lambda = 1.3$ μm.

The two-interface coupling loss of an $H = 5$ μm, $W = 5$ μm single waveguide is about 0.8 dB, in agreement with experiments. Figure 8.5 shows that the two-interface coupling loss is less than 0.25 dB if the waveguide end is expanded to $H = 7$ μm and $W = 9$ μm without diluting the delta. This loss value is satisfactory for most applications. The end waveguide is

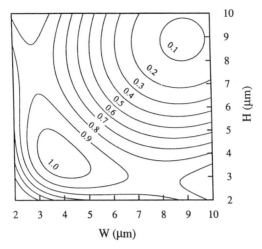

Fig. 8.5 Contour plot of calculated fiber–waveguide–fiber (two interfaces) mode field mismatch loss (in decibels) at $\lambda = 1.3$ μm. H and W are the height and width of the core, respectively. The waveguide has $\Delta \approx 0.63\%$, and the fiber is single-mode AT&T/Lucent 5D.

multimode for $H = 7$ μm and $W = 9$ μm, and the light can couple to the higher order modes in the presence of misalignment. However, the higher order modes are radiated if the expanded waveguide is adiabatically tapered to a single-mode waveguide (with $W = 4$ μm).

At Bell Laboratories, we use a mode taper of $H = 6.7$ μm and $W = 9$ μm with our P-doped ($\Delta \approx 0.6\%$) waveguides and have achieved 0.3-dB fiber–waveguide–fiber insertion loss for a 6-cm-long straight waveguide. This is comparable to the insertion loss achievable with low-delta (e.g., 0.4 %) waveguides.

B. PROPAGATION LOSS

Even the best optical waveguides have propagation losses that are many orders of magnitude greater than losses of optical fibers, despite the fact that the silica-based waveguides are similar in composition and cross section to optical fibers. One reason for this is that the glass in the fiber is much better flowed during the drawing at about 2000°C than the glass in the waveguide. Another reason is that drawing decreases the area and increases axial dimensions by more than four orders of magnitude, demagnifying any inhomogeneities existing in the preform and stretching them out so that they can be adiabatically followed by the mode. This remarkable smoothing process is not available in waveguide fabrication, where low loss is achieved by homogeneous depositions, by densification and flowing, and by patterning and etching the waveguide as smoothly as possible.

Because OICs are of order 10 cm in length and because it is worthwhile to remove every 0.1 dB of excess insertion loss, propagation loss should be reduced to about 1 dB/m, if possible. Losses in this range require special methods of measurement, because the attenuation length is much longer than the wafer size. Two successful methods have been devised.

Adar, Serbin, and Mizrahi (1994) observed the dips in transmission occurring when light of a narrow-linewidth laser is passed through a waveguide coupled to a ring resonator and the laser wavelength is tuned through several resonances. This method was used to evaluate the waveguides made at Bell Laboratories by CVD, which had P-doped cores with $\Delta = 0.70\%$. These researchers found losses of 4.72, 1.22, and 0.85 dB/m for radii of 10, 20, and 30 mm, respectively, at a wavelength of 1.55 μm.

Hida et al. (1995) of NTT used transmission in a 10-m-long waveguide with 15-mm radius bends to evaluate their waveguides made by FHD, which had $\Delta = 0.45\%$. They observed a loss of 1.7 dB/m at both 1.3 and 1.55 μm. Earlier, Hibino et al. (1993) of the same group observed a loss of about 3 dB/m for $\Delta = 0.75\%$. Suzuki and Kawachi (1994) summarized the available data for waveguide loss. High-delta waveguides have been demonstrated with Ge doping of the core by FHD (Suzuki et al. 1994). Unfortunately, waveguide loss is observed to increase with increasing Δ and, of course, coupling loss to a standard fiber also increases. The lowest loss reported for high-delta waveguides ($\Delta = 2\%$) is 7 dB/m, measured by Suzuki et al. (1994) of NTT. Suzuki and Kawachi (1994) associated the increased loss of high-delta waveguides with sidewall roughness.

C. BEND RADIUS AND COMPONENT SIZE

Unlike electronic integrated circuits, OICs are compact in only one dimension. The long length of OIC components is a result of the large bend radii required for low-loss transmission. The previously mentioned measurements of Adar, Serbin, and Mizrahi (1994) indicate that for medium-delta waveguides, bend radii of 10 mm or greater are required. For high-delta waveguides ($\Delta \gtrsim 1.2\%$), it is expected that bend radii will drop to 2–3 mm. Because waveguide bends are nearly parabolic in shape, the waveguide length will decrease approximately as the square root of the radius. This is still a significant improvement. For example, consider a 100-Å free spectral range MZ interferometer. For radii of 3 and 14 mm, it will be about 9 and 22 mm long, respectively. At radii of 2–3 mm, the turning diameter becomes compatible with a reasonable chip width. Thus, high-delta waveguides both shrink component size and make the folding of optical paths on relatively small chips possible.

V. Couplers and Splitters

A. SIDE-BY-SIDE COUPLING

The waveguide coupler is one of the basic elements in integrated optics. It splits lightwaves coherently in a manner similar to a beam splitter in bulk optics. A waveguide coupler has two (or more) closely separated side-by-side waveguides. The evanescent tail of the lightwave in one waveguide extends to a neighboring waveguide and induces an electric polarization. The polarization generates a lightwave in the second waveguide, which also couples back to the first waveguide. This coupling phenomenon is extensively studied using the coupled mode theories that can be found in the literature (see, for example, Marcuse 1982, 1991; Snyder and Love, 1983). In this section, we first discuss two special cases that are not only important in practical applications, but also illustrative of the basic physical principles involved. Then we summarize the principles governing more general couplers.

1. Two-by-Two Coupler with Identical Waveguides

In the 2×2 coupler with identical waveguides, the two single-mode waveguides are identical and parallel in the coupling region, and they bend away (and decouple) from each other gradually at both ends. The two coupled waveguides support a symmetrical and an asymmetrical guided supermode, represented by ψ_s and ψ_a, respectively. At the input, light launched into either of the two nearly uncoupled waveguides, represented by ψ_{1i} and ψ_{2i}, respectively, excites ψ_s and ψ_a with equal amplitude. That is,

$$\begin{bmatrix} A_s \\ A_a \end{bmatrix} = \frac{1}{\sqrt{2}} \begin{bmatrix} 1 & 1 \\ 1 & -1 \end{bmatrix} \begin{bmatrix} A_{1i} \\ A_{2i} \end{bmatrix}, \tag{8.10}$$

where the As are the amplitudes of the normalized mode fields. This is referred to as *mode projection* at a waveguide–coupler junction. In the input bend region, because of the symmetrical geometry, there is no mode conversion between ψ_s and ψ_a, and, provided the bends are gradual enough, the amplitudes of A_s and A_a remain the same and no higher order supermode is excited. In the coupling region, ψ_s and ψ_a propagate at different velocities, and after a distance l_{par} along the parallel waveguides, have a phase difference:

$$2\phi_{par} = 2\pi\frac{l_{par}(n_s - n_a)}{\lambda} = \pi\frac{l_{par}}{L_c}, \tag{8.11}$$

where n_s and n_a are the effective indices of the two supermodes, and $L_c \equiv \lambda/2(n_s - n_a)$ is the coupling length. The bend regions at the input and output also contribute a phase difference $2\phi_{ends}$, which is given later, and the total phase difference at the output is $2\phi = 2(\phi_{par} + \phi_{ends})$. At the output, A_s and A_a are related to the lightwave amplitudes in the two uncoupled waveguides through mode projection (ignoring a common phase factor):

$$\begin{bmatrix} A_{1o} \\ A_{2o} \end{bmatrix} = \frac{1}{\sqrt{2}}\begin{bmatrix} 1 & 1 \\ 1 & -1 \end{bmatrix}\begin{bmatrix} A_s e^{i\phi} \\ A_a e^{-i\phi} \end{bmatrix}. \tag{8.12}$$

Therefore,

$$\begin{bmatrix} A_{1o} \\ A_{2o} \end{bmatrix} = T_\phi\begin{bmatrix} A_{1i} \\ A_{2i} \end{bmatrix}, \quad \text{and} \quad T_\phi = \begin{bmatrix} \cos\phi & i\sin\phi \\ i\sin\phi & \cos\phi \end{bmatrix}, \tag{8.13}$$

where T_ϕ is the transfer matrix of the coupler.

The coupling length can be calculated by solving the eigenmodes of the coupled waveguides. It has an approximately exponential dependence on the separation d of the two waveguides:

$$L_c \approx L_0 e^{d/D_0}, \tag{8.14}$$

where L_0 and D_0 are phenomenological constants. Assuming that the waveguides bend away with a radius R at the input and output regions, the phase contribution from both ends is

$$2\phi_{ends} \approx 2\int_0^\infty \frac{\pi}{L_c(z)}dz \approx \int_0^\infty \frac{2\pi}{L_c e^{z^2/RD_0}}dz = \frac{\pi}{L_c}\sqrt{\pi D_0 R}, \tag{8.15}$$

i.e., the bend regions have a coupling contribution equivalent to an extra length of $l_{ends} \approx \sqrt{\pi D_0 R}$ in the parallel coupling region. In a full coupler, the total coupler length ($l = l_{par} + l_{ends}$) equals odd multiples of the coupling length, and light transfers completely from one waveguide to the other. In a 3-dB coupler, the total coupler length equals half multiples of the coupling length, and the optical power in one waveguide splits equally into both waveguides.

The coupling length L_c and coupler ends contribution l_{ends} can be obtained experimentally by measuring the fraction of crossed power in two couplers of different l_{par}. Figure 8.6 shows wavelength and polarization dependence of L_c and l_{ends}, along with the bar-state and cross-state transmis-

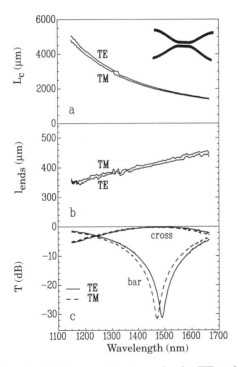

Fig. 8.6 Properties of a 2 × 2 symmetrical coupler for TE and TM polarizations. The waveguide has $H \approx 5$ μm, $W \approx 5$ μm, and $\Delta \approx 0.63\%$. The center-to-center separation of the two waveguides in the parallel region is 9.5 μm, and the bend radius is 15 mm. (a) Coupling length (*inset*, geometric layout). (b) Contribution of the coupler ends. (c) Bar- and cross-state transmission of a coupler with the length of the parallel waveguides $l_{par} \approx 1000$ μm.

sion of a typical coupler. The coupling length is larger at small wavelengths because the waveguide confinement is stronger. The wavelength-dependent nature of the coupling makes it possible to use couplers as WDMs. The minimum in the bar-state transmission corresponds to the wavelength at which the coupling length equals the total coupler length. The minimum transmission (-20 to -30 dB) is limited by irregularities of the waveguides and excitations of high-order modes in the bend regions. The strain in the silica layers results in a more confined TE mode than the TM mode; therefore, the coupling properties are polarization dependent (see Sections IX.A and IX.B for details).

2.   Two-by-Two Adiabatic 3-dB and Full Couplers

In an adiabatic coupler, the two waveguides are not identical and can change in width and separation along the propagation direction. Unlike the symmetrical coupler discussed previously, symmetry no longer exists to prevent mode conversion between the supermodes of the coupled waveguides. However, if the change in the two waveguides is gradual enough, mode conversion is effectively eliminated (see also Section III.B).

   An approximate formulation of the condition for adiabatic change was found by Landau and Zener in quantum mechanics (Landau and Lifshitz 1958). They treated an idealized time-dependent crossing of two energy levels. Their problem is equivalent to propagation in an adiabatic full coupler (Fig. 8.3a), which has constant mutual coupling between the waveguides. In the absence of mutual coupling, the waveguide's effective indices in this model change linearly with axial position and cross. The mutual coupling removes the crossing of the supermodes, shown in Fig. 8.7. The cross-talk power (transitions between supermodes) associated with a nonadiabatic transition is

$$\text{cross talk} = \exp\left(-\frac{2\pi\delta^2}{d\,|\,k_1 - k_2\,|\,/dz}\right) = \exp\left(-\frac{\pi^2 L}{4L_c}\right), \quad (8.16)$$

where $2\delta = \pi/L_c$ is the separation of supermode propagation constants at the crossing, L_c is the coupling length, k_1 and k_2 are the propagation constants of the uncoupled waveguides, and L is the characteristic length of the crossing, defined in Fig. 8.7. When the exponent becomes large, the coupler is adiabatic and the cross talk becomes negligible. This occurs when the splitting is large and the crossing is gradual, which is achieved when L is more than $2L_c$ or $3L_c$.

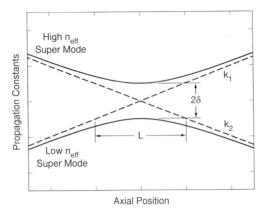

Fig. 8.7 Idealized crossing on two coupled waveguides. The propagation constants of the waveguides in the absence of coupling are k_1 and k_2, 2δ is the splitting of supermode propagation constants at the crossing due to mutual coupling, and L is the characteristic length of the crossing.

The length of this idealized device is infinite. In practice, it must be large compared with L_c to approximate the previous description. Thus, the coupler must be many coupling lengths long to effect an adiabatic crossing. However, apart from this requirement, an adiabatic coupler has no critical dimensions. (For example, a symmetrical full coupler must have a length equal to the coupling length, whereas an adiabatic coupler has no such specified length.) Furthermore, the transmission characteristics of adiabatic couplers are insensitive to wavelength, polarization, and waveguide irregularities.

In an adiabatic full coupler (Fig. 8.3a), if light is launched in the input waveguide with the larger effective index, the fundamental mode is initially excited. The light remains in the fundamental mode throughout the device because of the adiabatic condition, and thus exits in the output waveguide with the larger effective index. Similarly, light launched in the narrower input waveguide excites and remains in the first-order mode, and exits in the narrower output waveguide. An adiabatic 3-dB coupler (Fig. 8.3b), consists of half an adiabatic full coupler followed by a region in which the two waveguides are identical and bend away from each other symmetrically. Either the fundamental or the first-order supermode, corresponding to the wider or narrower input waveguide, respectively, excites an equal amount of light in the two symmetrical output waveguides.

The transmission of an adiabatic 3-db coupler of Adar, Henry, Kistler, *et al.* (1992) is shown in Fig. 8.3c. At the shorter wavelengths, the coupling between guides is insufficient to separate adequately the values of $\bar{n}$, and there is a departure from adiabatic behavior and equal splitting of the optical power.

3. General Coupling and Propagation

The symmetrical coupler and the adiabatic coupler are two special cases of couplers. In general, a coupler can be neither symmetrical nor adiabatic, such as the wavelength-insensitive coupler (WINC) reported by NTT (Takagi, Jinguji, and Kawachi 1992), in which the wavelength dependence of transmission is reduced by changing the waveguide widths and gap nonadiabatically. There is currently no simple and general method for visualizing the physical process and predicting the transmission properties of such general couplers. They are usually modeled by beam propagation or coupled mode theory. Furthermore, couplers are not limited to two waveguides, but multiwaveguide couplers have not yet found significant applications and have not been studied extensively. An exception is the case of a periodic structure of coupled identical waveguides, which is the basis of star couplers (discussed in Section VII.A). Furthermore, multimode interference devices (Oddvar and Sudbo 1995) can be regarded as couplers in which the coupled waveguide section is replaced by a multimode waveguide. The basic principles introduced in the previous discussion are general and applicable to other waveguide devices. These principles include mode projection at a waveguide junction, propagation of supermodes, mode conversion forbidden by symmetry and by the adiabatic condition.

An additional general property of propagation, in the absence of magnetic fields (and Faraday rotation), is reciprocity. The power transmission between specific modes at two ports does not depend on the direction of transmission. Furthermore, the phase change of such transmission is independent of the direction of propagation. These results hold even in the presence of attenuation. Reciprocity can generally be established by applying Green's theorem to a pair of fields obeying the scalar wave equation, whose sources are the modes at the two ports. Reciprocity is obvious for a component with axial inversion symmetry, such as the symmetrical coupler (see Section V.A.1). It is not obvious for devices of lower symmetry, such as the adiabatic 3-dB coupler shown in Fig. 8.3b or a Y branch, which is discussed next.

B. Y-BRANCH SPLITTER

A Y branch consists of a single waveguide splitting symmetrically into two waveguides and is commonly used as a 1×2 power splitter. Cascading N stages of Y branches gives a 1×2^N power splitter (Hibino *et al.* 1995). The geometric symmetry prevents conversion of the fundamental mode into any asymmetrical mode. Ideally, the waveguide-splitting transition is adiabatic, which also prevents conversion of the fundamental mode into higher order symmetrical modes. These two conditions ensure an equal power split with little excess loss. Compared with a symmetrical 3-dB coupler, a Y branch is usually shorter in dimension and the equal power split is much less wavelength and polarization dependent.

In practical fabrication, however, the minimum width of the gap between two waveguides is limited by the resolution of the photolithography and chemical etching processes. The sudden opening of the gap violates the adiabatic condition, resulting in an extra insertion loss. Some fabrication processes, such as interdiffusion of the core and cladding materials, effectively smear the abrupt gap opening and reduce the insertion loss, but they are not available for all fabrication systems. Hibino *et al.* (1995) reported 0.13-dB extra insertion loss per Y branch fabricated with low-delta (0.3%) waveguides. Unfortunately, the loss associated with the sudden opening of the gap increases with delta. At Bell Laboratories, an excess loss of 0.2 dB per Y branch has been achieved with $\Delta = 0.6\%$ waveguides.

VI. Mach–Zehnder and Fourier Filter Multiplexers

A. MACH–ZEHNDER INTERFEROMETER

The MZ interferometer is a basic waveguide interference device. It consists of two couplers (or Y branches) connected by two waveguides of different length (see Fig. 8.8). We refer to these two waveguides as a *differential delay* because the common phase delay does not contribute to the interference.

The transfer matrix of an MZ interferometer is given by

$$T_{MZ} = T_{\phi_2} T_{\theta} T_{\phi_1}, \qquad (8.17)$$

where T_{ϕ_1} and T_{ϕ_2} are given by Eq. (8.13), and

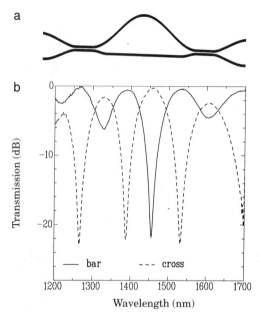

Fig. 8.8 Mach–Zehnder interferometer. (a) Geometric layout expanded in the vertical dimension. (b) Bar- and cross-state transmission spectra. The length difference of the two delay waveguides is 10 μm. The couplers are 1000 μm long in the parallel regions and are 3 dB at a wavelength of 1.46 μm.

$$T_\theta = \begin{bmatrix} e^{i\theta} & 0 \\ 0 & e^{-i\theta} \end{bmatrix} \tag{8.18}$$

is the transfer matrix of the differential delay, where $\theta = \pi \bar{n} s / \lambda$, and s is the length difference of the two delay waveguides. The cross-state field transmission through the MZ interferometer is the sum of contributions associated with two paths for the lightwave, as shown in Fig. 8.9b. For 2×2 devices, the bar- and cross-transmission states correspond to the input and output ports being on the same or different waveguides, respectively.

Figure 8.8b shows the transmission spectra of a typical MZ interferometer. Because θ is proportional to the optical frequency ν, the transfer function is ideally sinusoidal in ν. However, a null in the bar transmission requires both couplers to be 3 dB (i.e., $\phi_1 = \phi_2 = \pi/4$), whereas a null in the cross transmission requires only $\phi_1 = \phi_2$. In practical devices, the coupling length is wavelength dependent. As a result, good bar nulls are observed only at wavelengths for which the couplers are close to 3 dB, whereas the cross nulls are observed in a much broader wavelength range. The wavelength-dependent property of the transfer function makes MZ interferometers suitable as simple optical filters and WDMs.

B. FOURIER FILTER PRINCIPLE AND APPLICATIONS

The Fourier filter (Li, Henry, Laskowski, Mak, *et al.* 1995; Li, Henry, Laskowski, Yaffe, *et al.* 1995; Li *et al.* 1996) is a generalization of the MZ interferometer. It consists of a chain of N ($N > 2$) optical couplers of different coupling ratios linked by $N - 1$ differential delays of different lengths. If the differential delays are multiples of a fundamental delay, the transfer function is periodic. Fourier expansion becomes a natural method for analyzing and synthesizing such periodic optical filters. Compared with other types of multisection filters, such as optical lattice filters (Moslehi *et al.* 1984; Kawachi and Jinguji 1994) and resonant couplers (Yaffe, Henry, Serbin, *et al.* 1994), the Fourier filter is directly based on Fourier expansion and is more general in structure.

1. Principle of the Sum of All Optical Paths

The transfer matrices of a filter with couplers of $\phi_1, \phi_2, \ldots, \phi_N$ and differential delays of $\theta_1, \theta_2, \ldots, \theta_{N-1}$ are

$$T_{\phi\theta} = T_{\phi_N} T_{\theta_{N-1}} \ldots T_{\theta_2} T_{\phi_2} T_{\theta_1} T_{\phi_1}. \tag{8.19}$$

From Eq. (8.19), we see that the transfer function from any input port to any output port consists of a sum of the form

$$t_{\phi\theta} = \sum f(\phi_1, \phi_2, \ldots, \phi_N) e^{i(\pm\theta_1 \pm \theta_2 \cdots \theta_{N-1})}. \tag{8.20}$$

By examining transfer functions, we obtain the *principle of the sum of all optical paths* of a chain of N couplers and $N - 1$ differential delays: the transfer function from any input port to any output port consists of the unweighted sum of contributions of all (2^{N-1}) distinct optical paths. The contribution of each path is a product of $2N - 1$ factors: traversing a coupler without crossing gives $\cos\phi$ and with crossing $i\sin\phi$; traversing the longer arm of a differential delay gives $e^{i\theta}$ and the shorter arm $e^{-i\theta}$.

This principle is illustrated with the MZ interferometer and the simple examples in Fig. 8.9. A negative θ corresponds to an interchange of the longer and shorter delay arm. The MZ is a special case of $N = 2$ having two optical paths (Fig. 8.9b). For $N = 3$ (Fig. 8.9c), there are four distinct optical paths from any input port to any output port, and the transfer function is a sum of four terms.

2. Filter Synthesis Using Fourier Expansion

If the θs have integral ratios, the sum in Eq. (8.20) becomes a truncated Fourier series in optical frequency. To synthesize a filter, the coupler lengths and differential delays are optimized so that this Fourier series best approximates the desired filter response. The optimization is usually done numerically. The wavelength dependence of the couplers is included for broadband filters. The differential delays are not limited to equal lengths with the same sign as in lattice filters (Moslehi *et al.* 1984; Kawachi and Jinguji 1994) and can be negative.

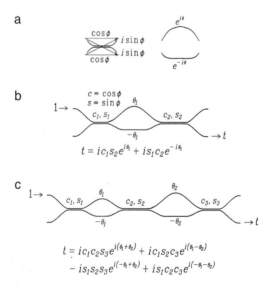

Fig. 8.9 Illustration of the principle of the sum of all possible optical paths for (a) a coupler and a differential delay, and for filters with (b) two and (c) three couplers. See text for details.

3. Applications

The Fourier filter principle discussed in the previous two sections is applicable to optical waveguide filters with arbitrary amplitude or phase response. In this section, we demonstrate a 1.31/1.55-μm WDM with a rectangular response (Li, Henry, Laskowski, Yaffe, *et al.* 1995) and an erbium-doped fiber amplifier (EDFA) gain equalization filter (Li, Henry, Laskowski, Mak, *et al.* 1995). Other devices, such as multichannel WDMs and wavelength-independent splitters, can also be made using concatenated Fourier filters. These devices are fabricated with doped silica planar waveguides.

Figure 8.10 shows the schematic layout and measured spectral response of our 1.31/1.55-μm WDM. Three stages of Fourier filters are cascaded to

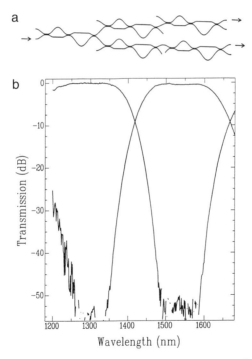

Fig. 8.10 1.3/1.55-μm WDM with a nearly rectangular passband and low cross talk. (a) Layout. *Arrows* represent the input and output ports. The total length of the device is 60 mm, and the width is 0.6 mm. The vertical scale has been expanded 20 times for clarity. (b) Measured spectral response. (After Li, Henry, Laskowski, Yaffe, and Sweatt 1995).

reduce the cross talk by triple filtering. The WDM has low cross talk of about −50 dB over rejection bands of 100 nm wide centered around both 1.31 and 1.55 μm. We have achieved a total fiber-to-fiber insertion loss of less than 1 dB. Such optical performance was achievable earlier only with hybrid devices employing thin film filters, which significantly complicated the fabrication and increased the cost of the WDM.

The solid curve in Fig. 8.11 shows the schematic layout and the measured spectral response of our EDFA filter. The circles represent the required filter response, optimized for maximum end-to-end flatness of an EDFA system over its full gain bandwidth, from 1.53 to 1.56 μm. The measured spectral response closely follows the design values, except the measured spectrum has an extra insertion loss of about 1.5 dB. The response outside

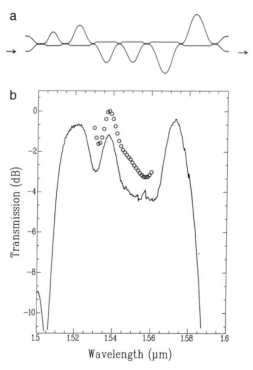

Fig. 8.11 EDFA gain equalization filter. (a) Layout. *Arrows* represent the input and output ports. The total length of the device is 90 mm, and the width is 1.2 mm. The vertical scale has been expanded 20 times for clarity. (b) Designed (*circles*) and measured (*solid curve*) spectral responses. (After Li, Henry, Laskowski, Mak and Yaffe 1995).

the 1.53- to 1.56-μm window is not important. In this case, it is designed to cut off out-of-band noise and for testing purposes.

VII.   Array Waveguide Devices

A.   STAR COUPLER

An $N \times N$ star coupler couples the lightwave from any input to all the outputs without wavelength selectivity. It is widely used as a basic cross-connect element in passive optical networks (Agrawal 1992). Ideally, the optical power from any input splits evenly into all the outputs without power loss, so that each output receives $1/N$ of the input power. Because $1/N$ is the maximum power transmission efficiency of an $M \times N$ coupler with $M < N$, an efficient $N \times N$ coupler can also be used as an efficient $M \times N$ coupler. When N is a power of 2, an efficient $N \times N$ star coupler can be implemented as a network of 2×2 couplers, but for large and arbitrary N the array waveguide star coupler (Dragone *et al.* 1989) is a more compact and wavelength-independent implementation. Figure 8.12 illustrates the schematic layout and measured coupling ratios of a star coupler.

The array waveguide star coupler consists of an array of input waveguides and an array of output waveguides connected by a two-dimensional waveguide slab in a normally center symmetrical geometry (see Fig. 8.12). The waveguides in each array have a constant angular separation and point radially to a focal point. Dummy waveguides are added beyond the input and output waveguides at the center, so that the center waveguides appear periodic. These waveguides are far apart and uncoupled initially, but they become close together and strongly coupled when they approach the slab. The two focal points are separated by the focal distance. The input and output boundaries of the slab are arcs centered at focal points and with radii equal to the focal distance.

The even power split is a result of the properties of a coupled periodic waveguide array (Dragone 1989, 1990). The optical field in such an array is analogous to the electron wave function in a crystal. The supermodes are Bloch functions. Just as with energy levels in a crystal, the supermode effective indices divide into bands. If the isolated waveguides are single mode, the highest effective index band corresponds to the guided supermodes of the waveguide array.

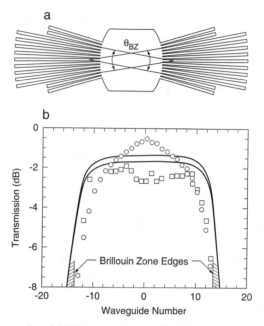

Fig. 8.12 Star coupler. (a) Schematic layout. In this diagram, θ_{BZ} is the Brillouin zone angle. (b) Measured power transmission of a 19×19 device. The *circles* are transmissions from a central input waveguide, and the *squares* are transmissions from the ninth (outermost) waveguide. (After Dragone *et al.* 1989).

Far away from the slab, the array waveguides are weakly coupled. In the weak coupling limit, the Bloch functions are linear combinations of the modes $\Phi(x)$ of the isolated waveguides, with a constant phase difference between neighbors. The band of Bloch functions corresponds to a range of this phase from $-\pi$ to π. The Bloch function has the form of a Fourier series:

$$\psi_{k_x}(x) \sim \sum_m \Phi(x - ma)e^{ik_x ma}, \tag{8.21}$$

where a is the period of the waveguide array, and $-\pi/a < k_x \leq \pi/a$ is the Bloch wave vector in the first Brillouin zone. The inverse Fourier transform,

$$\Phi(x - ma) \sim \int_{-\pi/a}^{\pi/a} \psi_{k_x}(x)e^{-ik_x ma}dk_x, \tag{8.22}$$

implies that light launched into waveguide m excites Bloch supermode k_x with a complex amplitude proportional to $e^{-ik_x ma}$. Therefore, light launched into the fundamental mode of any input waveguide excites all the Bloch modes in the highest index band with equal power. As the array waveguides come closer together adiabatically, the periodic symmetry prevents mode conversion within the highest index band, and the adiabatic condition prevents excitation of modes in other index bands. Thus, the lightwave remains in the Bloch modes of the highest index band with equal amplitudes throughout the input waveguide array.

As the waveguides approach the slab, they gradually merge, and $\psi_{k_x}(x)$ evolves into a circular wave front at an angle $\theta = \arctan(k_x/k_z)$ relative to the arc of the slab, where k_z is the propagation constant of the Bloch supermode. Each of these waves propagates across the slab, coming to focus at the arc at the other side of the slab. Because the Bloch waves are excited with equal amplitudes, the arc subtending the Brillouin zone angle θ_{BZ} is uniformly illuminated. The angle θ_{BZ} is given by

$$\theta_{BZ} \equiv 2 \arcsin(\pi/ka) \approx \frac{\lambda}{n_s a}, \tag{8.23}$$

where $k = \sqrt{k_x^2 + k_z^2} = 2\pi n_s/\lambda$ is the propagation constant of light in the slab, and n_s is the effective refractive index of the slab. The star coupler is designed such that all the output waveguides cover the illuminated region — i.e., $Na/R = \theta_{BZ}$. Ideally, the uniform illumination of the output waveguides at the slab would transfer the light evenly in the output waveguides without power loss as they separate and become uncoupled.

In practice, the waveguides near the slab do not fully merge. This results in a nonuniform illumination of the output waveguides and a fall off of light intensity in the output waveguides near the zone edges. There is also a power loss due to the light illuminating beyond the Brillouin zone edges. The coupling of power from the illuminated to the unilluminated waveguides is unavoidable in an unending periodic array. It can be avoided only in a circularly periodic structure, which is not realizable in planar structures. This nonuniform power split is illustrated in the theoretical curves and measurement results in Fig. 8.12, and can be improved by optimizing the array design (Okamoto, Takahashi, et al. 1992). Star couplers with large splitting ratios — e.g., 144 × 144 with an average excess insertion loss of 2.0 dB and uniformity 1.3 dB — have been demonstrated (Okamoto, Okazaki, et al. 1992).

B. WAVEGUIDE GRATING ROUTER AND APPLICATIONS

The WGR is widely used as an $M \times N$ WDM (Dragone 1990, 1991; Dragone, Edwards, and Kistler 1991; Smit 1989; Vellekoop and Smit 1991; Takahashi et al. 1990). It consists of two star couplers connected by an array of waveguides that act like a grating. The waveguide grating has a constant length difference between adjacent waveguides (Fig. 8.13). The two star couplers are mirror images, except the number of inputs and outputs can be different.

The principle of the WGR can be described as follows. The lightwave from an input waveguide couples into the grating waveguides by the input star coupler. If there were no differential phase shift in the grating region, the lightwave propagation in the output star would appear as if it were the reciprocal propagation in the input star. The input waveguide would thus be imaged at the interface between the output slab and the output waveguides. The imaged input waveguide would be butt coupled to one of the

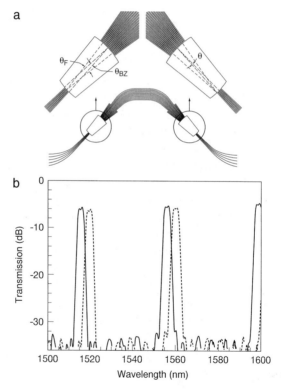

Fig. 8.13 Waveguide grating router (WGR). (a) Schematic layout. The far-field angle of the input/output waveguide is θ_F, θ_{BZ} is the Brillouin zone angle, and θ is the deflection angle of the image of the center input waveguide. (b) Graph of the measured power transmission of a 1×10 device showing the transmission of two adjacent outputs. The flat-top transmission is achieved with a Y-branch input.

output waveguides. The linear length difference in the waveguide grating results in a wavelength-dependent tilt of the wave front of the lightwave in the grating waveguides and thus shifts the input waveguide image to a wavelength-dependent position. As the wavelength changes, the input waveguide image sweeps across and couples light into different output waveguides.

We use this notation in the following discussion: l = length difference between adjacent grating waveguides; R = focal length of the input/output slab; a, a_i, a_o = separation of grating, input, and output waveguides, respectively, at the slab interfaces; $\alpha = a/R$, $\alpha_i = a_i/R$, $\alpha_o = a_o/R$ are the angular separations; and $\bar{n}$ and n_s are the effective refractive indices of the grating waveguides and input/output slabs, respectively.

The array waveguides are excited by a circular wave front at angle θ_i and contribute to a circular wave front at the output at angle θ_o, where both angles deflect downward from the center, when positive. The condition that the output wave front has a common phase is that neighboring paths differ by $2\pi m$:

$$k\bar{n}l + kn_s a(\theta_i + \theta_o) = 2\pi m, \tag{8.24}$$

where m is the integer order of the grating and $k = 2\pi/\lambda = 2\pi\nu/c$.

The WGR transmission spectrum is ideally periodic in optical frequency. The period — i.e., the free spectral range $\Delta\nu_{FSR}$ — corresponds to a change of grating order by 1 — i.e., $\nu_1\bar{n}(\nu_1)l/c + \nu_1 n_s(\nu_1)a(\theta_i + \theta_o)/c = m$, $\nu_2\bar{n}(\nu_2)l/c + \nu_2 n_s(\nu_2)a(\theta_i + \theta_o)/c = m + 1$. Therefore, at the center of the input and output, $\theta_i = \theta_o = 0$,

$$\Delta\nu_{FSR} = \nu_1 - \nu_2 = \frac{c}{n_g l}, \tag{8.25}$$

where $n_g = \bar{n} + \nu(d\bar{n}/d\nu)$ is the group index of the grating waveguides.

The channel spacing at the array center, $\theta_i + \theta_o = 0$, is found by differentiating Eq. (8.24) with respect to θ_o:

$$\Delta\nu = \alpha_o \frac{d\nu}{d\theta_o} = \frac{\alpha_o \nu^2 n_s a}{mc}. \tag{8.26}$$

The WGR output ports "wrap around"; i.e., wavelength channel p goes to the same output port as channel $p + N$, if $\Delta\nu_{FSR}/\Delta\nu = N$. Furthermore, for an $N \times N$ WGR, if $\alpha_i = \alpha_o$, input/output combinations (p,q) and $(p \pm 1, q \mp 1)$ have the same transmission spectrum. If the wraparound condition is also satisfied, $(1,q)$ and $(N, q + 1)$ have the same transmission spectrum, and so do $(p,1)$ and $(p + 1,N)$. Note that whereas the transmission spectrum of each output port is periodic in ν, the channel spacing $\Delta\nu$ is wavelength dependent. As a result, the wraparound condition is wavelength dependent and is normally satisfied for only one $\Delta\nu_{\text{FSR}}$.

The change of the transverse offset between the input waveguide image and the output waveguide as a function of wavelength determines the shape of the passband, which is approximately Gaussian as indicated by Eq. (8.7). A double waveguide input, in the form of a Y branch, with a spacing comparable to the mode size of the output waveguide, can be used to produce a flat-top passband (Dragone 1995). The Y branch introduces about a 3-dB insertion loss. We have also achieved a flattened passband by modification of the waveguide grating (Dragone 1995).

The transfer function of the WGR from input port p to output port q can be obtained by the principle of the sum of all optical paths (each of which corresponds to a waveguide in the grating array; see also Section VI.B.1):

$$
\begin{aligned}
T_{pq} &= \sum_m f_{pm} e^{i2\pi(m\bar{n}l/\lambda)} g_{mq} \approx \\
&\sum_m |f_{pm}g_{mq}| \exp\left(i2\pi m \frac{\bar{n}l + n_s a(p\alpha_i + q\alpha_o)}{\lambda}\right),
\end{aligned}
\tag{8.27}
$$

where f_{pm} and g_{mq} are the coupling coefficients of the input star and output star, respectively, and can be calculated using beam propagation. We have used the paraxial approximation in the last part of Eq. (8.27) for the phase of f_{pm} and g_{mq} — i.e., relative to the center axial waveguide, the phase of f_{pm} is determined by the distance between input waveguide p and array waveguide m in the input slab, and similarly for g_{mq}. The amplitudes of f_{pm} and g_{mq} are approximately given by the far-field pattern of the input and output waveguides, respectively. Thus, $T_{pq}(\nu)$ is approximately a Fourier transform of $|f_{pm}g_{mq}|$, and a change in (p,q) mainly shifts the spectrum as discussed previously.

To design a WGR, the free spectral range is first determined from the channel spacing and the number of channels. Then l is determined from the free spectral range using Eq. (8.25). The passband width to channel spacing ratio is used to determine the output waveguide spacing relative to the waveguide width. To ensure low adjacent channel cross talk, the waveguide spacing is usually large enough, so that the output waveguides are nearly uncoupled, which is different from the arrangement in a regular star coupler for uniform power division. The slab radius is chosen so that all the output waveguides fit in the first Brillouin zone — i.e.,

$(a_o/R)(\Delta\nu_{FSR}/\Delta\nu) = \theta_{BZ}$, where θ_{BZ} is given by Eq. (8.23). The number of grating wave-guides is chosen so that they cover the far-field distribution of the input and output waveguides. The number of grating waveguides should also be about $3N$ or more to ensure an efficient and clean imaging of the input waveguides in the output slab. The grating waveguides are closely spaced at the slab interface to minimize insertion loss. The device transmission can then be simulated using Eq. (8.27).

In practice, several factors degrade the device's performance. First, the fluctuations in effective index of the grating waveguides cause random phase errors in the grating and degrade the cross talk. These phase errors have been measured using incoherent-light interference techniques (Takada *et al.* 1994). A simulation based on Eq. (8.27) shows that for the cross talk to be less than -25 dB, the random phase errors must be less than $10°$ root mean square (rms), which corresponds to less than about $\frac{1}{40}$ μm in path length. Second, nonideal effects in the star couplers result in insertion loss and nonuniform power transmission in the output ports, as discussed in Section VII.A. Currently, a WGR made with $\Delta \sim 0.6\%$ waveguides has an insertion loss of 2–3 dB from a center input port and a center output port, and 4–6 dB from an edge input port to an edge output port. Flattening the passband by using a Y-branch input or by modifying the array waveguide transmission results in an extra loss of 2–3 dB. Third, although the accuracy of channel spacing is relatively easy to control, that of the absolute channel position and free spectral range is more difficult. To meet the requirement of accurate channel position, we have used vernier inputs with $\alpha_i = [(n \pm 1)/n]\alpha_o$ (where, e.g., $n = 10$). This enables shifting the channel position by $(1/n)\Delta\nu$ by postmanufacturing selection of the input port (Dragone 1995). Furthermore, stress in the waveguide structure causes birefringence, which manifests as a relative shift of the transmission spectrum of the two polarizations (2–3 Å for silica-on-silicon waveguides). A detailed discussion on the birefringence compensation can be found in Section IX.A.

The WGR is considered the most promising integrated WDM for multiwavelength-channel networks (Glance, Kaminow, and Wilson 1994; Agrawal 1992). Its advantages, including compactness and accurate channel spacing, are more prominent for dense WDM applications with a large number of channels. For broadband WDMs and a small number of channels, Fourier filter-based devices (see Section VI.B) are more practical, but broadband WGRs (Adar *et al.* 1993) have also been demonstrated. Much work has been devoted to improving the WGR performance (see, for example, Takahashi *et al.* 1995). Figure 8.13 shows the transmission spectrum of a 1×10 WGR with a flattened passband, channel spacing of 40 Å, cross talk less than -30 dB, and an insertion loss of less than 7 dB (not including fiber–waveguide coupling loss). A 64×64 router with cross talk less than -27 dB has been demonstrated (Okamoto, Moriwaki, and Suzuki 1995). WGRs have also been integrated with thermooptic switches on a single chip to make reconfigurable add–drop filters (Okamoto, Takiguchi, and Ohmori 1995) and wavelength routers (Ishida *et al.* 1994).

VIII. Bragg Reflection

Bragg reflectors couple forward and backward propagating modes. The difference of wave vectors is provided by a periodic refractive index change in the form of a finely corrugated grating running along the waveguide. A good derivation of the formulas for the reflection and transmission of Bragg reflectors is given by McCall and Platzman (1985). If a Bragg reflector is sufficiently long, it reflects all the power within the Bragg reflection band. This is a distinct advantage over the side-by-side couplers discussed previously, which transfer power completely only if the coupler is one coupling length long. A problem associated with Bragg reflection filters is that they couple the fundamental mode of a single-mode waveguide into backward propagating cladding modes, which results in undesired loss on the short-wavelength side of the Bragg reflection band; see Section VIII.B.

Traditionally, the grating was made by exposure of photoresist and etching a periodic pattern into the core. First-order Bragg reflection requires a period of about 0.5 μm. At Bell Laboratories, this was done by deep UV photolithography (Henry *et al.* 1989). In addition to simple Bragg reflectors, this method was used to form narrow band resonant optical reflectors, made by coupling a waveguide to a resonant cavity formed with Bragg reflectors (Olsson *et al.* 1987) and multiplexers made from Bragg reflectors with elliptical shaped lines (Henry *et al.* 1990).

A. UV-INDUCED BRAGG GRATINGS

Much work has been done since the late 1980s in forming Bragg reflectors in optical fibers by deep UV-induced changes in refractive index (Meltz, Morey, and Glenn 1989). The magnitude of the refractive index change is greatly enhanced by loading the optical fiber with hydrogen prior to UV exposure (Lemaire *et al.* 1993). This method has been used to demonstrate very high-quality Bragg reflection filters in optical fibers. Bragg reflectors formed by UV-induced Bragg gratings can also be made in silica-based optical waveguides by the same methods. This has been demonstrated in both Ge-doped waveguides (Bilodeau *et al.* 1993; Kashyap, Maxwell, and Ainslie 1993; Strasser *et al.* 1994) and P-doped waveguides (Erdogan *et al.* 1996). The UV-induced gratings have several advantages over etched gratings. The UV-induced grating is formed throughout the core, which reduces the coupling to cladding modes and the associated transmission loss on the short-wavelength side of the Bragg reflection band. The grating corrugations are very smooth compared with those of etched gratings, which reduces loss. The grating may be fabricated in finished waveguides, which should increase the yield and makes possible the monitoring of the Bragg reflection as the grating is being made. The gratings are stabilized by annealing after fabrication, which removes the short-lived UV-induced changes. The annealing slightly alters the position and width of the Bragg reflection band, and these effects have to be taken into account when the grating is made.

B. BRAGG REFLECTING ADD–DROP FILTERS

An important potential application of integrated optic Bragg reflection is the fabrication of narrow band add–drop filters (Kashyp, Maxwell, and Ainslie 1993). The method is illustrated in Fig. 8.14a (Erdogan *et al.* 1996). The waveguides form an MZ geometry with two 3-dB couplers separated by equal-length arms. Identical Bragg reflectors are made in each arm. The initial traffic enters port 1, carrying several wavelengths. The dropped wavelength is Bragg reflected and leaves through port 2. The remaining wavelengths pass through the arms and leave through port 4. An added wavelength enters port 3, is Bragg reflected and leaves through port 4. The advantage of this arrangement is that, ideally, no energy is lost. However, this requires making identical Bragg reflectors in each arm at equal distances to the 3-dB couplers. This is feasible with OICs, but difficult with optical fiber components. A simple add–drop filter made with a single Bragg reflector and two 3-dB couplers, which can be made with optical fiber components, has an inherent 6-dB loss. Figures 8.14b and 8.14c show the transmission spectra of the added and dropped signals of an add–drop filter made by Erdogan *et al.* (1996) using P-doped waveguides and UV-induced gratings.

   Figure 8.15 shows the polarization dependence of the transmitted signal found by Erdogan *et al.* (1996). It illustrates a number of problems that

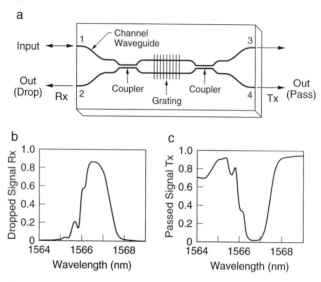

Fig. 8.14 Bragg reflecting add–drop multiplexer. (a) Layout. (b) Spectrum of the dropped signal Rx. (c) Spectrum of the passed signal Tx. (After Erdogan *et al.* 1996.)

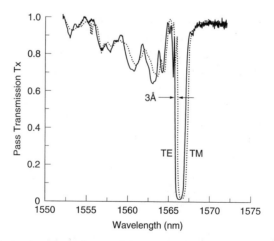

Fig. 8.15 Polarization dependence of the passed signal of a Bragg reflecting add–drop filter. TE polarization has the E field parallel to the substrate, and TM polarization has the E field normal to the substrate. The waveguide has $H \approx 5 \ \mu$m. (After Erdogan *et al.* 1996.)

must be solved before IOC Bragg reflecting add–drop filters are suitable for applications. One problem is the polarization splitting of the Bragg reflection band. This occurs in all types of narrow band filters and is discussed in Section IX.A. In addition, there is a polarization-dependent loss on the short-wavelength side due to the contradirectional coupling of the fundamental mode into cladding modes. These problems appear solvable. This unwanted loss can be shifted away from the Bragg reflection band by using high-delta waveguides. It also can be reduced in magnitude in designs for which the Bragg grating has a more complete overlap with the fundamental mode. If the Bragg grating completely covers the fundamental mode, mode orthogonality prevents contradirectional coupling. Recently, Kohnke *et al.* (1996) achieved an add–drop filter with suppression of cross talk to less than −40 dB, by doubly filtering both the passed and the dropped signals.

IX.   Wavelength and Polarization Control

A.   *POLARIZATION-DEPENDENT SPLITTING AND ITS CONTROL*

The most widely used optical fibers do not preserve polarization. Thus, optical components used with optical fibers must be made polarization independent. The polarization dependence of waveguides is mainly due to strain. The additional polarization dependence resulting from boundary conditions of the vector field is negligible. The strain is caused by the different thermal expansion coefficients of the two materials. It is relieved during anneals at 1000°C, which densify and flow the glass. But when the wafers cool to room temperature, the silicon substrate contracts more than the doped silica, and this results in an in-plane compressive strain of the waveguide layers. The strain causes a polarization-dependent change in refractive index that is larger for the TM mode (E normal to substrate) than for the TE mode. The relative change is about $2–3 \times 10^{-4}$ in effective index for silica waveguides on silicon. The polarization dependence of the mode effective refractive index results in a TE–TM wavelength splitting of narrow band filters of 2–3 Å. The strain is also different in the base, core, and cladding layers, which causes a larger delta for the TE mode than for the TM mode; therefore, TM has more bend loss than TE. Depending on the compensation method, these two effects may not be canceled simultaneously.

Many approaches have been taken to remove this unwanted polarization-dependent splitting. Adar, Henry, Kazarinov, *et al.* (1992) etched a trench on either side of the Bragg reflector to reduce the lateral strain field and reduced the splitting to about 0.3 Å (see also Kawachi 1990). Inoue *et al.* (1994) removed the strain splittings of WGRs by inserting a thin half-wave plate in the center of the WGR waveguide array so that propagation in each array waveguide occurs equally in both polarizations. Another method is depositing an amorphous silicon film on the top surface of the waveguide. The film properties and thickness are adjusted so that the film generates a strain field sufficient to cancel the average compressive strain over the waveguide mode (Kawachi 1990). Suzuki, Inoue, and Ohmori (1994) have also demonstrated WGRs of only 0.1-Å polarization-dependent splitting formed with FHD Ge-core waveguides deposited on fused silica substrates. Yaffe, Henry, Kazarinov, *et al.* (1994) have compensated the polarization dependence of narrow band MZ interferometers by inserting a thin layer of silicon nitride underneath the waveguide.

B.   POLARIZATION- AND WAVELENGTH-INDEPENDENT TAP

Taps made from parallel waveguide couplers are important in monitoring applications. They can remove a small portion of the transmitted signal with little excess loss. However, the tapped signal is both wavelength and polarization dependent. The wavelength dependence occurs because mode confinement decreases with increasing wavelength, which results in increased side-by-side coupling. Similarly, the polarization dependence results because confinement changes with polarization. Strain alters the refractive indices of the base, core, and cladding layers in a polarization-dependent manner, as discussed previously, which makes the depth of the potential well of Fig. 8.2 polarization dependent. These changes result in a weaker confinement for the TM polarization than for the TE polarization, which causes the tapped signal to be stronger for TM polarization. This is illustrated in Fig. 8.16a and in Fig. 8.6.

The dependence of the tap on wavelength and polarization follows from the dependence of the coupling length L_c on these quantities. Both dependencies are removed by the compensated tap shown in Fig. 8.16b, which consists of two couplers in series and is designed so that the transmission of the tapped signal is independent of L_c in the first order (Henry and Scotti 1996). The tapped signal is the product of the cross-state transmission of the first coupler and the bar-state transmission of the second coupler.

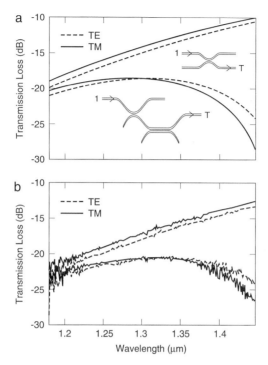

Fig. 8.16 TE and TM tapped signals versus wavelength for a conventional parallel waveguide tap and a compensated tap. (a) Calculated tapped signals (*insets*, waveguide layouts of a conventional tap [*upper*] and a compensated tap [*lower*]). (b) Measured tapped signals.

These transmissions have opposite dependencies on L_c, which makes cancellation possible. The observed and expected transmissions are shown in Fig. 8.16. The cost of compensation is at most a 3.7-dB decrease in the strength of the tapped signal, because part of the tapped power is not detected.

C. WAVELENGTH CONTROL

The mass fabrication of narrow band devices requires making filtering elements with controlled wavelengths over a series of wafers. This can be achieved only if the spread of filter wavelengths associated with waveguide nonuniformity is small. Figure 8.17 shows the results of an experiment to measure the distribution of filter wavelengths (Henry and Yaffe 1994).

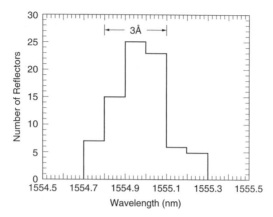

Fig. 8.17 Wavelength distribution of identical Bragg reflectors across a set of five wafers. (After Henry and Yaffe 1994.)

Identical narrow band Bragg reflectors were formed across a set of five wafers, and the distribution of Bragg wavelengths was measured. The spread was about 3 Å. If the wavelength change is attributed solely to an inhomogeneity in core thickness, it corresponds to a variation of 10%. Thus, relatively small wavelength spreads should be possible with increased control of the material properties.

X.   Amplification in Erbium-Doped Waveguides

The development of EDFAs in optical fiber has been a revolutionary step in optical communications. These amplifiers can also be made as OICs. As the performance of planar amplifiers improves and hybrid-optical integration makes it possible to assemble an entire amplifier consisting of a semiconductor pump laser, a WDM combining pump and signal wavelengths, an amplifying waveguide, and other components in a single compact package, this amplifier will become attractive and competitive with the optical fiber amplifier for some applications. Camy *et al.* (1995) of Corning Europe have demonstrated a lossless Y-branch splitter in which the splitting loss is compensated for by amplification in glass doped with Er and Yb. Hattori *et al.* (1994) of NTT have demonstrated a planar amplifier integrated with a pump–signal combiner having a threshold of 55 mW and a gain of 27 dB at 264 mW of pump power. The same group earlier packaged a planar optical amplifier pumped having four laser diodes (Kitagawa *et al.* 1993).

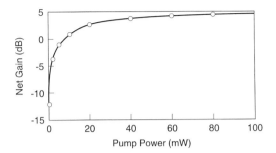

Fig. 8.18 Amplification versus pump power of a planar optical amplifier with a 6-cm-long waveguide composed of an Er-doped soda lime glass core. (After Ghosh *et al.* 1996.)

Optical fiber amplifiers are of order 20–40 m in length. The length of a planar amplifier is limited by the optical attenuation length of the material to tens of centimeters. To achieve significant gains in distances of order 10 cm requires a concentration of Er that is two orders of magnitude greater than that used in optical fibers. At higher concentrations, the fraction of closely spaced Er ions increases. Unfortunately, at close spacing, Er ion pairs exhibit nonradiative up-conversion when both ions have populated upper levels. One electron decays nonradiatively to the ground state, transferring its energy to the other electron, which is promoted to a higher level. This problem is accentuated by the clustering of Er ions in silica doped to high Er concentrations. Up-conversion is less pronounced in soda lime glass than in silica because the cagelike structure of soda lime glass tends to separate the Er ions and reduce clustering. Using a core of soda lime glass, Ghosh *et al.* (1996) have demonstrated an amplification with a threshold of 8 mW and a 4-dB gain at 80 mW in a 6-cm-long amplifier (Fig. 8.18). Such an amplifier can be pumped with a single diode laser.

XI. Integrated Optical Switches

MZ interferometers with thin film heaters above the delay waveguides have been widely used as a basic 2 × 2 thermooptic switch element. By applying an electric current in the heater to raise the temperature of the adjacent waveguide, we can change the phase of the interferometer by π, thereby switching the MZ from bar to cross transmission, or from cross to bar transmission. There are no moving parts in the switch; this feature makes it highly reliable. However, because of the small temperature coefficient of the refractive index of silica ($\sim 1 \times 10^{-5}$/°C), the required heater power is relatively large (~ 0.5 W per MZ, which depends on the thickness of the waveguide layers but is almost independent of the length of the heater). The silicon substrate is used not only to confine the temperature change locally, but also as a heat sink. To reduce the required heater power, the silica waveguide can be etched into strips along the heater (Moller *et al.* 1993), or the silicon substrate directly underneath the heated waveguide can be removed (Sugita *et al.* 1990). Because of the sinusoidal spectral response of the MZ, clean switching occurs only in a narrow wavelength range and requires accurate control of the heater power level. In order to have a clean initial transmission state, a bias power is usually applied to the heater, or the MZ initial phase is actively tuned (Moriwaki *et al.* 1995). The currently achievable extinction ratio is -20 to -30 dB. The switching time is approximately 1 ms, largely determined by the heat capacity and heat conductivity of silica, which is adequate only for low-speed switching applications. MZ-based thermooptic switches have been widely used in implementing compact $M \times N$ matrix switches (see Okuno *et al.* 1993, 1994) and reconfigurable IOCs — e.g., reconfigurable wavelength add–drop filters (see Okamoto, Takiguchi, and Ohmori 1995) and wavelength routers (see Ishida *et al.* 1994).

Poling waveguides may be a promising method for making better integrated optical switches. UV-assisted poling in Ge-doped silica fibers has been demonstrated to produce a permanent electrooptic coefficient close to 1 pm/V (Fujiwara, Wong, and Fleming 1995; Myers *et al.* 1995). The large electrooptic effect is induced by treating Ge-doped silica with a large electric field. If the same technique is developed with Ge-doped silica waveguides and the electrooptic coefficient is increased by a factor of 5–20, compact electrooptic switches similar to those using lithium niobate waveguides can be realized.

Besides thermooptic and electrooptic switches, waveguide switches based on other mechanisms have been investigated. Capillary mercury mirrors manipulated by the electrocapillary effect have been demonstrated (Sato 1994). A bistable optical switch based on frustrated total reflection from electrochemically generated bubbles has been demonstrated by Jackel, Johnson, and Tomlinson (1990). Optical switches involving deflectable suspended waveguides have also been investigated. However, these switches involve moving parts and are considerably more complicated to fabricate than other waveguide devices (Ollier, Labeye, and Mottier 1995).

XII.   Hybrid Integration

The functionality of passive OICs can be greatly increased in some applications by adding lasers and detectors to the OIC chip by hybrid integration. This is illustrated by the transceiver of of Bell Laboratories (J. V. Gates, private communication), shown schematically in Fig. 8.19, which sends and receives signals on a single fiber. The OIC has a WDM to separate received signals at 1.3 and 1.55 μm, a coupler to separate the transmitted and received signals at 1.3 μm, a laser for transmitting 1.3 μm, two detectors for receiving the two wavelengths, and one for monitoring the laser power.

A.   *TURNING MIRRORS AND DETECTOR INTEGRATION*

A small etched mirror intercepts light leaving the waveguide and directs it upward to a surface-mounted detector (Fig. 8.20). The surface of the mirror is metallized, and metallization and solder are patterned on the top

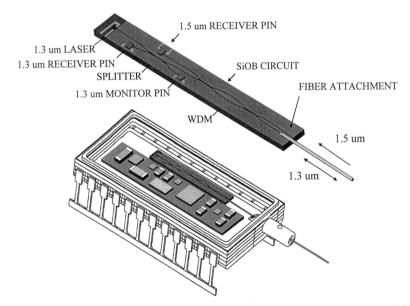

Fig. 8.19   Transceiver for two-way transmission on a fiber suitable for transmitting and receiving audio at 1.3 μm and receiving video at 1.55 μm. (J. V. Gates, private communication.)

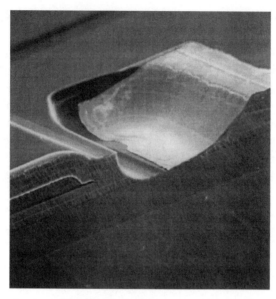

Fig. 8.20 Termination of an optical waveguide by an etched and metallized turning mirror. (J. V. Gates, private communication.)

surface so that the detectors can be bonded in place. About 85–90% of the light in the waveguide is detected (J. V. Gates, private communication).

B. LASER INTEGRATION

Gates *et al.* (1995) developed a laser subassembly consisting of a chip of silicon to which an edge-emitting laser, a back-face monitor detector, and a lensed optical fiber are bonded. The alignment of the optical fiber to the laser is done by building a miniature strip heater that melts the solder used to anchor the metallized fiber into place. The fiber is aligned to maximize the received spontaneous emission from the laser. About 75% of the light emitted from the laser facet is coupled into the optical fiber. This fiber, in the form of a short stub, is attached to the OIC; thus, the laser light is coupled into the waveguide of the transceiver (Fig. 8.19).

The difficulty of aligning a conventional edge-emitting laser to the waveguide prohibits bonding the laser directly to the silicon chip. This difficulty has been overcome at NTT, where an expanded beam laser was used

(Okamoto 1995). The expanded beam structure was also used to efficiently couple the waveguide to photodiodes (Yamada *et al.* 1995).

XIII. Conclusion

Silica-based OICs have undergone major development since the mid-1980s. Coupling and propagation losses have been reduced to the point that the fiber–waveguide–fiber insertion loss of a 6-cm straight waveguide is as low as 0.3 dB. The fusion of optical fibers to silica waveguide chips has also been demonstrated. WGRs and Fourier filters are examples of achievable new multiplexing devices that make use of the excellent control of couplers and delays. UV irradiation has led to improved methods of forming Bragg reflectors in waveguides. Low-threshold traveling-wave optical amplifiers have been demonstrated.

Despite this progress, various problems remain to be solved. A preferred method of silica deposition has not yet emerged; CVD, FHD, and electron beam deposition are being used by major laboratories. The mechanisms that limit waveguide loss are not well clarified and high-delta waveguides with negligible loss are yet to be demonstrated. Low cost and reliable methods of optical fiber attachment remain the greatest challenges for successful commercial application of this technology. The WGR has found application in dense WDM applications, but add–drop OICs need additional refinement. A low-power alternative to the thermooptic switch is desirable. This technology is relatively new, and important new inventions are likely to be forthcoming.

Acknowledgments

We have greatly benefited from discussing these issues with our colleagues. We particularly want to thank G. Barbarosa, A. J. Bruce, L. G. Cohen, G. F. DeVeau, C. Dragone, J. V. Gates, B. I. Greene, A. Kilian, G. Kohnke, E. J. Laskowski, J. B. MacChesney, C. Narayanan, H. Presby, R. E. Scotti, M. R. Serbin, J. Shmulovich, R. K. Smith, Y. Wang, A. E. White, and H. H. Yaffe. We thank H. H. Yaffe and C. Dragone for valuable suggestions on the manuscript. We are indebted to M. A. Cappuzzo, E. J. Laskowski, D. Mozer, and M. R. Serbin for processing our wafers.

References

Adar, R., C. H. Henry, C. Dragone, R. C. Kistler, and M. A. Milbrodt. 1993. Broadband array multiplexers made with silica waveguides on silicon. *J. Lightwave Tech.* 11:212–219.

Adar, R., C. H. Henry, R. F. Kazarinov, R. C. Kistler, and G. R. Weber. 1992. Adiabatic 3-dB couplers, filters and multiplexers made with silica waveguides on silicon. *J. Lightwave Tech.* 10:46–50.

Adar, R., C. H. Henry, R. C. Kistler, and R. F. Kazarinov. 1992. Polarization independent narrow band Bragg reflection gratings made with silica on silicon waveguides. *Appl. Phys. Lett.* 60:1779–1781.

Adar, R., M. R. Serbin, and V. Mizrahi. 1994. Less than 1 dB per meter propagation loss of silica waveguides measured using a ring resonator. *J. Lightwave Tech.* 12:1369–1372.

Agrawal, G. P. 1992. *Fiber-optic communication systems.* New York: Wiley.

Bilodeau, F., B. Malo, J. Albert, D. C. Johnson, K. O. Hill, Y. Bibino, M. Abe, and M. Kawachi. 1993. *Opt. Lett.* 18:953.

Camy, P., A. Beguin, C. Lerminiaux, C. Prel, J. E. Roman, M. Hempstead, J. S. Wilkinson, J. C. Vad der Plaats, W. Willems, and A. M. J. Koonen. 1995. Diode-pumped planar lossless splitter at 1.5 μm for optical networks. In *Proceedings of the 21st European Conference on Optical Communications (ECOC'95)*, Belgium: IMEC, 1067–1070.

Chiang, K. S. 1994. Review of numerical and approximate methods for modal analysis of general optical dielectric waveguides. *Opt. Quantum Electron.* 26:S113–S134.

Day, S., R. Bellerby, R. Cannell, and M. Grant. 1992. Silicon based fibre pigtailed 1 × 16 power splitter. *Electron. Lett.* 28:920–922.

Dragone, C. 1989. Efficiency of a periodic array with a nearly ideal element pattern. *IEEE Photon. Tech. Lett.* 1:238–240.

Dragone, C. 1990. Optimum design of a planar array of tapered waveguides. *J. Opt. Soc. Am. A* 7:2081–2093.

Dragone, C. 1991. An $N \times N$ optical multiplexer using a planar arrangement of two star couplers. *IEEE Photon. Tech. Lett.* 3:812–815.

Dragone, C. 1995. Frequency routing device having a wide and substantially flat passband. *U.S. patent* no. 5412744, issued May 2.

Dragone, C., C. A. Edwards, and R. C. Kistler. 1991. Integrated optics $N \times N$ multiplexer on silicon. *IEEE Photon. Tech. Lett.* 3:896–899.

Dragone, C., C. H. Henry, I. P. Kaminow, and R. C. Kistler. 1989. Efficient multichannel integrated optics star coupler on silicon. *IEEE Photon. Tech. Lett.* 1:241–243.

Erdogan, T., T. A. Strasser, M. A. Milbrodt, E. J. Laskowski, C. H. Henry, and G. E. Kohnke. 1996. Integrated-optical Mach–Zehnder add–drop filter fabricated by a single uv-induced grating exposure. *IEEE Photon. Tech. Lett.* (submitted for publication).

Farin, G. 1988. *Curves and surfaces for computer aided geometric design.* New York: Academic Press.

Findakly, T. 1985. Glass waveguides by ion-exchange: A review. *Opt. Eng.* 24:244–250.

Fujiwara, T., D. Wong, and S. Fleming. 1995. Large electro-optic modulation in thermally-poled germanosilicate fiber. *IEEE Photon. Tech. Lett.* 7:1177–1179.

Gates, J. V., G. Henein, J. Shmulovich, D. J. Muehlner, W. M. MacDonald, and R. E. Scotti. 1995. Uncooled laser packaging based on silicon optical bench technology. *Proc. SPIE Photon. East* 2610:127–137.

Ghosh, R. N., J. Shmulovich, C. F. Kane, M. R. X. DeBarros, G. Nykolak, A. J. Bruce, and P. C. Becker. 1996. 8 mW Threshold Er^{3+}-doped planar waveguide amplifier. *IEEE Photon. Tech. Lett.* 8:518–520.

Glance, B., I. P. Kaminow, and R. W. Wilson. 1994. Applications of the integrated waveguide grating router. *J. Lightwave Tech.* 12:957–962.

Hadley, G. 1992. Transparent boundary condition for the beam propagation method. *IEEE J. Quantum Electron.* 28:363–369.

Hattori, K., T. Kitagawa, M. Oguma, Y. Ohmori, and M. Horiguchi. 1994. Erbium-doped silica-based waveguide amplifier integrated with a 980/1530 nm WDM coupler. *Electron. Lett.* 30:856–857.

Henry, C. H., and R. E. Scotti. 1996. Polarization and wavelength independent optical tap. U.S. Patent no. 5539850, issued July 23.

Henry, C. H., G. E. Blonder, and R. F. Kazarinov. 1989. Glass waveguides on silicon for hybrid optical packaging. *IEEE J. Lightwave Tech.* 7:1530–1538.

Henry, C. H., R. F. Kazarinov, Y. Shani, R. C. Kistler, V. Pol, and K. Orlowsky. 1990. Four-channel wavelength division multiplexers and bandpass filters based on elliptical Bragg reflectors. *J. Lightwave Tech.* 8:748–755.

Henry, C. H., and Y. Shani. 1991. Analysis of mode propagation in optical waveguide devices by Fourier expansion. *IEEE J. Quantum Electron.* 27:523–530.

Henry, C. H., Y. Shani, R. C. Kistler, T. E. Jewell, V. Pol, N. A. Olsson, R. F. Kazarinov, and K. J. Orlowsky. 1989. Compound Bragg reflection filters made by spatial frequency lithography doubling. *J. Lightwave Tech.* 7:1379–1385.

Henry, C. H., and B. H. Verbeek. 1989. Solution of the scalar wave equation by two dimensional Fourier analysis. *J. Lightwave Tech.* 12:309–313.

Henry, C. H., and H. H. Yaffe. 1994. Characterization of silicon optical bench waveguides. *NIST'94 Tech. Dig. National Institute of Standards and Technology Special Publication* 864:105–108.

Hibino, Y., F. Hanawa, H. Nakagome, M. Ishii, and N. Takato. 1995. High reliability optical splitters composed of silica based planar lightwave circuits. *J. Lightwave Tech.* 13:1728–1735.

Hibino, Y., Y. Hida, H. Okazaki, and Y. Ohmori. 1993. Propagation loss characteristics of long silica-based optical waveguides on 5 inch Si wafers. *Electron. Lett.* 29:1847–1848.

Hida, Y., Y. Hibino, H. Okazaki, and Y. Ohmori. 1995. In *NTT Opto-electronics Laboratories annual report*. Japan: NTT publication.

Hunsperger, R. G. 1991. *Integrated optics: Theory and techology*. 3d ed. Berlin: Springer-Verlag.

Imoto, K. 1994. Progress in high silica waveguide devices. *Integr. Photon. Res. 1994 Tech. Dig.* 3:62–64.

Inoue, Y., Y. Ohmori, M. Kawachi, S. Ando, T. Sawada, and H. Takahashi. 1994. Polarization mode converter with polyimide half-wave plate in silica-based planar light-wave circuits. *IEEE Photon. Tech. Lett.* 6:626–628.

Ishida, O., H. Takahashi, S. Suzuki, and Y. Inoue. 1994. Multichannel frequency-selective switch employing an arrayed-waveguide grating multiplexer with fold-back optical paths. *IEEE Photon. Tech. Lett.* 6:1219–1221.

Jackel, J. L., J. J. Johnson, and W. J. Tomlinson. 1990. Bistable optical switching using electrochemically generated bubbles. *Opt. Lett.* 15:1470–1472.

Kashima, N. 1995. *Passive optical components for optical fiber transmission*. Boston: Artech House.

Kashyap, R., G. D. Maxwell, and B. J. Ainslie. 1993. Laser-trimmed four-port bandpan filter fabricated in single-mode photo-sensitive Ge-doped planar wave-guide. *IEEE Photon. Tech. Lett.* PTL-5:191.

Kato, K., K. Okamoto, H. Okazaki, Y. Ohmori, and I. Nishi. 1993. Packaging of large-scale integrated-optic $N \times N$ star couplers. *IEEE Photon. Tech. Lett.* 4:348–351.

Kawachi, M. 1990. Silica waveguides on silicon and their application to integrated-optic components. *Opt. Quantum Electron.* 22:391–416.

Kawachi, M., and K. Jinguji. (1994). Planar lightwave circuits for optical signal processing. *OFC'94 Tech. Dig.* 4:281–282.

Kawachi, M., M. Yasu, and M. Kobayashi. 1983. *Jpn. J. Appl. Phys.* 22:1932.

Kitagawa, T., K. Hattori, K. Shuto, M. Oguma, J. Temmyo, S. Suzuki, and M. Horiguchi. 1993. Erbium-doped silica-based planar amplifier module pumped by laser diodes. In *ECOC'93, 19th European Conference on Optical Communications, September 12–16, Montreux, Switzerland.* (Vol. 3). Zurich: Swiss Electrotech Association, pp. 41–44.

Kohnke, G. E., T. A. Strasser, A. E. White, M. A. Milbrodt, C. H. Henry, and E. J. Laskowski. 1996. Planar waveguide Mach–Zehnder bandpass filter fabricated with single exposure UV-induced gratings. *OFC'96 Tech. Dig.* 2:277.

Kominato, T., Y. Ohmori, H. Okazaki, and M. Yasu. 1990. Very low-loss GeO_2-doped silica waveguides fabricated by flame hydrolysis deposition method. *Electron. Lett.* 26:327–329.

Koren, U. 1994. Waveguide based photonic integrated circuits. In *Optoelectronic integration*, ed. O. Wada, chapter 7. Kluwer Academic Publishers.

Korotky, S. K., and R. C. Alferness. 1987. $Ti:LiNbO_3$ integrated optic technology. In *Integrated optical circuits and components*, ed. L. D. Hutcheson, 169–228. New York: Marcel Dekker.

Landau, L. D., and E. M. Lifshitz. 1958. *Quantum mechanics*. Reading, MA: Addison-Wesley. (See section on the Landau–Zener transition.)

Lemaire, P. J., R. M. Atkins, V. Mizrahi, and W. A. Reed. 1993. High pressure H_2 loading as a technique for achieving ultrahigh UV photosensitivity and thermal sensitivity in GeO_2 doped optical fibres. *Electron. Lett.* 29:1191–1193.

Leonberger, F. J., and J. P. Donnelly. 1990. Semiconductor integrated optic devices. In *Guided-wave optoelectronics*, 2d ed., ed. T. Tamir, 317–390. Berlin: Springer-Verlag.

Li, Y. P., and C. H. Henry. 1996. Silica-based optical integrated circuits. *IEE Proc.-Optoelectron.* 143:263–280.

Li, Y. P., C. H. Henry, E. J. Laskowski, C. Y. Mak, and H. H. Yaffe, 1995. A waveguide EDFA gain equalization filter. *Electron. Lett.* 31:2005–2006.

Li, Y. P., C. H. Henry, E. J. Laskowski, and H. H. Yaffe. 1996. Fourier transform based optical waveguide filters and WDMs. *OFC'96 Tech. Dig.* 2:97–98.

Li, Y. P., C. H. Henry, E. J. Laskowski, H. H. Yaffe, and R. L. Sweatt. 1995. A monolithic optical waveguide 1.31/1.55 μm WDM with -50 dB crosstalk over 100 nm bandwidth. *Electron. Lett.* 31:2100–2101.

Marcuse, D. 1982. *Light transmission optics*. 2d ed. New York: Van Nostrand-Reinhold.

Marcuse, D. 1991. *Theory of dielectric optical waveguides*. 2d ed. New York: Academic Press.

Marz, R. 1994. *Integrated optics: Design and modeling*. Boston: Artech House.

McCall, S. L., and P. M. Platzman. 1985. An optimized $\pi/2$ distributed feedback laser. *IEEE J. Quantum Electron.* 21:1899–1904.

Meltz, G., W. W. Morey, and W. H. Glenn. 1989. Formation of Bragg grating in optical fibers by a transverse holographic method. *Opt. Lett.* 14:823–825.

Moller, B. A., L. Jensen, C. Laurent-Lund, and C. Thirstrup. 1993. Silica waveguide thermo-optic phase shifter with low power consumption and low lateral heat diffusion. *IEEE Photon. Tech. Lett.* 5:1415–1418.

Moriwaki, K., M. Abe, Y. Inoue, M. Okuno, and Y. Ohmori, 1995. New silica-based 8 × 8 thermo-optic matrix switch on Si that requires no bias power. *OFC'95 Tech. Dig.* 8:211–212.

Moslehi, B., J. W. Goodman, M. Tur, and H. J. Shaw. 1984. Fiber-optic lattice signal processing. *Proc. IEEE* 72:909–911.

Myers, R. A., X. C. Long, S. R. J. Brueck, and R. P. Tumminelli. 1995. Effect of hydrogen loading on temperature/electric-field poling of SiO_2-based thin films on Si. *Electron. Lett.* 31:1604–1606.

Neyer, A., T. Knoche, and L. Muller. 1993. Fabrication of low loss polymer waveguides using injection moulding technology. *Electron. Lett.* 4:790–796.

Nolting, H. P., and R. Marz. 1995. Results of benchmark tests for different numerical BPM algorithms. *J. Lightwave Tech.* 13:216–224.

Oddvar, L., and A. Sudbo. 1995. 8-Channel wavelength division multiplexer based on multimode interference coupler. *IEEE Photon. Tech. Lett.* 7:1034–1036.

Okamoto, K. 1995. Application of planar lightwave circuits to optical communications systems. In *Proceedings of the 21st European Conference on Optical Communications (ECOC'95)*, Belgium: IMEC, 75–82.

Okamoto, K., H. Takahashi, M. Yasu, and Y. Hibino. 1992. Fabrication of wavelength-insensitive 8 × 8 star coupler. *IEEE Photon. Tech. Lett.* 4:61–63.

Okamoto, K., K. Moriwaki, and S. Suzuki. 1995. Fabrication of 64 × 64 arrayed-waveguide grating multiplexer on silicon. *Electron. Lett.* 31:184–185.

Okamoto, K., H. Okazaki, Y. Ohmori, and K. Kato. 1992. Fabrication of large scale N × N star couplers. *IEEE Photon. Tech. Lett.* 4:1032–1035.

Okamoto, K., K. Takiguchi, and Y. Ohmori. 1995. 16-Channel optical add/drop multiplexer using silica-based arrayed-waveguide gratings. *Electron. Lett.* 31:723–724.

Okuno, M., K. Kato, Y. Ohmori, and T. Matsunaga. 1994. Improved 8 × 8 integrated optical matrix switch using silica-based planar lightwave circuits. *J. Lightwave Tech.* 12:1597–1606.

Okuno, M., A. Sugita, T. Matsunaga, M. Kawachi, Y. Ohmori, and K. Kato. 1993. 8 × 8 Optical matrix switch using silica-based planar lightwave circuits. *IEICE Trans. Electron.* E76-C:1215–1223.

Ollier, E., P. Labeye, and P. Mottier. 1995. A micro-opto-mechanical switch for optical fiber networks. *SPIE Proc.* 2401:116–124.

Olsson, N. A., C. H. Henry, R. F. Kazarinov, H. J. Lee, B. H. Johnson, and K. J. Orlowsky. 1987. Narrow linewidth 1.5 micron semiconductor laser with a resonant optical reflector. *Appl. Phys. Lett.* 51:1141–1142.

Presby, H. M., and C. A. Edwards. 1992. Packaging of glass waveguide silicon devices. *Opt. Eng.* 31:141–143.

Presby, H. M., S. Yang, A. E. Willner, and C. A. Edwards. 1992. Connectorizing integrated star couplers on silicon. *Opt. Eng.* 31:1323–1327.

Ramaswamy, R. V., and R. Srivastava. 1988. Ion-exchanged glass waveguides. *J. Lightwave Tech.* 6:984–1002.

Sato, M. 1994. Electrocapillary optical switches. *IEICE Trans. Commun.* E77-B:197–203.

Shani, Y., C. H. Henry, R. C. Kistler, R. F. Kazarinov, and K. J. Orlowsky. 1992. Adiabatic 3-dB couplers, filters, and multiplexers made with silica waveguides on silicon. *J. Lightwave Tech.* 1:46–50.

Singer, P. 1995. CVD technology, trends. *Semiconductor Intl.* 18:58–62.

Smit, M. K. 1989. New focusing and dispersive planar component based on an optical phase array. *Electron. Lett.* 24:385–386.

Smith, B. T., J. M. Boyle, J. J. Dongarra, B. S. Garbow, Y. Ikebe, V. C. Klema, and C. B. Moler. 1976. *Matrix eigensystem routines — EISPACK guide.* New York: Springer-Verlag.

Smith, R. K. 1995. Stable finite element solutions to the vector wave equation. *Integr. Photon. Res. 1995 Tech. Dig.* 209–210.

Snyder, A. W., and J. D. Love. 1983. *Optical waveguide theory*. London: Chapman & Hall.

Strasser, T. A., T. Erdogan, A. E. White, V. Mizrahi, and P. J. Lemaire. 1994. Ultraviolet laser fabrication of strong, nearly polarization-independent Bragg reflectors in germanium-doped silica waveguides on silica substrates. *Appl. Phys. Lett.* 65:3308–3310.

Sugita, A., K. Jinguji, N. Takato, K. Katou, and M. Kawachi. 1990. Bridge-suspended silica-waveguide thermo-optic phase shifter and its application to Mach–Zehnder type of optical switch. *Trans. IEICE* E73:105–109.

Sugita, A., K. Onose, Y. Ohmori, and M. Yasu. 1993. Optical fiber coupling to single-mode silica-based planar lightwave circuits with fiber-guiding grooves. *Fiber Integr. Opt.* 12:347–354.

Suzuki, S., Y. Inoue, and Y. Ohmori. 1994. Polarization-insensitive arrayed-waveguide grating multiplexer with SiO_2-on-SiO_2 structure. *Electron. Lett.* 30:642–643.

Suzuki, S., and M. Kawachi. 1994. Planar lightwave circuits based on silica waveguides on silicon. Part 2. *Electron. Commun. Jpn.* 77:25–36.

Suzuki, S., M. Yanagisawa, Y. Hibino, and K. Oda. 1994. High-density integrated planar lightwave circuits using SiO_2–GeO_2 waveguides with a high refractive index difference. *J. Lightwave Tech.* 12:790–796.

Takada, K., Y. Inoue, H. Yamada, and M. Horiguchi. 1994. Measurement of phase error distributions in silica-based arrayed-waveguide grating multiplexers by using Fourier transform spectroscopy. *Electron. Lett.* 30:1671–1672.

Takagi, A., K. Jinguji, and M. Kawachi. 1992. Silica-based waveguide-type wavelength-insensitive couplers (WINC's) with series-tapered coupling structure. *J. Lightwave Tech.* 10:1814–1824.

Takahashi, H., K. Oda, H. Toba, and Y. Inoue. 1995. Transmission characteristics of arrayed waveguide $N \times N$ wavelength multiplexer. *J. Lightwave Tech.* 13:447–455.

Takahashi, H., S. Suzuki, K. Katoh, and I. Nishi. 1990. Arrayed waveguide grating for wavelength division multi/demultiplexer with nanometer resolution. *Electron. Lett.* 26:87–88.

Tamir, T. E. 1990. *Guided-wave optoelectronin*. (2nd ed.). Berlin: Springer-Verlag.

Vellekoop, A. R., and M. K. Smit. 1991. Four-channel integrated-optic wavelength demultiplexer with weak polarization dependence. *J. Lightwave Tech.* 9:310–314.

Yaffe, H. H., C. H. Henry, R. F. Kazarinov, and M. A. Milbrodt. 1994. Polarization-independent silica-on-silicon Mach–Zehnder interferometers. *J. Lightwave Tech.* 12:64–67.

Yaffe, H. H., C. H. Henry, M. R. Serbin, and L. G. Cohen. 1994. Resonant couplers acting as add–drop filters made with silica-on-silicon waveguide technology. *J. Lightwave Tech.* 12:1010–1015.

Yamada, Y., F. Hanawa, T. Kitoh, and T. Maruno. 1992. Low-loss and stable fiber-to-waveguides connection utilizing UV curable adhesive. *IEEE Photon. Tech. Lett.* 4:906–908.

Yamada, Y., S. Suzuki, K. Moriwaki, Y. Hibino, Y. Thomori, Y. Akatsu, Y. Nakasuga, T. Hashimoto, M. Yanagisawa, Y. Inoue, Y. Akahori, R. Nagase, and H. Terui. 1995. A hybrid integrated optical WDM transmitter/receiver module for optical subscriber systems utilizing a planar lightwave circuit platform. *OFC'95 Tech. Dig.* 8:370–373.

Yasu, M., M. Kawachi, and M. Kobayashi. 1985. Fabrication of SiO_2–TiO_2 glass planar optical waveguides by soot deposition. *Trans. Inst. Electron. Commun. Eng. Jpn.* J68-C:454–461.

Yevick, D. 1994. A guide to electric field propagation techniques for guided-wave optics. *Opt. Quantum Electron.* 26:S185–S197.

Chapter 9 | Lithium Niobate Integrated Optics: Selected Contemporary Devices and System Applications

Fred Heismann
Steven K. Korotky
John J. Veselka

Lucent Technologies, Bell Laboratories, Holmdel, New Jersey

9.1 Introduction

9.1.1 INTRODUCTION AND SCOPE

Recently, the first examples of an emerging family of optoelectronic devices referred to as *integrated optic devices,* or alternatively as *photonic integrated circuits (PICs),* made their debut in commercial long-distance optical fiber communications systems. From their beginnings in research, these components have been advanced and refined through the development and engineering process and have been, or are very soon to be, deployed in both submarine and terrestrial lightwave applications. Among the types of photonic circuits that have found use in these demanding areas are lithium niobate electrooptic amplitude, phase, and polarization modulators; InGaAsP integrated laser and electroabsorption intensity modulators; and glass-on-silicon waveguide grating router multiplexers. In addition, although not yet ready for field use, two other categories of photonic circuits that can be used to implement a wide variety of desirable functions, especially in wavelength-division multiplexed systems, and that have been used in laboratory system demonstrations, are lithium niobate tunable wavelength filters and polarization controllers.

Significant advances have been made in the design, performance, and use of these most basic of integrated optic components since the publication of *Optical Fiber Telecommunications II* in 1988 (Miller and Kaminow 1988).

377

Copyright © 1997 by Lucent Technologies.
All rights of reproduction in any form reserved.
ISBN: 0-12-395171-2

For this reason, it is worthwhile to review these results in this volume. An introduction and survey of the entire field of integrated optics is, however, beyond the scope of this chapter. Several books (Tamir 1979; Hunsperger 1982; Hutcheson 1987; Tamir 1988; Marz 1995) and special issues of archival journals (Alferness and Walpole 1986; Boyd 1988; Korotky, Campbell, and Nakajima 1991) dealing with these broader aspects of the subject are available to the reader. Rather, in this chapter we consider only a subset of integrated optics and focus on its role in contemporary optical fiber telecommunications. Also, we specifically confine ourselves to photonic circuits based on the lithium niobate material.

The evolution of fiber optic technology and that of integrated optic technology have been intertwined from the beginning. And, with the recent trend toward the increased use of wavelength-division multiplexing in long-distance fiber systems, there is every reason to believe that integrated optics will play an ever-increasing role in lightwave communication systems and networks. To understand better the role that integrated optics may have in future systems, and to set the remainder of this chapter into context, it is worthwhile in this introduction first to summarize briefly the interplay that has transpired to bring us to the present. After that and a short overview of this chapter, we delve into the details of our specific subjects.

9.1.2   BACKGROUND AND MOTIVATION

Time and time again, advances in fiber optics and integrated optics, driven by the compelling economics of ever higher capacity and longer distances between regenerators, have created new possibilities and opportunities for each other. We may begin by citing, for example, the complementary breakthroughs in room-temperature laser diodes and low-loss optical fibers, which occurred in the early 1970s and set the stage for revolutionizing long-distance telecommunications. These pioneering works and their theoretical foundations were the subject of the first volume of this book, *Optical Fiber Telecommunications I* (Miller and Chynoweth 1979). It was not long after the deployment of the first multimode fiber transmission systems in the late 1970s that the demand for longer distances and higher bit rates necessitated lower dispersion and lower loss solutions. This sped the development of single-transverse-mode, zero-dispersion fiber and single-longitudinal- and single-transverse-mode injection lasers working in the 1.3-μm wavelength region. With the switch to single-mode fibers beginning in the early 1980s came a resurgence in integrated optics, which performed best in a single-transverse-mode environment. In addition to waveguide lasers, research on single-mode modulators, switches, and wavelength filters was vigorously undertaken. However, except for the lasers, integrated optics was not yet mature enough to have a significant impact on systems being installed in the field.

The unrelenting demands of long-distance telecommunications next pushed the primary transmission wavelength from 1.3 μm to the lower loss window near 1.5 μm. The systems were now no longer loss limited, but again dispersion limited. To push back the dispersion limitation, work then intensified on lower chirp lasers and integrated optic external modulators. It was not long before advances on both types of transmitters returned us to the situation where the systems were loss limited. The last volume of this book was written near the end of that period (Miller and Kaminow 1988). Because the performance of the directly modulated lasers was adequate to reach the new loss limit, the newer external modulator technology would not be deployed then, but continued to remain the subject of laboratory hero experiments.

Since *Optical Fiber Telecommunications II* was written in 1988, the mutual bootstrapping between fiber and integrated optics to bring down the loss, dispersion, and capacity barriers has continued unabated. Without a doubt, the most significant event to reshape the fiber transmission landscape since then has been the advent of a practical optical amplifier. Almost overnight the loss limit of fiber systems operating near a wavelength of 1.5 μm was completely removed. With that event, fiber dispersion, optical amplifier noise, and fiber nonlinearities became the limits to the performance of fiber optic transmission systems. This situation created many unique opportunities for photonic technologies. Integrated optic external modulators, for example, with their low, easily controlled, and reproducible chirp characteristics, were immediately in high demand to push the dispersion barrier to its theoretical limit. Perhaps not surprising, with the demand created by the optical amplifier, external modulator prototypes made their way from research benches into operating commercial systems within a few years. Today, external-modulator-based transmitters are the option of choice for optically amplified, long-distance digital systems and are the standard by which other modulation techniques are judged. Interestingly, the availability of high-speed phase and amplitude modulators has created a situation where they have begun to find use in a number of unexpected ways to circumvent system limitations that could not have been predicted in the early stages of either device or system research. We also discuss these in this chapter.

In addition to external amplitude modulators, integrated optics has provided several other unique devices and capabilities to capitalize on the system possibilities created by the erbium-doped fiber amplifier (EDFA), such as the devices mentioned earlier in this introduction. With the introduction of multiple-wavelength systems, and the prospects of optical networking, there is little doubt that integrated optics technology will be central to the future of fiber telecommunications. It is our goal then, when this chapter is taken with its companions that compose this book, for the reader to be presented with a panoramic picture of the dramatic change and progress that have occurred in fiber telecommunications since the years just prior to the EDFA's arrival on the scene, and, it is hoped, a glimpse of what may soon follow.

9.1.3   CHAPTER OVERVIEW

In this chapter, we survey the design and performance of devices based on the lithium niobate technology that have found application in high-speed, long-distance, fiber optic communication systems. We review the advances that have occurred to make their deployment possible, give examples of the unique roles that they serve in contemporary systems, summarize the state-of-the-art device demonstrations in these areas, and explore examples of other types of devices that have reached very promising stages of research and development and examine how they may be used in future generations of lightwave systems and networks.

This chapter includes three major sections that cover the subjects of (1) high-speed external phase and amplitude modulators and time-domain switches; (2) polarization mode converters, controllers, and modulators; and (3) tunable wavelength filters. Another key function addressed by the lithium niobate technology, space switching, is the subject of Chapter 10 of Volume IIIB.

9.1.4   RELATIONSHIP TO OTHER CHAPTERS

The material presented in this chapter does not stand in isolation; rather, it is closely related to many of the topics covered in other chapters of this book. In the areas of motivation, complementary technologies, and applications, we call particular attention to the chapters listed next.

Motivation
   EDFA (Volume IIIB, Chapter 2)
   Polarization phenomena (Volume IIIA, Chapter 6)
Complementary Technologies and Devices
   Electroabsorption modulators (Volume IIIB, Chapter 4)
   InP-based devices (Volume IIIB, Chapter 5)
   Glass-on-silicon devices (Volume IIIB, Chapter 8)
   Space switches (Volume IIIB, Chapter 10)
Applications
   Optical amplifier systems (Volume IIIA, Chapters 9 and 10)
   High bit-rate transmission (Volume IIIA, Chapter 11)
   Soliton transmission (Volume IIIA, Chapter 12)
   Analog systems (Volume IIIA, Chapter 14)

A potentially important class of integrated optic devices that is absent from the discussions in this book are those based on polymers. This technology has not yet reached the stage where it has an impact on commercial telecommunications; however, its prospects for large and low-cost optical back planes, for example, suggest that it should not be overlooked by the reader (Hornak 1992).

9.2 High-Speed Phase and Amplitude Modulators and Switches

9.2.1 INTRODUCTION

Since the mid-1980s, significant advances in the understanding, design, and performance of high-speed lithium niobate phase and amplitude waveguide electrooptic modulators have been made in the areas of (1) chirp performance; (2) modulation bandwidth and switching voltage optimization; (3) DC bias stability, environmental stability, and packaging; and (4) device demonstrations and system applications. In this section we summarize the principles of operation of these high-speed devices and describe the significant results in each of these four important areas.

9.2.1.1 Background and Motivation

As discussed in the introduction to this chapter, if we consider the waveguide injection laser as the first application of integrated optics to commercial long-distance fiber optic transmission systems, then waveguide modulators are arguably the second. The primary application of the waveguide modulators, which have recently been deployed for the first time, is as the data-encoding element of the optical transmitter. We illustrate the difference between how the information, which is in the form of an electrical signal, is encoded onto the optical carrier in the earlier and present approaches in Fig. 9.1.

The earlier approach employed in digital telecommunication systems was *direct* modulation of the laser, in which the perturbation that causes the modulation is produced *internal* to the laser cavity. In the typical configuration for direct modulation, which is depicted in Fig. 9.1a, the electrical signal is combined with a bias current and applied to the terminals of the laser diode. The electrical signal thereby *directly* modulates the laser gain, and consequently the optical intensity at the laser chip's output facet. A basic advantage of this approach is its inherent simplicity. However, when attempts are made to modulate at data rates beyond about 1–2 Gb/s, the complex dynamics of the coupling of the electron and photon densities within the laser cavity cause undesirable optical frequency variations, referred to as *chirp*, during and following the transitions in the driving data stream. Although examples of low-chirp lasers have been demonstrated, attaining adequate yields for system use has been a very challenging endeavor.

An alternative approach to the direct modulation of the laser is external modulation. This technique is illustrated in Fig. 9.1b. With external modulation, only a constant bias current is applied to the laser diode, which in this case produces a narrow line-width, continuous optical wave or carrier.

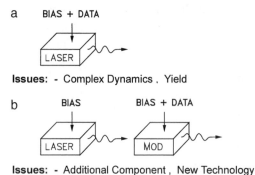

Fig. 9.1   Methods of information encoding in lightwave systems. (a) Direct modulation of the laser. (b) External modulation of the laser.

Data are then encoded onto the optical carrier by applying the modulating signal to the modulator that is *external* to the laser cavity. The advantage of this approach is that by decoupling the processes of generating and modulating the light, the overall performance of the transmitter may be improved. Key features of external modulators that make them attractive for this application are their broad modulation bandwidth, large extinction ratio, excellent spectral purity of the transmitted signal, high optical power handling capability, low modulation distortion, and fabrication yield and reproducibility. Of course, to make use of the improved performance, the added complexity and reliability issues of using an additional component and a new technology must be addressed. In particular, the improved performance is expected to be achieved at a lower total cost. Today, both discrete external modulators and external modulators integrated onto the same chip as the laser are available. Both the hybrid and integrated types may be fabricated in LiNbO$_3$ (Suche *et al.* 1993) and InGaAsP technologies (Chapter 4, Volume IIIB) and consequently provide system designers with a wide selection of options.

The application areas where external modulators can find use may be classified by market segments and modulation formats. In addition to long-distance base-band digital telecommunication applications, which are the subject of this chapter, other potential communication uses include the local loop, the trunk and distribution networks of analog and digital subcarrier-multiplexed community-antenna television (CATV) systems (Childs and O'Byrne 1990), analog RF subcarrier links (Stephens and Joseph 1987; Cox, Betts, and Johnson 1990), and base-band digital computer data communication. Other key markets include analog instrumentation (Jungerman and Dolfi 1991) and sensors–transducers (Bulmer and Burns 1984; Leonberger, Findakly, and Suchoski 1989; Ezekiel and Udd 1991; Tsang and Radeka 1995). All these applications are important and worthy of further consideration in their own right.

9.2.1.2   Present and Near-Term Telecommunication Applications

In Table 9.1 we list some of the present and very near-term fiber communication systems being engineered and deployed. The systems are categorized according to the modulation format, frequency or bit rate of operation, and typical transmission distances. It may be surmised from this table, and the other chapters of this book, that the trends in these systems are toward ever higher capacity, lower cost per unit information bandwidth, graceful

Table 9.1   **Present and Near-Term Communications Applications**

Application (Digital/Base Band)	Bit Rate or Frequency (Gb/s)	Typical Distance (km)
Transoceanic	2.5–10	10,000
Repeaterless	2.5–10	250–500
Terrestrial long haul	2.5–10	100–1000
Terrestrial short haul	2.5–10	50
(Analog/Subcarrier)	**(GHz)**	**(km)**
Community-antenna television (AM-VSB)[a]	1	5
Antenna remoting	10–50	10

[a] AM-VSB, amplitude modulation–vestigial sideband.

growth scenarios, and reliability. The modulation data rates (or frequencies) and transmission distances being sought for all these systems push the limits of photonic technologies. The photonic technology solutions that are making these systems possible are low-loss fibers, low-noise optical amplifiers, high-speed external modulation, narrow-linewidth lasers, low cross-talk wavelength-division multiplexers, and high-speed receivers.

9.2.1.3   System Challenges and Solutions

Initially considered for their innate ability to attain high-modulation speeds (Kaminow, Bridges, and Pollack 1970), external electrooptic modulators were also recognized for their disposition toward low modulation chirp and, consequently, as a means of minimizing the degradations arising from dispersion in the transmission fiber (Korotky, Eisenstein, Gnauck, *et al.* 1985). The difficulties with direct laser modulation arise because the change in charge carrier density associated with the transition in the input current not only changes the optical gain of the material, but also changes the real part of the index of refraction within the laser cavity. The latter change causes the lasing frequency to shift — i.e., chirp — and thereby to broaden the optical spectrum of the transmitted pulses to a width greater than the minimum set by the Fourier transform limit (Koch and Bowers 1984; Kobayashi and Mito 1989). Unfortunately, when combined with the group velocity dispersion of standard single-mode silica fibers, the sign (direction) and magnitude of the frequency chirp for typical lasers result in an acceleration of the rate at which the pulses spread and interfere as they propagate along the fiber. External modulators based on changes of the real part of the index of refraction, on the other hand, theoretically may be operated without producing any chirp (Koyama and Iga 1988).

To illustrate the nature of the limitations caused by the combination of fiber dispersion and modulation chirp, in Figs. 9.2 and 9.3 we show the results of numerical calculations that simulate the behavior of direct laser amplitude modulation and external amplitude modulation. First, in Fig. 9.2, we graph the transmitted optical spectra from a typical directly modulated laser and from a typical externally modulated laser. The model assumes that in both cases the electrical drive signal is a sinusoidal waveform having a magnitude that produces 100% optical modulation depth, and that the devices may be characterized by a modulation chirp parameter, which is commonly denoted as α. As we elaborate later, the α parameter characterizes the relative amounts of amplitude and phase modulation present in the transmitted signal. Ideal amplitude modulation corresponds to $\alpha = 0$. The value of the α parameter for a typical directly modulated laser is +4. Virtually all types of external modulators can provide $|\alpha| < 1$, and some, such as the lithium niobate Mach–Zehnder modulator, can readily obtain $|\alpha| < 0.1$ (Korotky et al. 1991).

Visible in the optical spectra are the discrete modulation sidebands and the optical carrier frequency at the center of each group. It is immediately clear from this figure that the spectral width of the transmitted signal is significantly broader for the case corresponding to the directly modulated

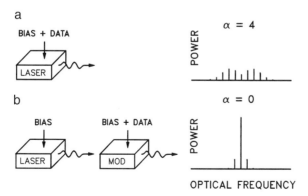

Fig. 9.2 Comparison of transmitted spectra for (a) direct and (b) external modulation.

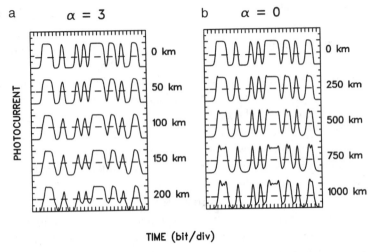

Fig. 9.3 Propagation of 2.5-Gb/s 1.5-μm non-return-to-zero (NRZ) data through conventional fiber. (a) Direct modulation. (b) External modulation. Note that the propagation distance scales for the two panels differ.

laser. In the case of an isolated, linearly chirped, Gaussian-shaped intensity pulse, it may be shown that the optical spectrum is broader than the Fourier-transform-limited value (corresponding to $\alpha = 0$) by a factor of $\sqrt{(1 + \alpha^2)}$. When the signal is passed through the fiber, the wider the optical spectrum, the greater the spread in arrival time of the various frequency components; that is, the pulse broadens.

The consequences of increased pulse broadening introduced by chirp for a 2.5-Gb/s digital transmission system are immediately evident from Fig. 9.3. In this figure, we again compare the situation for a directly modulated laser with that for an externally modulated laser. For the simulation we have assumed a chromatic dispersion coefficient for the fiber of 17 ps/nm-km, which is typical of the single-mode fiber deployed extensively in the United States at the low-loss wavelength of 1.5 μm. Plotted are snapshots of the evolution of a set of $2^5-1 = 31$ pseudo-random optical data bits encoded in the non-return-to-zero (NRZ) format as they would appear having traveled various distances in the fiber. The photocurrent that a detector would generate, which is proportional to the optical intensity, is plotted along the vertical axis, and time is plotted along the horizontal axis. The dashed horizontal lines indicate the position of the decision threshold for determining whether the signal voltage should be interpreted as a zero or one digital level. In the presence of noise, the closer the signal to the decision threshold, the greater the probability of a decision error.

The theoretical dispersion limit for such an on–off-keyed system is defined as the distance for which the reduction in the signal margin for a pure amplitude-modulated waveform (i.e., $\alpha = 0$) corresponds to a 1-dB optical power penalty for a predetermined bit error rate (BER), which is typically assumed to be 10^{-9} (Henry 1986). For the present purposes this may be interpreted as the distance at which the worst case difference between the high and low optical intensity levels has been reduced to 80% of the initial peak-to-peak difference. Thus, by examining Fig. 9.3b, we note that the theoretical dispersion limit for the system under consideration, which may be reached using external modulation, is about 750 km. However, as may be seen from Fig. 9.3a, a similar reduction of the optical margin for a typical directly modulated laser occurs for a propagation distance of only approximately 75 km. Thus, for systems that operate at high bit rates and over large cumulative dispersion, externally modulated laser transmitters are essential.

Another aspect of external modulators that has proven valuable in improving the performance of the latest generation of fiber transmission systems, which include optical amplification, is the nature of their modulation response. Unlike lasers, which exhibit overshoot–undershoot ringing (referred to as relaxation oscillations) because of the coupling of the gain population inversion and cavity photon densities, the response of external modulators, such as the electrooptic modulators, is parametric — i.e., essentially prompt and proportional to the drive signal. This allows external modulators also to attain the high extinction ratio that they are capable of without sacrificing low-chirp performance or digital waveform fidelity. A high extinction ratio is desirable in optically amplified systems to minimize the noise level in the zero states, which will increase through the process of stimulated emission in the amplifiers in response to the background light level. Ideally, if the transmitter launches the optical waveform with an infinite extinction ratio, the noise in the zeros will increase only because of the spontaneous emission component. Numerical simulations indicate that the signal-to-noise ratio (SNR) at the receiver in transoceanic optically amplified digital systems, which include hundreds of optical amplifiers, increases by approximately 1 dB for each 1-dB increase of the transmitted extinction ratio when the extinction ratio is near 15 dB. Each decibel improvement in the SNR is extremely valuable because this margin may be used, for example, to increase significantly the distance between optical amplifiers, which has a direct impact on the cost of the system.

9.2.2 *MODULATOR FUNDAMENTALS*

9.2.2.1 Modulator Specifications

The characteristics, and hence the specifications, of modulator performance may be divided into several categories. When we treat a modulator as an object that has both inputs and outputs, it is convenient to classify its specifications according to the following categories: electrooptic characteristics, optical characteristics, electrical characteristics, mechanical and environmental characteristics, and, finally, financial characteristics. Clearly, to determine the suitability of a given technology to a given task, all these categories and their trade-offs should be assessed. As an example with which to illustrate these categories in greater detail, we consider the specifications of an amplitude modulator in Table 9.2.

Space does not permit us to address all these characteristics and their interrelationships in this chapter. Instead, our emphasis is primarily on the electrooptical characteristics of the devices because these are essential functional features of modulators. Researchers and devleopers of devices should not assume, however, that these are always the most important characteristics that system designers use in selecting among competing technologies. In some cases system designers may choose to sacrifice electrooptic performance to achieve other goals, such as the perceived ease of system integration. Also, risk management, market pressures, and contractual arrangements act to create a preference for technologies that are available from multiple suppliers over those that simply offer higher raw performance.

Already, in the beginning of this section, we have introduced several of the concepts that fall under the category of electrooptic characteristics, such as chirp and distortion (e.g., ringing). To complete this subsection we define some of the other key modulator characteristics. Perhaps the most basic characteristic of any electrooptic intensity modulator is the relationship between the input voltage and the output light level at low frequency, which is often referred to as the *switching curve*. In Fig. 9.4 we illustrate the switching curve for a generic electrooptic intensity modulator. Plotted is the transmitted — i.e., output — light intensity I versus the input drive voltage V. Except in special cases, the modulator is used in a region of the switching curve where the relationship between output and input is single valued and monotonic. Note that the switching curve has come to be defined and measured as a low-frequency characteristic of modulators; i.e., it is

Table 9.2   **Amplitude Modulator Characteristics**

Electrooptic Characteristics
Switching voltage–current
Extinction ratio
Frequency response (amplitude, phase, ripple)
Chirp
Distortion
Bias voltage stability

Optical Characteristics
Level of integration
Operating wavelength range
Polarization dependence
Input/output fibers and connectors
Insertion loss
Return loss
Maximum input power

Electrical Characteristics
Number of inputs
Input impedance
Connector type(s)
Return loss
Bias voltages/currents and controls

Mechanical and Environmental Characteristics
Size
Temperature sensitivity
Humidity sensitivity
Shock and vibration sensitivity
Fiber pigtail pull strength
Failure rate

Financial Characteristics
Research and development time and resource costs
Manufacturing supply, tooling, and yield costs
User installation, maintenance, and inventory costs
Availability of second source

determined under conditions where the rate that the applied voltage is swept to map out the switching curve is very small compared with the modulation bandwidth of the modulator.

Again referring to Fig. 9.4, we can see that if the input optical intensity

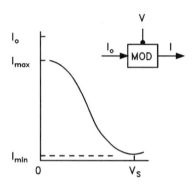

Fig. 9.4 External modulator switching curve. The curve illustrates light output versus applied voltage for a generic modulator.

is I_o and the maximum of the output light intensity is I_{max}, then the transmittance of the modulator is usually defined as the ratio I_{max}/I_o, and the insertion loss (in decibels) is defined as

$$loss\ (\text{dB}) = 10\ \log_{10}\ (I_{max}/I_o). \tag{9.1}$$

Similarly, the extinction ratio of the modulator is usually defined in relation to the I versus V switching curve. If the maximum light intensity output is I_{max} and the minimum intensity that the optical output may be extinguished to is I_{min}, then the extinction ratio is $ER = I_{min}/I_{max}$, or

$$ER\ (\text{dB}) = -10\ \log_{10}\ (I_{min}/I_{max}). \tag{9.2}$$

The switching voltage V_S is defined as the difference (absolute value) between the voltages that correspond to the maximum and minimum output intensities, I_{max} and I_{min}. Again, it is standard and accepted practice to measure and quote the switching voltage at virtual DC for base-band modulators. (*Note:* Unless explicitly stated otherwise, throughout this chapter the use of the terms *switching voltage* and *extinction ratio* imply this operating condition. The qualifier *intrinsic* may occasionally be used to emphasize that these conditions apply.) It should be clear that if one is willing to operate with a worse extinction ratio, the required amplitude of the modulation voltage is also reduced. Finally, note that when the DC switching voltage and the extinction ratio are taken together with a model of the device dynamics and independent measurement of the modulation bandwidth, it is possible to deduce the behavior for any modulation frequency or waveform. At the end of this subsection we address the question of defining and measuring the dynamic switching voltage and extinction ratio for high-speed modulation conditions without reference to the modulation bandwidth or a device model.

Of course, a key characteristic of external electrooptic modulators is their modulation frequency response, which is often summarized by a single number, the modulation bandwidth. The most common definition of the modulation frequency response for electrooptic modulators is specified in relation to how these modulators are normally used in systems. In Fig. 9.5 we provide a schematic of a typical lightwave communication link based on an external electrooptic intensity modulator. This link also serves to define the frequency response and indicates one method of how to measure it. The link consists of a continuous-wave (CW) optical laser source (e.g., a laser diode), the electrooptic modulator, and a square-law photodetector, such as a p-i-n photodiode. The light from the optical source is coupled into the modulator, and the modulated lightwave signal is coupled onto the photodetector. The electrical signal to be transmitted is applied to the electrooptic modulator, optically transported to the detector, and converted back to an electrical signal by the detector.

To define and measure the modulator's frequency response, the input signal is chosen to be a sinusoidal waveform, the frequency of which we denote as ν. Because all intensity modulators will exhibit some degree of nonlinearity in their switching curves, the magnitude of the drive waveform is chosen to be small compared with the switching voltage. This ensures that

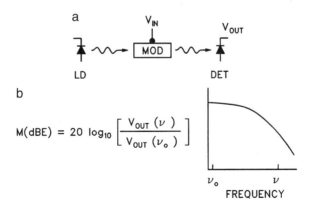

Fig. 9.5 Modulation bandwidth of electrooptic modulators. (a) Typical measurement arrangement. (b) Schematic modulation response. DET, photodetector; LD, laser diode; MOD, modulator.

the linear approximation is valid and hence that the modulation frequency response is uniquely specified. In the small-signal, or linear, regime, the electrical signal produced by the photodetector will predominantly be a sinusoidal waveform at the same frequency as the modulator drive signal, ν. If $V_{out}(\nu_o)$ represents the amplitude of the output of the detector at a reference frequency ν_o, then the magnitude of the modulation frequency response $M(\nu)$ expressed in electrical decibels (dBE) may be defined as

$$M(\nu) \ (\text{dBE}) = 20 \log_{10} [V_{out}(\nu)/V_{out}(\nu_o)]. \tag{9.3}$$

Thus, $M(\nu)$ represents the relative efficiency of transferring RF power over the optical link. The modulation bandwidth $\nu_{-3\,\text{dBE}}$ is defined as the contiguous frequency range over which the modulation response is greater than -3 dBE relative to the maximum response. This definition is consistent with the definition used for directly modulated lasers and thus allows a direct comparison of the two. In some cases, a modulator's response may be quoted in terms of the frequency at which the optical modulation depth decreases to 50% of its maximum value. This situation corresponds to the frequency at which the received electrical RF power has fallen by 6 dBE. For consistency, throughout this chapter *all* references to the modulation bandwidth, both theoretical and experimental, conform to the usage implied by Eq. (9.3) and implicitly refer to ν at -3 dBE. Finally, note that by using phase-lock detection of the signal at frequency ν at the receiver, both the magnitude and the phase of the electrooptic modulator frequency response may be determined. The use of phase-lock detection of the fundamental component also ensures that the measured response corresponds to the linear modulation response, even when the drive signal is not, strictly speaking, small.

The reader has likely already realized that in the procedure that we have just described to determine the electrooptic modulator's frequency response, we have assumed that the detector's frequency response could either be ignored because it is nearly constant or be calibrated out. This is indeed true; however, for some types of electrooptic modulators that exhibit well-understood and frequency-independent nonlinearities, such as the Mach–Zehnder interferometric modulator, which we discuss in greater detail later, it is possible to measure the modulation response without knowledge of the detector's response (Uehara 1978; Jungerman *et al.* 1990). Also, optical spectral techniques may be used to determine the frequency response of electrooptic modulators, including optical phase modulators (Kaminow 1974).

Having introduced the concept of the modulation frequency response, we now address the subject of the dynamic switching voltage and extinction ratio. The latter characteristics are clearly large-signal parameters, whereas the former is deliberately constructed as a small-signal parameter in acknowledgment of the nonlinear nature of the switching curve of all known external modulators. Therein lies the practical dilemma, because neither the dynamic switching voltage nor the dynamic extinction ratio may be uniquely defined, and thus measured, independent of a detailed knowledge of the drive waveform conditions. A consequence, for example, is that by the very nature of the fact that modulators have a finite bandwidth, the observed dynamic extinction ratio will degrade as the bandwidth of the modulation waveform is increased — that is, unless the modulation waveform is preemphasized to account for the modulator's frequency response. With proper preemphasis, such as may be accomplished with electrical equalization, an extinction ratio equal to the DC value may be achieved for very high modulation rates. For purposes of creating and confirming large-signal specifications, there is clearly one type of waveform that is universally reproducible and easily equalizable, namely a single-frequency tone. Therefore, often a modulator intended for digital NRZ modulation is required to meet a dynamic switching voltage specification for a sinusoidal waveform having a frequency equal to approximately one-half the clock rate.

9.2.2.2   Waveguide Electrooptic Phase Modulation

In previous texts we and other authors have described the basic principles of optical waveguides, electrooptic phase and amplitude modulation, high-speed modulator design, and device fabrication (Kaminow 1974; Tamir 1979; Hunsperger 1982; Hutcheson 1987; Korotky and Alferness 1988). Consequently, in this section we give only a brief summary of these essentials and devote the main portion of the discussion to the advances that have occurred in the state of the art since the publication of the previous volume in this series.

Both the phase and amplitude modulators that we describe in this section are based on electrooptically induced phase changes to the optical signal as it propagates along a waveguide formed in the lithium niobate crystalline substrate. The details of the perturbation caused by the electrooptic effect to the polarizability ΔP of the electric field E^o of the propagating optical wave in general depend upon the relative orientations of the polarization state of the input optical signal, the crystalline axes and crystal cut, and the applied electric field (Kaminow 1974; Korotky and Alferness 1987). Consequently, the electrooptic effect may be formally described as a change to the dielectric tensor $\Delta\varepsilon$ of the substrate induced by the applied electric field E^a. Mathematically, this may be expressed as

$$\Delta P_i = \sum_{j=1}^{3} \Delta\varepsilon_{ij} E_j^o, \tag{9.4}$$

where the subscripts denote indices corresponding to the Cartesian coordinate basis vectors. For lithium niobate, which is a uniaxial crystal, the electrooptically induced change to the dielectric tensor may be written as

$$\Delta\varepsilon_{ij} = \begin{bmatrix} -r_{22}E_y^a + r_{13}E_z^a & -r_{22}E_x^a & r_{51}E_x^a \\ -r_{22}E_x^a & r_{22}E_y^a + r_{13}E_z^a & r_{51}E_y^a \\ r_{51}E_x^a & r_{51}E_y^a & r_{33}E_z^a \end{bmatrix}, \tag{9.5}$$

where $r_{\alpha\beta}$ are the nondegenerate electrooptic coefficients specific to lithium niobate. The diagonal terms in this matrix represent electrooptically induced changes of the phase of the propagating optical signal, and the off-diagonal terms represent polarization converison. Measured values of the electrooptic coefficients for LiNbO$_3$ and other crystals have been tabulated (Kaminow 1974; Kaminow 1987). The largest of the electrooptic coefficients of the diagonal tensor elements for LiNbO$_3$ is r_{33}, which has a value of approximately 31×10^{-12} m/V.

When we consider the electrooptically induced changes of the phase of the optical carrier, it is convenient to reformulate the modification of the dielectric tensor in terms of changes to the index of refraction for the orthogonal input linear polarization states. This is because, assuming that the device is uniform along its length, the optical phase shift $\Delta\Phi$ that is produced is proportional to the product of the electrooptically induced change in the index of refraction Δn_{eo} and the active length of the device L. Explicitly,

$$\Delta\Phi = \Delta\beta_o L = k_o \Delta n_{eo} L, \tag{9.6}$$

where $\Delta\beta_o$ is the change in the optical propagation constant, $k_o = 2\pi/\lambda_o$ is the optical wave number, and λ_o is the free-space optical wavelength of the light source. By considering the perturbation of the dielectric tensor, we can show that the local change in the index of refraction Δn_{eo} for light linearly polarized along a principal axis of the crystal may be written as

$$\Delta n_{eo} = \frac{-n^3 r}{2} E^a, \tag{9.7}$$

where n is the unperturbed index of refraction of the material, E^a is the relevant component of the local applied electric field, and r is the relevant electrooptic coefficient. In Fig. 9.6 we illustrate a common configuration for a single-transverse-mode waveguide electrooptic phase modulator fabricated in LiNbO$_3$ and list typical values of the parameters that determine the magnitude of the electrooptic index change. Note that in such an optical waveguide device, it is the change of the effective index of refraction of the guided optical mode, denoted ΔN_{eo}, that is of interest. In this case the perturbation is a spatial average of the index change Δn_{eo} over the modal optical intensity profile. As a method of summarizing these details and the overall electrooptic efficiency of the design, often the average over the spatially varying optical and applied electric fields is expressed in the form of the product of a prototypical electric field strength and an overlap factor. Thus, on the basis of perturbation theory, the change of the effective index is frequently written as

$$\Delta N_{eo} = \frac{-n^3 r}{2} \Gamma \frac{V}{G}, \tag{9.8}$$

where V is the differential voltage applied to the electrodes, G is the gap between the electrodes, and Γ represents the spatial overlap of the optical intensity profile and the applied electric field. At an optical wavelength of 1.5 μm, the value of the index of refraction for LiNbO$_3$ is approximately

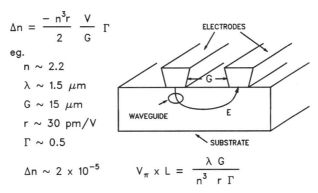

$$\Delta n = \frac{-n^3 r}{2} \frac{V}{G} \Gamma$$

eg.

$n \sim 2.2$

$\lambda \sim 1.5\ \mu m$

$G \sim 15\ \mu m$

$r \sim 30\ pm/V$

$\Gamma \sim 0.5$

$\Delta n \sim 2 \times 10^{-5}$

$$V_\pi \times L = \frac{\lambda\ G}{n^3\ r\ \Gamma}$$

Fig. 9.6 Electrooptic modulation (Pockel's effect). The parameters correspond to a prototypical Ti:LiNbO$_3$ optical waveguide device.

2.2, and typical values for the electrode gap and overlap factor are $G = 15\ \mu m$ and $\Gamma = 0.5$, respectively. From this the reader may infer that the change in the effective index of refraction caused by the electrooptic effect is approximately $\Delta N_{eo} = 2 \times 10^{-5}$. Another figure of merit that may be deduced is the voltage–length product $V_\pi \times L$ required to induce a π phase shift of the optical carrier wave. On the basis of the aforementioned relations and parameters, we derive

$$V_\pi \times L = \frac{\lambda G}{n^3 r\ \Gamma}, \tag{9.9}$$

and we estimate that $V_\pi \times L = 14$ V-cm for $\lambda = 1.5\ \mu m$ wavelength.

9.2.2.3 Waveguide Electrooptic Amplitude Modulation

Electrooptic devices operate by an index change produced by an electric field, so the modulation of the optical phase velocity is the single ingredient, albeit a versatile one, that other optical functionality must be fashioned from. Amplitude modulation in the form of on–off-keying is currently the preferred method of transmitting data optically in long-haul systems, and many optical circuit configurations are known for converting the fundamental electrooptic phase modulaton into amplitude modulation. Among these switch–modulator structures are a differential polarization phase retarder combined with polarizers (Kaminow and Turner 1966), a phase modulator combined with a Fabry–Perot spectral filter (Kaminow 1963), a phase modulator within a Mach–Zehnder interferometer serving as a spatial filter (Ramaswamy and Standley 1976), spatial directional couplers (Miller 1954), and spatial modal evolution junctions (Burns, Lee, and Milton 1976), to name a few.

Of the various possibilities, the waveguide interferometric and directional coupler switch modulators have been the more popular options because of their simplicity and versatility (Korotky and Alferness 1988). Although the first gigabit-per-second digital fiber optic transmission experiments investigating external modulation used directional coupler switch modulators (Korotky, Eisenstein, Gnauck, *et al.* 1985), interferometric modulators have come to be preferred because they permit higher modulation speed for a given drive voltage and a larger extinction ratio at high speed (Korotky and Veselka 1990). The better bandwidth–voltage performance of the interferometric modulator compared with that of the directional coupler switch is not because of a fundamental difference between the two, however (Korotky and Alferness 1983; Korotky 1986). Rather, it is a consequence of the difference between the electrical signal loss that can be attained on the drive electrodes of the two structures and the subtleties of how the variation of the amplitude of the electrical signal, caused by attenuation, affects the switching curve of the directional coupler. In a waveguide electrooptic interferometric modulator, the interwaveguide separation is adjustable independent of the width and the gap of the electrode. This additional degree of freedom results in an optimization having a larger electrode width and gap than is possible for a typical directional coupler, and therefore a lower electrical attenuation and a higher bandwidth.

That the interferometric modulator is sensitive only to the integral of the electrooptically induced phase-velocity difference between the waveguide arms along the interaction length L implies that a high extinction ratio may be attained independent of the modulation drive frequency. This is not the case for the directional coupler, which has an extinction ratio that is sensitive to local variations in the ratio of the phase-velocity difference to the coupling coefficient in addition to the global difference (McCaughan and Korotky 1986). For these reasons we restrict the following discussions on high-speed amplitude modulation and switching to the interferometric structures.

Waveguide electrooptic Mach–Zehnder interferometric switch–modulators may be realized in a variety of configurations. All consist of single-transverse-mode optical waveguides, a means of splitting the input optical wave between a pair of decoupled waveguides (often referred to as the *arms* of the interferometer), a set of electrodes for controlling the relative optical phase between the optical fields in the two arms, a means of combining the phase-modulated optical fields, and a means of spatially filtering the resultant of the interfered waves. The most basic and easiest to fabricate variant is the Y-branch structure, which is illustrated in Fig. 9.7. This configuration takes its name from the Y-shaped junctions that are used to split and recombine the optical power into and from the waveguide arms. Ideally, each Y branch produces balanced power splitting and combining. The device then operates as follows: In the absence of any applied voltages to the electrodes, light entering the input single-mode waveguide is divided equally into the two identical waveguide arms with zero relative phase difference between the split fields. The separated optical fields propagate along the arms and are then combined in the output junction. Because the arms have equal lengths, the two fields arrive at the output junction in phase, and the total power is combined into the output waveguide, which may be deduced by the consideration of time-reversal symmetry.

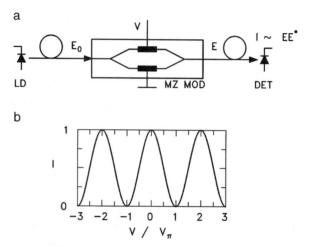

Fig. 9.7 Waveguide Mach–Zehnder (MZ) interferometer. (a) Schematic of the device (top view). (b) Theoretical switching curve.

The situation just described corresponds to the ON state of the modulator. The OFF state is obtained by introducing a π phase shift between the separated fields, which is accomplished by applying a voltage to the electrodes and using the electrooptic effect. Now, the combined field that impinges on the output waveguide is antisymmetrica — i.e., its transverse mode profile has negative (odd) symmetry (parity) upon spatial inversion about the center of the output waveguide. Because by construction the output waveguide is a single-transverse-mode waveguide and this single mode has even-parity symmetry, the odd-parity light field from the branches cannot be coupled into the output waveguide. Instead, the light is radiated into the substrate. This fascinating behavior is illustrated in Fig. 9.8, which is the output from a beam-propagation-method calculation. (*Note*: This OFF state differs from the case for hollow metallic waveguides, in which there are not radiation modes. Consequently, the antisymmetrical wave, which cannot exit the output, is reflected back to the source.) If the input and output branches are ideal 50:50 splitter–combiners, the transmitted optical intensity is described by the relation

$$\cos^2\left(\frac{\pi \Delta N_{eo} L}{\lambda}\right) = \cos^2\left(\frac{\pi}{2} \cdot \frac{V}{V_s}\right). \tag{9.10}$$

For interferometers it is common to refer to the switching voltage V_s as the *half-wave voltage,* or V_π.

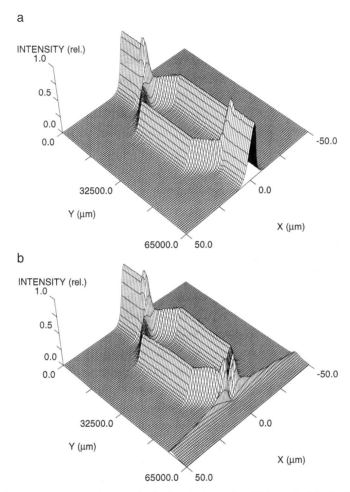

Fig. 9.8 Beam-propagation-method simulation of a Y-branch Mach–Zehnder modulator. The figure depicts (a) the ON state and (b) the OFF state of the modulator. The x axis represents the spatial coordinate that is transverse to the direction of optical propagation, and the y axis represents the spatial coordinate that is along the direction of propagation. The optical intensity is plotted along the z axis.

Besides being designated as a *Y-branch modulator,* the structure just described is also known as a 1×1 Mach–Zehnder modulator, in reference to the number of its input and output optical waveguides. Other commonly used interferometric structures replace one or both of the 1×1's Y branches with a 3-dB directional coupler. In the case where the output Y branch is replaced, the device is referred to as a *1 × 2 Mach–Zehnder switch–modulator*. This configuration has several attractive features; the main one is that the switching action causes the light to be directed from one output waveguide to the other, as opposed to being lost in the substrate. Consequently, the 1 × 2 Mach–Zehnder modulator is an extremely versatile device. In addition to digital amplitude modulation, the 1 × 2 configuration finds use where having two complementary output signals is necessary or of increased value. Examples include broadcasting in CATV systems, diverse routing for protection (e.g., optical ring networks), optical time-division demultiplexing, and polarization scrambling. Of course, the input Y branch may also be replaced with a 3-dB directional coupler to form a four-port 2 × 2 Mach–Zehnder switch–modulator, which is also known as a *balanced-bridge Mach–Zehnder interferometer* (Ramaswamy, Divino, and Standley 1978). Interestingly, as we discuss in the following section, carefully designed Mach–Zehnder modulators may be used for phase modulation in addition to amplitude modulation.

9.2.3 EXTINCTION RATIO AND CHIRP

Two of the extremely important characteristics of external modulators for telecommunication applications are the extinction ratio and chirp, as was discussed previously. The relations necessary to calculate both of these may be derived for the waveguide electrooptic Mach–Zehnder modulator by keeping account of the optical field as it propagates through the device. A main factor limiting the intrinsic extinction ratio for interferometers can be the power splitting ratio in the Y branches or directional couplers. To model imperfect power splitting, we suppose that the input Y branch divides the power in the ratio $a^2/(1 - a^2)$ and that the output Y branch divides power in the ratio $b^2/(1 - b^2)$, as indicated in Fig. 9.9. The factor that

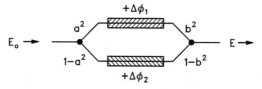

Fig. 9.9 Waveguide branching and modulation ratios for an interferometric modulator.

determines the chirp produced by the Mach–Zehnder modulator is the ratio of the two electrooptically induced phase shifts, $\Delta\Phi_1(t)$ and $\Delta\Phi_2(t)$, produced in the two arms of the interferometer (Koyama and Iga 1988; Korotky *et al.* 1991). If we denote the guided-wave optical field that propagates in the input waveguide of the interferometer by E_o, then the output field E is given by

$$E = E_o\{abe^{-\sigma_1 L}\ \exp[i\Delta\Phi_1(t)] \\ + \sqrt{1-a^2}\sqrt{1-b^2}e^{-\sigma_2 L}\ \exp[i\Delta\Phi_2(t)]\}, \quad (9.11)$$

with $0 \leq a, b \leq 1$. The factors $e^{-\sigma_1 L}$ and $e^{-\sigma_2 L}$ account for the optical propagation loss that occurs in each of the waveguide arms. In the absence of optical reflections, Eq. (9.11) implicitly represents a complete model of a Mach–Zehnder modulator transmitter, which may be used for system-level simulations. (*Note*: The input field may be taken as $E_o = \psi(x, y)e^{+i(\beta_o z - w_o t)}$, where x and y are the transverse coordinates, z is the direction of propagation, ω_o is the optical angular frequency, β_o is the optical propagation constant, and $\psi(x, y)$ represents the transverse mode of thel field.)

To illustrate the effect that the splitting ratio has on the extinction ratio, we may consider the case when the input splitting is imperfect with $a^2 = \frac{1}{2} + \varepsilon$, the output splitting is ideal with $b^2 = \frac{1}{2}$, and there is no differential waveguide propagation loss. In this situation the extinction ratio is $ER = \varepsilon^2/2$. Experimentally, intrinsic extinction ratios greater than 30 dB are routinely attainable, which indicates that the power splitting in the branches is better than the ratio of $\frac{46}{54}$. If an interferometric modulator is driven with a single-frequency source, the peak dynamic extinction ratio can approach the intrinsic extinction ratio (Korotky and Veselka 1990). In the case of pseudo-random NRZ modulation at 5 Gb/s, instrument-limited dynamic extinction ratios of 15–17 dB over the central 100 ps of the bit time slot have been observed (N. R. Dietrich, personal communication, 1993). As a final note relating to the modulator extinction ratio, we point out that the phase of the background light level, which can influence how pulses propagate in a nonlinear medium, depends on the mechanism that limits the extinction ratio. For example, the phase of the background level will have opposite signs in the cases where the modulation voltage swing either under- or overdrives the device.

Next we consider the modulation chirp characteristics of optical transmitters, which is summarized by the chirp parameter, α. The dynamic chirp $\alpha(t)$ represents the instantaneous ratio of the phase modulation to amplitude modulation of the transmitted optical field and may be expressed as (Koyama and Iga 1988)

$$\alpha(t) = \frac{\dfrac{d\phi}{dt}}{\dfrac{1}{2I} \cdot \dfrac{dI}{dt}}, \tag{9.12}$$

where ϕ and I are, respectively, the phase and intensity of the optical field E, and t denotes time. Clearly the chirp is a function of the type of optoelectronic transducer, as well as the particular operating point and modulation depth. Like the modulation bandwidth, the chirp parameter is uniquely defined only in the limit of small-signal modulation. To illustrate the basics of the chirp behavior of the Mach–Zehnder modulator, we examine the 1×1 Y-branch modulator for the case where the splitting ratios of both branches are ideal — i.e., $a^2 = b^2 = \frac{1}{2}$ — and there is no differential waveguide propagation loss. In this situation, Eq. (9.11) can be rewritten as

$$E = E_o \exp[i\{\Delta\Phi_1(t) + \Delta\Phi_2(t)\}/2] \cos[\{\Delta\Phi_1(t) - \Delta\Phi_2(t)\}/2]. \tag{9.13}$$

In this case it is evident that in general the output field contains both phase (exp) and amplitude (cos) modulation contributions. Also, we observe that if the two waveguide arms are driven by complementary waveforms — i.e., $\Delta\Phi_2(t) = -\Delta\Phi_1(t)$ — then the phase modulation term is identically zero, and pure amplitude modulation without chirp is obtained for an arbitrarily large modulation depth.

We may derive a concise expression for the small-signal modulation chirp parameter, α, for Mach–Zehnder modulators based on the linear electrooptic effect for the typical operating conditions where the modulator is biased at the midpoint of its switching curve and the two arms, 1 and 2, are driven by synchronized signals having the same waveform shape. Under these circumstances α is independent of time and may be written uniquely in terms of the peak-to-peak voltages, ΔV_1 and ΔV_2, applied to the two arms of the interferometer, viz.

$$\alpha = \frac{\Delta V_1 + \Delta V_2}{\Delta V_1 - \Delta V_2}. \tag{9.14}$$

Although this expression applies rigorously only in the small-signal regime, it is accurate to relatively large values of the modulation depth because of the shape of the Mach–Zehnder switching curve. We note too that by choosing the relative magnitude and sign of the drive amplitudes, values of α in the range $-\infty \leq \alpha \leq +\infty$ may be selected. Ti:LiNbO$_3$ waveguide electrooptic Mach–Zehnder modulators having independent drive electrodes for each arm of the interferometer have been constructed to take advantage of this capability (Korotky et al. 1991). Chirp parameter magnitudes smaller than 0.1 and as large as 4 have been demonstrated using dual-drive Mach–Zehnder modulators.

The ability to control and engineer the chirp of optical transmitters can be a benefit to the performance of optical fiber transmission systems. In the case of systems that must operate far from the zero-dispersion wavelength, low chirp is necessary to ensure low dispersion penalties. Interestingly, optimum performance for such systems is obtained for slightly negative values of the modulation chirp parameter, which together with the sign of the dispersion coefficient of the fiber introduces some amount of pulse compression (Koyama and Iga 1988). This behavior is illustrated in Fig. 9.10. In this figure, the experimentally observed dispersion penalty is plotted versus the modulation chirp parameter for the transmission of 5-Gb/s NRZ pseudo-random data, which were generated using a dual-drive Mach–Zehnder modulator (Gnauck et al. 1991). For transmission through

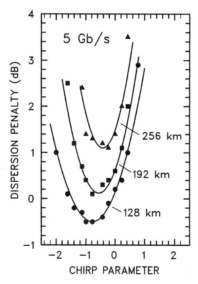

Fig. 9.10   Transmission dispersion penalty versus chirp using a Ti:LiNbO$_3$ modulator.

128 km of fiber having a dispersion coefficient of 17 ps/km-nm, employing the optimum chirp value of $\alpha = -0.8$ is observed to reduce the dispersion penalty by 0.7 dB, which is a significant improvement in the system margin.

9.2.4 HIGH-SPEED OPTIMIZATION

9.2.4.1 Design Strategies and Models

Also among the more important characteristics of external modulators for nearly all applications is the modulation bandwidth. The phenomena that limit the modulation bandwidth may be classified into several categories, such as those listed as follows.

Resistance-capacitance time constant

Transit time

Velocity mismatch

Frequency-dependent loss

Parasitics

Driver current

Extrinsic dynamics

Intrinsic process

In the case of optoelectronic effects, the intrinsic response time is exceedingly fast ($\sim 10^{-13}$ s); therefore, one or more of the other considerations will dominate. To achieve a high modulation bandwidth and a low drive voltage simultaneously, electrooptic modulators are most often designed as traveling-wave modulators. In this configuration the electrical drive signal propagates on a miniature transmission line along the direction of the optical waveguide to permit an extended interaction (hence the designation *traveling-wave*). An example of a dual-drive traveling-wave Y-branch Mach–Zehnder modulator is illustrated in Fig. 9.11. Whereas the long interaction length permits the drive voltage to be kept low, the traveling-wave interaction achieves a higher bandwidth than either the total capacitance or optical transit time would otherwise allow. Under these circumstances the bandwidth is limited by the difference between the microwave and optical velocities, as well as frequency-dependent microwave attenuation (Korotky and Alferness 1983). Next we discuss the advances in modulator optimization that have taken place since the last volume of this book was published and that have provided extraordinary reductions in both the velocity mismatch and microwave attenuation.

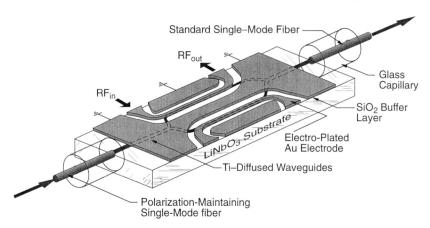

Fig. 9.11   Schematic of a dual-drive Ti : LiNbO$_3$ Y-branch Mach–Zehnder modulator.

To understand the fundamentals of the contemporary approach to modulator design, it suffices to consider the modulation response to a sinusoidal drive waveform for the case where both the source and termination impedances are matched to the modulator transmission line impedance and the electrical and optical wave propagate in the same direction. Under these conditions there is no backward-propagating electrical wave, and the voltage experienced by a photon as it traverses the modulator may be written as (Korotky and Alferness 1987)

$$V(z, t) = \frac{V_o e^{+i(\eta z - \Omega t)}}{2}, \tag{9.15}$$

where Ω is the angular drive frequency, V_o is the peak-to-peak drive voltage, and $\eta = \beta_e - (\Omega/\omega_o)\beta_o$ is a function of the electrical and optical propagation constants, β_e and β_o. The time-dependent electrooptically induced phase shift $\Delta\Phi(t)$ and hence the complex filter function $F(\Omega)$ that the modulator represents are obtained by integrating Eq. (9.15) over z and using the proportionalities specified by Eqs. (9.6) and (9.8). The result is

$$F(\Omega) = \frac{-i}{\eta L}\left(e^{+i\eta L} - 1\right). \tag{9.16}$$

The electrical propagation constant may be written as $\beta_e = N_m k_m + i\,\alpha_A/2$, where k_m is the microwave wavenumber, N_m represents the real part of the microwave index, and α_A is the microwave power attenuation coefficient. Because the optical loss is low, the optical propagation constant may be written as $\beta_o = N_o k_o$, in which case N_o is the effective optical index and k_o is the optical wave number. If we use these relations, η may be cast into the form $\eta = \gamma + i\,\alpha_A/2$, where $\gamma = \Omega(N_m - N_o)/c$ is the microwave-optical walk-off constant that depends on the angular electrical drive frequency Ω and the speed of light c.

Next, we may derive an expression for the modulation frequency response $M(\nu)$. Assuming that the impedances continue to remain matched as the drive frequency is varied, so that there is no backward-propagating wave, the response is then

$$M(\nu) = FF^* = \frac{e^{-\alpha_A L}\sin^2\left(\dfrac{\gamma L}{2}\right) + (\tfrac{1}{4})(1 - e^{-\alpha_A L})^2}{\left(\dfrac{\gamma L}{2}\right)^2 + \left(\dfrac{\alpha_A L}{2}\right)^2}, \qquad (9.17)$$

where we have made use of the relational definition $\Omega = 2\pi\nu$. In this expression for $M(\nu)$, dependences upon the fundamental characteristics of the electrode transmission line — i.e., the active length L, the microwave index N_m, and the attenuation coefficient α_A — are now evident. Note that in the absence of microwave attenuation Eq. (9.17) again simplifies, and we find that the product of the modulation bandwidth and the interaction length depends only on the difference between the microwave and optical indices, viz.

$$\nu_{-3\,\text{dBE}}L \cong \frac{1}{2}\frac{c}{N_m - N_o}. \qquad (9.18)$$

This implies that shorter interaction lengths provide higher bandwidths; however, shorter lengths also require higher drive voltages, as may be inferred from the voltage–length product, Eq. (9.9). Thus, when we optimize and evaluate high-speed modulator performance, it is the bandwidth per unit drive voltage that is the relevant figure of merit that should be considered.

The cross section of a typical traveling-wave lithium niobate electrooptic phase modulator, which is also the heart of other high-speed devices such as amplitude and polarization modulators, is shown in Fig. 9.12. Visible are the lithium niobate substrate, the optical waveguide, the silicon dioxide buffer layer, and the gold electrode; the latter serves as the electrical transmission line. Also indicated are the four fabrication parameters that directly

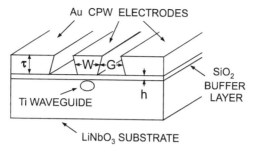

Fig. 9.12   Cross section of a traveling-wave lithium niobate integrated optic modulator. CPW, Coplanar Waveguide.

affect the microwave index, attenuation, and impedance of the electrical transmission line, as well as the switching voltage. These are the width W, gap G, and thickness τ of the coplanar waveguide (ground–signal–ground) transmission line and the height h of the SiO_2 buffer layer. Of course, the permittivities of the constituent materials also directly influence the operation. For completeness we note that the permittivities of the $LiNbO_3$ substrate, SiO_2 buffer layer, and air superstrate are 36, 3.8, and 1.0, respectively. (*Note*: $LiNbO_3$ is an anisotropic crystal; however, the electrodes usually operate in the regime of transverse electric and magnetic fields (TEM) as a consequence of the relatively small dimensions that are used for W, G, h, and τ. In this case the static approximation is valid and the permittivity may be taken as the geometric mean of the permittivities for the relevant axes.)

The goal of optimizing the modulator performance may be interpreted as identifying the values of W, G, h, τ, and the electrode length L that minimize the switching voltage V_π for a specified -3-dBE modulation bandwidth (Korotky 1989). Because for each set of values of W, G, h, and τ there is a unique value of L that yields the desired bandwidth, the functionals of interest are $N_m(W, G, h, \tau)$, $\alpha_A(W, G, h, \tau)$, and $V_\pi(W, G, h, \tau)$. Additionally, the functional for the electrode impedance $Z(W, G, h, \tau)$ is calculated because it influences the electrical return loss and the required drive power. Various models and methods exist for computing the characteristics of the transmission line, namely N_m, α_A, and Z, and for calculating the overlap of the electrical and optical fields in order to compute V_π. We refer the reader to the following references for an introduction to the subjects (Chipman 1968; Jordan and Balmain 1968; Gupta, Garg, and Bahl 1979; Marcuse 1989; Chung, Chang, and Adler 1991; Wadell 1991; Gopalakrishnan *et al.*

1994). In this section we note only that the complex electrical propagation constant and impedance may be written in terms of the distributed shunt capacitance $\mathcal{C}$, series inductance $\mathcal{L}$, series resistance $\mathcal{R}$, and shunt conductance $\mathcal{G}$ of the transmission line. We have

$$\beta_e = \sqrt{(\mathcal{R} + i\omega\mathcal{L})(\mathcal{G} + i\omega\mathcal{C})} \qquad (9.19)$$

and

$$Z = \sqrt{(\mathcal{R} + i\omega\mathcal{L})/(\mathcal{G} + i\omega\mathcal{C})} \qquad (9.20)$$

At gigahertz frequencies, the frequency-dependent terms will dominate and thus $Z = \sqrt{\mathcal{L}/\mathcal{C}} = 1/(c\sqrt{\mathcal{C}\mathcal{C}_o})$ and $N_m = c\sqrt{\mathcal{L}\mathcal{C}} = \sqrt{\mathcal{C}/\mathcal{C}_o}$, where $\mathcal{C}_o$ is the capacitance per unit length of the structure with all the dielectrics replaced by vacuum. Consequently, much intuition and fast computational results may be obtained by using the analytic conformal mapping solutions in combination with the method of partial capacitances to account for the thicknesses of the buffer layer and electrode (Gupta, Garg, and Bahl 1979; Kobota, Noda, and Mikami 1980; Chung, Chang, and Adler 1991).

9.2.4.2   Optimization and Performance

What are the ideal values of the electrode variables that should be used to achieve the optimum modulator performance? Suffice it to say that the answer is not obvious. More often than not, a change to one of the parameters that increases the bandwidth also increases the drive voltage. We have already observed that an increase of the active length L reduces not only the drive voltage, but also the bandwidth. An increase of the buffer layer thickness h has the positive effects of lowering the microwave index and attenuation to increase the bandwidth and the impedance, but it also results in an increase of the switching voltage. If the electrode thickness τ is increased, the optical-to-electrical velocity mismatch is reduced, and consequently the bandwidth is increased. An increase of τ also contributes to decreasing the attenuation; however, the influence is moderated by a reduction of the impedance owing to the increased capacitance. Similarly, modification of the electrode gap G produces conflicting outcomes. An increase of the gap decreases the electric field strength but improves the electrooptic overlap Γ. Moreover, increasing the electrode gap is beneficial in increasing the impedance and reducing the attenuation but results in diminishing the positive effect of a thick buffer layer, and so may lower the bandwidth.

As a consequence of these complex interdependencies, detailed model simulations and device prototyping are necessary to determine the four electrode and buffer layer parameters that simultaneously provide a high bandwidth-to-voltage ratio and an impedance near 50 Ω. Of course, they must also be practical for manufacturing. Before comprehensive device models were developed, available calculations and technology led designers to small electrode widths and gaps ($\sim$5 μm), thin buffer layers ($\sim$0.2 μm), and modest electrode thicknesses ($\sim$3 μm). These yielded a microwave index and an attenuation of $N_m \sim 4.0$ and $\alpha_A \sim 1.5$ dB/cm $-\sqrt{\text{GHz}} \times \sqrt{\nu(\text{GHz})}$, respectively, and consequently a bandwidth-to-voltage ratio of approximately 0.4 GHz/V (Izutsu, Yamane and Sueta 1977; Alferness *et al.* 1983; Gee, Thurmond, and Yen 1983; Becker 1984; Korotky *et al.* 1987). The improvements in design strategy, simulation tools, and fabrication techniques that have occurred since the last volume of this book was published have led to dramatic increases in the overall performance of high-speed modulators (Seino *et al.* 1987; Seino *et al.* 1989; Korotky 1989; Korotky *et al.* 1991; Chung, Chang, and Adler 1991; Gopalakrishnan *et al.* 1992; Gopalakrishnan *et al.* 1994). Contemporary designs use wide widths and gaps ($\sim$15 μm), thick buffer layers ($\sim$1 μm), and very thick metallization ($\sim$12 μm) to attain a microwave index of $N_m \sim 2.5$, which is very close to the optical index of $N_o \sim 2.2$; a low attenuation of $\alpha_A \sim 0.55$ dB/cm $-\sqrt{\text{GHz}} \times \sqrt{\nu(\text{GHz})}$; and a bandwidth-to-voltage ratio of about 2 GHz/V.

As a consequence of these improvements, LiNbO$_3$ Mach–Zehnder modulators are now used in deployed commercial fiber transmission systems operating at 2.5 and 5.0 Gb/s, are being engineered into 10-Gb/s systems, have been demonstrated in laboratory systems at 20 Gb/s and beyond, and have become the modulator technology most frequently used in experimental digital lightwave transmissions that define the state of the art. Already for several years, lithium niobate modulators with bandwidths more than adequate for 2.5 Gb/s and drive voltages low enough to be driven by commercially manufactured GaAs 2 V integrated multistage current–mirror–circuit drivers costing less than \$100 have been available in the marketplace. To illustrate the performance that is typical of contemporary LiNbO$_3$ Mach–Zehnder modulators, in Fig. 9.13 we plot the measured frequency response of a production device from Lucent Technologies that has an active length of 4 cm and a switching voltage of less than 4 V, and in Fig. 9.14 we display measured eye patterns for 5 and 10 Gb/s.

Still higher bandwidths and lower voltages have been achieved by using novel processing methods to attain nearly perfect velocity matching and

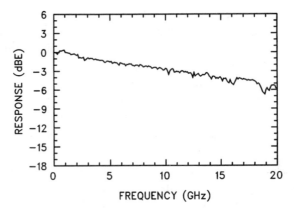

Fig. 9.13 Measured modulation frequency response. The data are for a Ti : LiNbO$_3$ traveling-wave Mach–Zehnder intensity modulator.

low attenuation. In one approach, a shielding cover located in proximity above the normal electrode structure has been used to obtain a bandwidth of 8 GHz and a switching voltage of 4.7 V (Kawano *et al.* 1989). Using an etched buffer layer and narrowed ground planes has resulted in a bandwidth of 44 GHz and a voltage of 12.8 V (Dolfi and Ranganth 1992). Grooving of the high-permittivity LiNbO$_3$ substrate has also been demonstrated to improve the velocity matching of the electrical and optical waves (Haga,

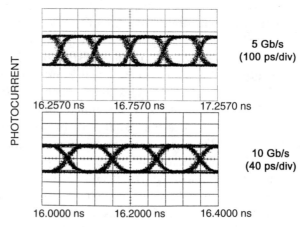

Fig. 9.14 Measured optical eye patterns. The waveforms are for a pair of Ti : LiNbO$_3$ traveling-wave Mach–Zehnder modulators.

Izutsu, and Sueta 1986). Recently, etching of the substrate has been used to achieve $N_m = N_o$ and $\alpha_A \sim 0.43$ dB/cm $-\sqrt{\text{GHz}} \times \sqrt{v(\text{GHz})}$, and consequently an incredible bandwidth of >60 GHz with a switching voltage of only 5 V (Noguchi *et al.* 1993; Noguchi *et al.* 1996). (We remind the reader that the bandwidths quoted here refer to v at -3 dBE, and the voltages are for a 1.5-μm wavelength.) Novel LiNbO$_3$ modulators for narrow band applications operating at nearly 100 GHz have also been reported (Sheehy, Bridges, and Schaffner 1993). Note that when working at modulation frequencies approaching 20 GHz and beyond, researchers have found it necessary to use thin substrates to minimize radiative losses and associated structure (Sueta and Izutsu 1982; Korotky *et al.* 1987; Dolfi and Ranganth 1992; Gopalakrishnan *et al.* 1994). We conclude from all these accomplishments that highly efficient base-band modulators having millimeter-wave bandwidths are feasible, and with investment and refinement in material processing techniques they will undoubtedly reach the marketplace and find application in systems.

9.2.4.3   Complete Circuit Model

With the advancements in the capability to achieve simultaneously high bandwidths and low drive voltages has come the introduction of other improvements in lithium niobate modulators. One example of another refinement that has become more common in the modeling and design of traveling-wave LiNbO$_3$ modulators is the use of a more complete description of the frequency-dependent modulator electrical circuitry. This circuitry includes the modulator transmission line, drive source, termination, and parasitics, such as bond wires. The increased realism that these refinements offer not only permits higher modulator performance to be attained, but also allows more accurate fiber transmission simulations. Because the circuit elements, such as those depicted in Fig. 9.15, and the electrooptic phase modulation are linear in voltage and current, the implementation of

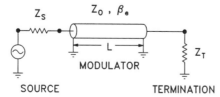

Fig. 9.15   Basic electric circuit model for a LiNbO$_3$ traveling-wave modulator.

the modulator electrical circuit model and its inclusion in the traveling-wave interaction are relatively straightforward. The circuit is normally analyzed in the frequency domain using the recursive voltage and current relations for two-port networks. The interested reader is referred to texts on standard linear circuit theory (Chipman 1968; Jordan and Balmain 1968) and, as an example, to the analytic solution for the optical response corresponding to the situation represented in Fig. 9.15 (Parsons, O'Donnell, and Wong 1986).

We note that a singularly important aspect of modulator circuit design is the consideration of the proper termination of the modulator transmission line, because an impedance discontinuity at the output end of the modulator electrode produces a backward-propagating component of the traveling-wave voltage. The velocity mismatch for the counterpropagating interaction is severe because it depends on the sum, rather than the difference, of the optical and microwave indices. Consequently, the counterpropagating contribution to the modulation efficiency is strongly peaked at low frequency, thus the backward wave can be the dominant factor determining the -3-dB bandwidth, even for a small impedance mismatch. To ensure flat and broad base-band performance, which is required for analog CATV and digital communication applications, traveling-wave modulators are therefore designed either to be matched to 50Ω within 1 or 2 Ω or to include a custom-matched termination. The match to the source is far less critical, because the electrical return loss for a traveling-wave modulator can be low even for a relatively large impedance difference. (*Note*: This differs markedly from the situation for a lumped element modulator, in which the termination usually consists of a resistor that is matched to the source impedance and placed in parallel with the capacitance of the modulator. In that case, the electrical return loss is -3 dB at the frequency corresponding to the -3-dBE bandwidth.)

9.2.5   *MODULATOR BIASING, ENVIRONMENTAL STABILITY, AND PACKAGING*

Optoelectronic transmitters, directly modulated and externally modulated alike, normally require a DC bias voltage or current to set the operating point of the device in order to obtain the optimum performance. Typically, an automatic biasing circuit (ABC), which can set the bias point dynamically, is used because there is always some degree of device-to-device variation, environmental sensitivity, or device aging that occurs. Lithium niobate electrooptic modulators are no exception, and an ABC is used to set the operating optical bias and to account for the variations of relative optical phase due to changes in the surroundings (e.g., temperature) and to the evolution of the voltage division within the highly insulating constituent dielectric materials. The latter behavior is well described as a multielement RC circuit effect (Yamada and Minakata 1981; Becker 1985; Korotky and Veselka 1994, 1996). We note that the configuration used to derive the feedback error signal for the ABC depends on the application. For digital transmitters, modulating the peak-to-peak amplitude of the drive waveform at low frequency is attractive and can be accomplished by dithering the power supply voltage of the output transistor of the drive amplifier (Kuwata et al. 1990).

Since the last volume in this series was published in 1988, significant progress has been made in reducing the rate of bias drift to extremely low levels and in demonstrating the long-term reliability of LiNbO$_3$ modulators for system applications. At the beginning of this period, early system experiments conducted both with and without ABC circuits had shown a modulator bias stability performance that allowed at least several days of continuous operation at room temperature (Giles and Korotky 1988). The direct proof of feasibility of continuous operation was later increased to many tens of days (Fishman, Nagel, and Bahsoun 1991). By the middle of this period, one group had succeeded in dramatically extending the time before which a modulator reset is required to longer than 15 years for operation at 70°C (Seino et al. 1992). This critically important advance was attained through improvements in the materials and processing of the SiO$_2$ buffer layer, which is located between the lithium niobate substrate and the electrode, and the preparation of the surface of the substrate itself. The progress toward these improvements was greatly aided by the experiments of the same group that established that the LiNbO$_3$ bias drift is temperature activated.

Today, lithium niobate waveguide electrooptic Mach–Zehnder modulators that can operate for longer than 20 years at 50°C without requiring the bias voltage to be reset are available from several manufacturers. An example of the bias stability typical of the latest generation of devices is graphed in Fig. 9.16 for an operating temperature of 80°C. For the measurement, an autobiasing circuit was used to maintain the optimum bias point. The voltage variation over the entire time line from 0.00015 to 300 h, which corresponds to approximately 30 years at room temperature, is approximately 20%. Nearly all this variation occurs within the first few hours of the initial application of the bias voltage, and at 100 hours the

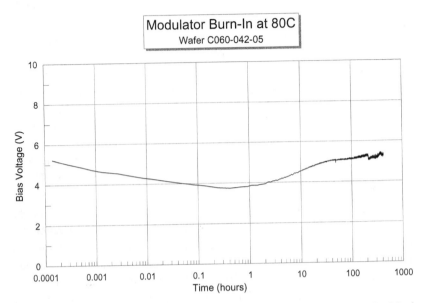

Fig. 9.16 Measured DC bias stability. The data are for a Ti:LiNbO₃ Mach–Zehnder modulator that includes a planar 1-μm-thick SiO₂ buffer layer and an operating temperature of 80°C.

bias point is observed to have returned to the value at which it had started. Although this performance is already very good, there is every reason to believe that further improvements will continue to be made in this key area.

Other areas of performance where there has been much progress on lithium niobate integrated optic modulators include environmental stability and packaging. Because LiNbO₃ is pyroelectric, there is the potential for large bias point shifts to occur under extreme and rapid temperature changes (Skeath *et al.* 1986). However, through the inclusion of charge dissipation layers (Sawaki *et al.* 1986) within the modulator structure, the use of symmetrically balanced modulator design principles (Veselka *et al.* 1992), and the attention to mechanical design, it has been possible to meet the demanding telecommunication standards required for optoelectronic components (Bellcore 1993). These standards call for the ability to operate indefinitely over the temperature range of 10–65°C and under cyclic temperature changes at rates of about 1°C/min. The devices must also be capable of being stored at temperatures of −10 to +85°C and operate after extended exposure to conditions of 85% relative humidity at 85°C. Lithium niobate waveguide electrooptic modulators, in fact, operate satisfactorily well beyond these temperature limits. For example, DC bias stability has now been improved to the point where experimental packaged devices are routinely operated at or above 125°C in order to accelerate the rate of drift so that it may be assessed within a few days. At the other extreme, fiber-pigtailed modulators have been operated at temperatures as low as 5°K (McConaghy *et al.* 1994). Packaged LiNbO$_3$ integrated optic circuits also meet the necessary shock and vibration specifications (Suchoski and Boivin 1992; O'Donnell 1995). Although hermeticity of the package is not always a requirement, some manufacturers of LiNbO$_3$ modulators already provide this additional feature.

Relative to the work on other areas of modulator performance, there has been less emphasis during recent years to improve further the optical insertion loss of fiber-pigtailed LiNbO$_3$ phase and amplitude modulators. This is because the insertion loss in a production environment was already typically in the range of 2.5–3.5 dB and because optical amplifiers had become readily available. Also, losses as low as 1 dB had been previously demonstrated, and losses consistently less than 2.5 dB are not out of the question (Alferness *et al.* 1982; Veselka and Korotky 1986). An extremely low optical return loss of less than −40 dB is achieved without difficulty by cutting and polishing the end faces of the chip at an angle of 6° from the normal to the waveguide axis (Kincaid 1988; Korotky *et al.* 1991). Alternatively, antireflection coating of the end faces, which permits a return loss of less than −30 dB, is used (Eisenstein *et al.* 1985).

9.2.6 APPLICATIONS AND DEPLOYMENT

The fourth area where there has been significant progress in the evolution of high-speed modulators and switches is system applications and deployment. In this section we highlight milestones in the use of high-speed lithium niobate waveguide modulators and switches.

9.2.6.1 Demonstration and Deployment in Digital Transmission Systems

Although lithium niobate intensity modulators had been used in research system experiments at data rates approaching 10 Gb/s as early as the mid-1980s (Korotky, Eisenstein, Gnauck, *et al.* 1985; Gnauck *et al.* 1986; Oki-yama *et al.* 1987), it was not until a practical optical amplifier — in the form of an erbium-doped fiber — was demonstrated in 1987 that there was a serious interest expressed by system developers. Perhaps the first long-distance field trials to use a LiNbO$_3$ integrated optic modulator were carried out at the AT&T Roaring Creek Earth Station in 1991 (Fishman, Nagel, and Bahsoun 1991). In those trials, which used an optical amplifier spacing of 70 km, system developers transmitted and received NRZ data at 1.7 Gb/s on each of four wavelengths over a total of 840 km of conventional fiber ($D = 17$ ps/nm-km). The results of that experiment are shown in Fig. 9.17, where the receiver sensitivity for a BER of 3×10^{-11} is plotted as a function of the transmission distance. A slightly negative dispersion penalty, indicating improved performance, was observed for all the distances that had been tested. Later, in the laboratory, the Bell Laboratories developers demonstrated error-free operation of a single-wavelength channel at 2.5 Gb/s over 617 km for a 60-day period (BER $< 7.7 \times 10^{-17}$) using a lithium-niobate-based transmitter (Fishman 1993).

More recently, lithium niobate Mach–Zehnder modulators have been deployed for the first time in long-distance telecommunication networks. Specifically, they are used in the land-based transmitters of the initial generation of optically amplified undersea cable systems (see Volume IIIA, Chapter 10). These systems operate at 1.5 μm over dispersion-shifted fiber and use NRZ data encoding. The lithium niobate modulators are used for both optical intensity modulation and, as is discussed in detail in Section 9.3, polarization scrambling. One of the first of these systems to be installed was a 2500-km, 2.5-Gb/s link that connects St. Thomas in the U.S. Virgin Islands with Florida. It was placed into commercial operation in September

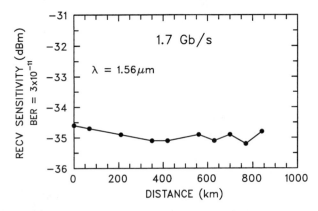

Fig. 9.17 Transmission results from the 1991 Roaring Creek field trials. The data are for a system using conventional fiber, a Ti : LiNbO$_3$ Mach–Zehnder modulator transmitter, and erbium-doped fiber amplifiers.

1995. Construction of several other much larger undersea cable systems that are based on optical amplifiers and lithium niobate modulators was also begun at about that time. These larger systems, which virtually span the globe, will operate at a line rate of 5 Gb/s over distances ranging from approximately 6000 to nearly 27,000 km (Barnett *et al.* 1996; Gunderson, Lecroat, and Tatekura 1996; Trischitta *et al.* 1996; Welsh *et al.* 1996). The trans-Atlantic cable (TAT-12/13) systems, the Trans-Pacific Cable (TPC-5) system, and the Asia-Pacific Cable Network (APCN) are planned to carry calls by the end of 1996. The longest system, which extends from the United Kingdom to Japan, is scheduled to be turned over to revenue-generating traffic by the end of 1997.

The relative ease with which lithium niobate modulators can encode data at high bit rates is evidenced by their ready availability in the marketplace and their ubiquitous use in research and development system experiments. Already for several years the overwhelming majority of the ground-breaking fiber optic transmission systems experiments reported at scientific and engineering conferences have relied on lithium niobate modulators. For example, soon after a section of the TPC-5 undersea system became functional, experiments were conducted at 10 Gb/s — twice the speed that the system was engineered for — using a lithium niobate modulator (Park *et al.* 1996). Also, the latest system results, which define the state of the art, use lithium niobate modulators operating at data rates of 10 and 20 Gb/s (Onaka *et al.* 1996; Gnauck *et al.* 1996; Morioka *et al.* 1996). To our knowledge, lithium niobate modulators have also been used in transmission experiments at data rates as high as 30 Gb/s (J. E. Bowers, personal communication, 1995).

With ready access to versatile high-speed lithium niobate modulators and switches, designers are exploring their use in a host of ways. To illustrate some of the many possibilities, we briefly review a few examples using our own recent work on soliton pulse generation, suppression of stimulated Brillouin scattering (SBS), and optical time-division multiplexing (OTDM).

9.2.6.2 Soliton Pulse Generation

A promising approach to ameliorating the effects of fiber dispersion and nonlinearities that are inherent in very long-distance, high bit-rate optical fiber transmission systems is the use of soliton pulse propagation (see Volume IIIA, Chapter 12). To help make such systems attractive to system designers, a practical source of soliton pulses is essential. The reproducible characteristics and flexible operation of lithium niobate integrated optic modulators make them ideally suited to the task. Various configurations based on LiNbO$_3$ Mach–Zehnder modulators that can produce nearly ideal secant-hyperbolic-squared soliton pulses have recently been demonstrated (Korotky and Veselka 1993; Veselka and Korotky 1996; Veselka et al. 1996). Among the features of these soliton generators that make them practical, and hence attractive, is that the one or two drive waveforms they require for operation are sinusoidal waveforms, which are easily produced and manipulated. Also, the selection of the pulse repetition rate is extremely flexible, and these generators work over a very wide wavelength range. Experimental soliton pulse subsystems for 2.5–15 Gb/s have been constructed and used in wavelength-division multiplexing soliton transmission system test-bed experiments. Using lithium niobate soliton generators, researchers have recently demonstrated the fiber transmission of eight 2.5-Gb/s wavelength channels over 10,000 km, eight 5-Gb/s wavelength channels over 9000 km, and eight 10-Gb/s wavelength channels over 10,000 km (Nyman, Evangelides, et al. 1996; Mollenauer, Mamyshev, and Neubelt 1996).

9.2.6.3 SBS Suppression

In many fiber optic system applications it is advantageous for the transmitter to launch very high optical power so as to obtain a large power budget. Examples include fiber optic community-antenna television distribution, remote sensing, and digital transmission. Unfortunately, nonlinearities in the fiber usually limit the optical power level that may be used. Most often it is SBS, which causes power-dependent backscattering, that is the first nonlinearity encountered. Because the threshold for SBS increases as the linewidth of the source is increased, an effective means of operating at higher powers is to broaden the spectrum of the source by an additional means of modulation. Introducing phase modulation via a lithium niobate phase modulator is a particularly attractive way of increasing the spectral width because it can be accomplished without producing spurious amplitude modulation, which may degrade system performance. In the case of repeaterless fiber optic transmission over low-dispersion fiber, very large increases in the threshold for SBS can be obtained using a novel phase modulation technique (Hansen et al. 1995). Using a phase modulator driven with four tones of selected amplitudes and frequencies, an increase of the SBS threshold of 17 dB at a wavelength of 1.5 μm was reproducibly obtained with a total RF drive power of less than 250 mW (Korotky et al. 1995).

9.2.6.4   High-Speed OTDM

Another area of application where there has been progress on lithium niobate devices is OTDM (Korotky, Eisenstein, Alferness, *et al.* 1985; Korotky and Veselka 1990). In OTDM, the burden of demultiplexing the high-speed time-multiplexed data channel into low-speed channels is moved from the electrical domain to the optical domain, which can permit significantly higher data rates to be attained (Tucker, Eisenstein, and Korotky 1988). For example, using OTDM, researchers have established the feasibility of single-wavelength systems at line rates approaching 100 Gb/s (Takada and Saruwatari 1988; Vasilev and Sergeev 1989; Korotky *et al.* 1990; Morioka *et al.* 1996). A critical component necessary to implement an OTDM system at these very high data rates is a fast optical switch for demultiplexing. In our recent OTDM experiments we used a velocity-matched, traveling-wave, lithium niobate, balanced-bridge 2 × 2 switch to demultiplex pairs of 5-ps optical pulses having a time separation corresponding to a data rate of 72 Gb/s (Korotky and Veselka 1990; Korotky *et al.* 1990). Using a streak camera having a resolution of approximately 3 ps to observe the optical waveforms, which are plotted in Fig. 9.18, we measured the cross talk in the demultiplexed channels and found it to be less than −19 dB for a drive power of approximately 400 mW. Integrating three high-speed switches in a binary tree configuration onto a single substrate to form a 4-to-1 demultiplexer for 50- to 100-Gb/s line rates is considered

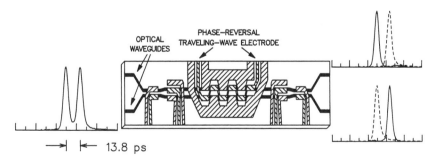

Fig. 9.18   72-Gb/s Ti:LiNbO₃ single-stage optical time-division demultiplexer (OTDM) switch. For the OTDM experiment, the switch was driven with a 36-GHz sinusoidal waveform. The measured optical waveforms at the input (*left*) and output (*right*) of the high-speed switch, which are depicted by the *solid curves*, were obtained with a streak camera. The *dashed curves* indicate where the optical bit that was switched to the complementary channel had been located.

feasible. In another experiment, the functions of modulating two 20 Gb/s channels and multiplexing them together were integrated onto a single substrate (Doi *et al.* 1996; Ishikawa *et al.* 1996).

9.3   Electrooptic Polarization Scramblers and Controllers

High-speed polarization scramblers and fast automatic polarization controllers have recently found important applications in long-distance lightwave systems. Polarization scramblers are now an essential component in optically amplified transoceanic communication systems, where they are used to eliminate anisotropic gain saturation (polarization hole burning) in the optical amplifiers by depolarizing the launched optical signal. Polarization controllers, on the other hand, allow automatic polarization demultiplexing in polarization-multiplexed transmission systems by transforming the unknown polarization states of the two multiplexed signals at the end of the transmission line into orthogonal linear states. Electrooptic polarization scramblers and controllers on LiNbO$_3$ have demonstrated superior performance over alternative modulation schemes and are therefore the preferred choice for practical systems. In the following sections we briefly discuss the operation and performance of electrooptic polarization modulators on LiNbO$_3$. These modulators are usually fabricated on z-cut or x-cut substrates, similar to conventional amplitude and phase modulators, and operate with low-loss single-mode waveguides that are directly coupled to single-mode fibers.

9.3.1 BACKGROUND AND CURRENT APPLICATIONS

It is well known that standard single-mode fibers do not preserve the launched state of polarization (SOP) of optical signals propagating through the fiber. Moreover, the SOP usually varies with time at any given point along the fiber because of small changes in the physical environment of the fiber or in the optical wavelength. These random polarization fluctuations, however, affect only those transmission systems that employ polarization-sensitive optical components, such as coherent optical receivers or polarization-dependent photonic switches and demultiplexers. Coherent receivers, for example, suffer from signal fading when the incoming optical signal is not in the same polarization state as that of the local oscillator laser. Even worse, the heterodyne signal can completely disappear if the two signals are in orthogonal polarization states. This undesired signal fading could be avoided, for example, by using polarization-maintaining fiber (PMF) throughout the entire transmission link. These fibers, however, tend to exhibit significantly higher propagation losses than those of standard single-mode fibers and are therefore rarely employed in long-distance lightwave systems. Nevertheless, polarization-induced signal fading can be eliminated — or at least substantially reduced — by applying one of the following three techniques:

(1) Continuous automatic control of the SOP of the local oscillator laser to match the SOP of the received optical signal (Okoshi 1985)

(2) High-speed polarization scrambling of either the transmitted optical signal or the local oscillator to effectively depolarize one of the two signals on the time scale of a bit period (Hodgkinson, Harmon, and Smith 1987)

(3) Polarization-diversity detection, where the incoming optical signal is decomposed into two orthogonally polarized components and then separately mixed with two local oscillator signals of matched polarization states (Glance 1987)

The recent advent of high-performance EDFAs, however, has diminished the interest in coherent optical communication systems because it was found that conventional intensity detectors in combination with optical preamplifiers can achieve similar or even better receiver sensitivities than those achieved by coherent detection — without incurring any of the adverse polarization effects discussed previously (Giles *et al.* 1989).

Polarization scramblers and controllers have nevertheless found other important applications in high-speed lightwave systems. As mentioned previously, polarization scramblers are currently employed in optically amplified transoceanic communication systems to remove transmission impairments resulting from anisotropic gain saturation (polarization hole burning) in the EDFAs, which occurs when an amplifier is saturated by a polarized optical signal. This otherwise desired saturation of the optical amplifiers gives rise to polarization-dependent gain in such a manner that the saturating optical signal always experiences less gain than the amplified spontaneous emission noise (ASE) in the orthogonal polarization state, which thus reduces the SNR in the system (Taylor 1993). Although the gain difference in a single EDFA is fairly small, typically of the order of 0.1 dB (Mazurczyk and Zyskind 1994), the resulting SNR impairments accumulate rapidly in

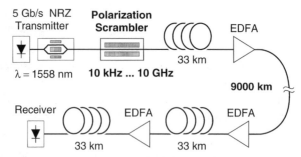

Fig. 9.19 Schematic diagram of a 9000-km-long all-optical communications system employing periodically spaced erbium-doped fiber amplifiers (EDFAs). A high-speed polarization scrambler depolarizes the 5-Gb/s NRZ optical signal before it is launched into the transmission fiber.

a long chain of concatenated amplifiers and can severely impair the signal transmission in ultralong transoceanic systems, such as the more than 6000-km-long trunks in the trans-Atlantic cable (TAT-12/13) network and the up to 9000-km-long Trans-Pacific Cable (TPC-5) system (see, for example, Trischitta *et al.* 1996; Barnett *et al.* 1996; Volume IIIA, Chapter 10).

The SNR impairments resulting from anisotropic gain saturation can be removed by depolarizing the launched optical signal on a time scale faster than the relatively slow gain recovery dynamics in the EDFAs, which is typically of the order of 0.1 ms (Bergano 1994). This can be accomplished with a fast polarization scrambler in the optical transmitter, as shown schematically in Fig. 9.19, where the scrambler modulates the launched SOP typically at rates between 10 kHz and 10 GHz. An example of the large performance improvements achieved with polarization scrambling is given in Table 9.3, which lists the time-averaged SNR in the received electrical

Table 9.3   **Transmission Performance of a 5900-km System**

Scrambling Frequency	*Average SNR[a] (Mean Q^2) (dB)*	*Corresponding Bit Error Rate*
0 Hz	19.2	4×10^{-20}
40 kHz	21.3	2×10^{-31}
5 GHz	24.9	2×10^{-69}
10 GHz	23.2	1×10^{-47}

[a] SNR, Signal-to-noise ratio.

signal (Q factor) of a 5900-km-long 5-Gb/s NRZ transmission test bed for various scrambling frequencies between 40 kHz and 10 GHz (Heismann 1995). The largest SNR improvements of more than 4 dB are obtained with high-speed scrambling at 5 and 10 GHz, where the detection of the high-speed digital data is not affected by undesired intensity modulation arising from interactions of the polarization scrambling with polarization-dependent loss in the system components (Taylor and Penticost 1994). However, the highest SNR improvement of about 5.7 dB is obtained when the SOP is modulated at 5 GHz phase synchronously with the high-speed digital data. The increased performance with bit-synchronous scrambling is attributed to nonlinear pulse compression in the transmission fiber that arises from a sinusoidal frequency chirp introduced by the scrambler (Bergano *et al.* 1996; Park *et al.* 1996).

The scrambling frequency for optimal transmission performance usually depends on the length of the system and its particular design. Electrooptic polarization scramblers offer the great advantage of an extremely broad range of modulation frequencies, from a few kilohertz to several gigahertz. Alternative schemes for fast polarization scrambling allow only a narrow range of modulation frequencies: scramblers based on piezoelectric fiber stretchers and acoustooptic polarization scramblers usually operate at fixed resonance frequencies of a few hundred megahertz (Kersey and Dandridge 1987; Taylor 1993; Noé *et al.* 1994), whereas depolarized light sources based on two polarization-multiplexed transmitters of slightly different optical wavelengths typically modulate the polarization state at rates of more than tens of gigahertz (Burns *et al.* 1991; Bergano, Davidson, and Li 1993).

Automatic polarization controllers are mostly used to transform the fluctuating output polarization at the end of a transmission fiber continuously into the preferred SOP of a polarization-sensitive optical component, such as a polarization-dependent optical switch or demultiplexer (Heismann, Ambrose, *et al.* 1993; Heismann, Hansen, *et al.* 1993). A particularly important application of such polarization controllers is found in polarization-multiplexed transmission systems, where they allow automatic demultiplexing of the two orthogonally polarized signals at the optical receiver, as shown schematically in Fig. 9.20. Polarization multiplexing is an attractive scheme for increasing the transmission capacity of lightwave systems by transmitting simultaneously two independent optical signals at the same wavelength in orthogonal polarization states (Hill, Olshansky, and Burns 1992; Evangelides *et al.* 1992). Although it is fairly easy to multiplex the two orthogonally polarized signals at the transmitter, it is far more difficult

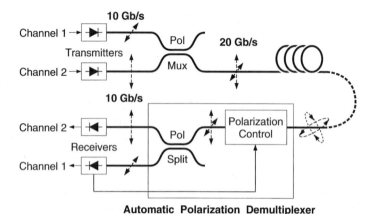

Fig. 9.20   Schematic diagram of a polarization-multiplexed 20-Gb/sNRZ transmission system with an automatic polarization demultiplexer, comprising a fast polarization controller and a conventional polarization splitter–combiner (PolSplit).

to demultiplex them at the receiver because their absolute polarization states are not preserved in the transmission fiber and may even change with time. This problem can be solved by transforming the unknown polarization states of the two signals simultaneously into two orthogonal linear states, which can then easily be separated by a conventional polarization splitter. Such automatic demultiplexing has indeed been successfully demonstrated in 20-Gb/s NRZ transmission systems as well as in ultralong 10-Gb/s soliton systems (Heismann, Hansen, *et al.* 1993; Mollenauer, Gordon, and Heismann 1995).

In these applications it is essential that the polarization controller does not interrupt, at any time, the continuous flow of the optical signals through the system. The controller therefore has to be fast enough to allow instantaneous compensation of the random and potentially large polarization fluctuations in the fiber. Moreover, to guarantee continuous adjustment of the output SOP, the controller has to have an effectively infinite transformation range. Once again, electrooptic polarization controllers on LiNbO$_3$ have shown superior performance over alternative schemes: polarization controllers using mechanically rotated bulk-optic or fiber optic wave plates are inherently slow and hence unsuitable for practical applications (Imai, Nosu, and Yamaguchi 1985; Okoshi 1985), whereas control systems based on electromagnetic fiber squeezers and liquid crystal polarization rotators are usually limited in speed by complicated drive algorithms that are required to reset the various control elements periodically (Noé, Heidrich, and Hoffmann 1988; Walker and Walker 1990; Rumbaugh, Jones, and Casperson 1990). Electrooptic polarization controllers based on the principle of cascaded, rotating wave plates exhibit an inherently unlimited transformation range and have demonstrated control speeds that clearly exceed the speeds of natural polarization fluctuations in long transmission fibers (Heismann and Whalen 1992).

9.3.2 POLARIZATION SCRAMBLERS

Electrooptic polarization modulators generally alter the polarization state by inducing a variable amount of linear birefringence in the waveguide, which may be any combination of TE–TM phase shifting and TE ↔ TM conversion. Circular birefringence, however, can also be produced by cascading several stages of linear phase retarders. Assuming that the input light is completely polarized on a time scale much shorter than the modulation period, we can represent a general input or output SOP by a normalized Jones vector of the form

$$\vec{\mathbf{J}} = \begin{bmatrix} a_{TM} \\ a_{TE} \end{bmatrix} = \begin{bmatrix} \cos\theta \exp(j\phi/2) \\ \sin\theta \exp(-j\phi/2) \end{bmatrix}, \tag{9.21}$$

whose two components describe the complex amplitudes of the TE- and TM-polarized modes in the waveguide, and where θ and ϕ characterize the angular orientation and the ellipticity of the polarization state. Moreover, neglecting optical losses, we can represent an arbitrary polarization transformation in the modulator by a unitary 2×2 matrix of the form

$$\underline{\mathbf{T}} = \begin{bmatrix} A - jB & -C - jD \\ C - jD & A + jB \end{bmatrix}, \tag{9.22}$$

which relates the Jones vector of the output SOP, $\vec{\mathbf{J}}_{out}$, to that of the input SOP, $\vec{\mathbf{J}}_{in}$, as $\vec{\mathbf{J}}_{out} = \underline{\mathbf{T}} \cdot \vec{\mathbf{J}}_{in}$. The matrix $\underline{\mathbf{T}}$ describes general elliptical birefringence where $2 \arcsin B$ is the amount of linear phase retardation at $0°$ (relative TE–TM phase shifting), $2 \arcsin D$ the amount of linear phase retardation at $45°$ (TE ↔ TM mode conversion), and $2 \arcsin C$ the amount of circular phase retardation (Heismann 1994). The total amount of induced elliptical phase retardation is given by $\psi = 2 \arccos(\pm A)$. Furthermore, rapid modulation of the amount or the orientation of the induced phase retardation allows depolarization of light in certain input SOPs. The output light of the scrambler is completely depolarized when, on average, all possible output polarization states are excited with equal probability — i.e., when the time averages of the three Stokes parameters, s_1, s_2, and s_3, vanish simultaneously, where $s_1 = |a_{TM}|^2 - |a_{TE}|^2$, $s_2 = 2\,\mathrm{Re}\,(a_{TM}a_{TE}^*)$, $s_3 = 2\,\mathrm{Im}\,(a_{TM}a_{TE}^*)$, and the asterisk denotes the complex conjugate (Born and Wolf 1975). Equivalently, this is the case when the average degree of polarization (DOP), defined as

$$DOP = \frac{\sqrt{\langle s_1 \rangle^2 + \langle s_2 \rangle^2 + \langle s_3 \rangle^2}}{s_0}, \qquad (9.23)$$

vanishes, where the angle brackets indicate time averaging over at least one full modulation period and $s_0 = |a_{TM}|^2 + |a_{TE}|^2$.

9.3.2.1 High-Speed Depolarizer

A conventional electrooptic phase shifter on z-cut LiNbO$_3$, as described in Section 9.2, acts as a polarization modulator if the input light is polarized in such a way that it equally excites the TE- and TM-polarized modes in the waveguide — e.g., if it is linearly polarized at 45°, as shown in Fig. 9.21 (Howerton and Burns 1994; Heismann et al. 1994). An external electric field, E_z^a, perpendicular to the crystal surface, generates an anisotropic index change in the waveguide by means of the r_{13} and r_{33} electrooptic coefficients ($r_{13} \approx 8.6 \times 10^{-12}$ m/V and $r_{33} \approx 30.8 \times 10^{-12}$ m/V), where the index change for TM-polarized light is given by

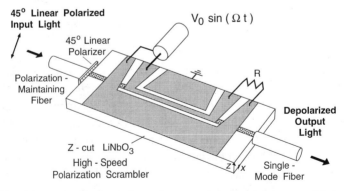

Fig. 9.21 Schematic diagram of an electrooptic polarization scrambler on z-cut lithium niobate (LiNbO$_3$) operating with a single-mode waveguide and a high-speed traveling-wave electrode. R, termination resistor.

$$\Delta n_{TM} = \Gamma_z \frac{n_e^3 r_{33} E_z^a}{2}, \tag{9.24}$$

and that for TE-polarized light by

$$\Delta n_{TE} = \Gamma_x \frac{n_o^3 r_{13} E_z^a}{2}. \tag{9.25}$$

In these two equations, $n_o \approx 2.207$ and $n_e \approx 2.135$ denote the ordinary and extraordinary indices of refraction, and Γ_z and Γ_x characterize the overlap integral of the external electric field with the electric fields of the optical modes. Note that both indices always increase or decrease in the same direction but at substantially different rates — i.e., $\Delta n_{TM} \approx 3.6 \, \Delta n_{TE}$. This differential index change gives rise to a phase retardation, $\Delta\Phi$, between the TE- and TM-polarized modes, with

$$\Delta\Phi = (\Delta n_{TM} - \Delta n_{TE}) \frac{2\pi}{\lambda_0} L, \tag{9.26}$$

where L is the length of the electrooptic interaction. In addition, the common change in the two indices modulates the common phase of the two modes by

$$\Delta\Psi = (\Delta n_{TM} + \Delta n_{TE}) \frac{\pi}{\lambda_0} L, \tag{9.27}$$

which is of the same order of magnitude as the TE–TM phase retardation — i.e., $\Delta\Psi \approx 0.8 \, \Delta\Phi$ — and hence may cause substantial frequency chirping, $d \, \Delta\Psi/dt$, at high modulation speeds. The overall transfer characteristic of this polarization modulator is therefore represented by the matrix

$$\underline{\mathbf{T}} = \begin{bmatrix} \exp(j \, \Delta\Phi/2) & 0 \\ 0 & \exp(-j \, \Delta\Phi/2) \end{bmatrix} \exp(j \, \Delta\Psi), \tag{9.28}$$

where we have ignored the large polarization mode dispersion (PMD) in the modulator waveguide of about 2.5 ps/cm, which arises from the large crystal birefringence. This PMD causes an additional offset in the relative TE–TM output phase, which depends strongly on wavelength and temperature, but usually does not affect the operation as a polarization scrambler (Heismann *et al.* 1994).

The output light can be depolarized by driving the scrambler, for example, with a sinusoidal voltage of the form $V(t) = V_0 \sin(\Omega t)$ at some modulation frequency $\Omega/2\pi$. In this case, the DOP varies as $DOP = |J_0(\pi V_0/V_\pi)|$

with drive voltage amplitude V_0, where J_0 denotes the zeroth-order Bessel function of the first kind, and V_π is the voltage for inducing a relative TE–TM phase shift $\Delta\Phi = \pi$. Complete depolarization is obtained at any root of J_0 — e.g., at $V_0 = 0.7655V_\pi$. Alternatively, one can drive the scrambler with a symmetrical sawtooth voltage of total peak-to-peak voltage swing $2V_p$, in which case $DOP = |\sin(\pi V_p/V_\pi)|/(\pi V_p/V_\pi)$, and complete depolarization is obtained at $V_p = mV_\pi$, with $m = 1, 2, \ldots$. In either case, the scrambler does not require any particular DC offset in the drive voltage, which is of great advantage for practical applications. The voltage–length product of this scrambler is typically around $V_\pi L \simeq 25$ V · cm at a 1.5-μm wavelength and, hence, comparable to that of conventional phase modulators. Thus, when equipped with traveling-wave electrodes, this scrambler is well suited for high-speed modulation at up to 10 GHz or even higher (Heismann et al. 1994).

An example of the performance of such a high-speed scrambler is shown in Fig. 9.22, which displays the residual DOP in the output light of a 4-cm-long device as a function of electric drive power. The scrambler is operated at a wavelength of 1.558 μm and modulated sinusoidally at frequencies of 10 kHz and 5 GHz. At the 10-kHz modulation frequency, the first minimum in the DOP, with $DOP \approx 0.02$, occurs at 210 mW of electric drive power. At 5 GHz, however, the scrambler requires about 380 mW of drive power for complete depolarization because of higher electrical attenuation in the coplanar electrodes and increased dephasing between the electrical and optical waves. Obtaining such a low residual DOP requires fairly precise adjustment of the input SOP to the scrambler. In the scrambler of Fig. 9.21

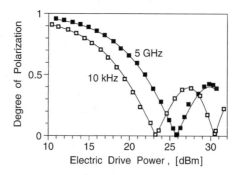

Fig. 9.22 Measured degree of polarization in the output light of a 4-cm-long high-speed polarization scrambler on z-cut LiNbO$_3$ as a function of electric drive power for modulation frequencies of 10 kHz and 5 GHz.

the input SOP is defined by a thin-film polarizer placed between the input fiber and the modulator waveguide, which is aligned at 45° relative to the crystal axes with an accuracy of less than ±1°. The overall fiber-to-fiber insertion loss of this scrambler is typically around 3 dB. The drive power for complete depolarization varies with optical wavelength, as expected from Eq. (9.26) for $\Delta\Phi$. Detuning the wavelength from its original value without readjusting the drive power usually increases the DOP in the output light, as shown in Fig. 9.23. In this example, the residual DOP stays at less than 0.05 over a total wavelength range of about 40 nm.

The additional common phase modulation in the output light, $\Delta\Psi$, can easily be observed at high modulation frequencies by analyzing the optical spectrum with a scanning Fabry–Perot interferometer, as shown in Fig. 9.24. In this instance, the scrambler is modulated at 7.5 GHz with 490 mW of electric drive power to obtain nearly complete depolarization. The top and middle diagrams display separately the output spectra of TM- and TE-polarized light, each being phase modulated with a peak phase deviation of 3.1 rad for TM-polarized light and 0.7 rad for TE-polarized light, such that their difference yields $\Delta\Phi \approx 2.4$ rad. The bottom diagram displays the spectrum of the entire depolarized output light of the scrambler. Comparison of this spectrum with one for pure polarization modulation with $\Delta\Phi = 2.4$ rad clearly shows that the output light of the z-cut scrambler is phase modulated with a peak phase deviation of about 1.9 rad. This fairly large phase modulation can be partially converted into amplitude modulation by chro-

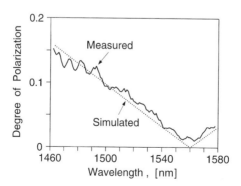

Fig. 9.23 Measured (*solid line*) and simulated (*dotted line*) wavelength dependence of the output degree of polarization of a z-cut polarization scrambler modulated sinusoidally at a constant drive power. The simulated curve assumes a wavelength dependence of $\Delta\Phi \sim \lambda^{-2}$.

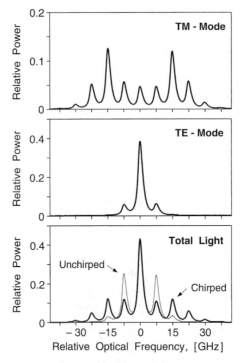

Fig. 9.24 Optical spectra of the depolarized output light of a z-cut polarization scrambler modulated sinuosoidally at 7.5 GHz. (*Top*) TM-polarized mode. (*Middle*) TE-polarized mode. (*Bottom*) The *bold line* displays the entire spectrum of the phase and polarization modulated (chirped) output light. For comparison, the *thin line* shows a spectrum of depolarized light without any additional common phase modulation.

matic dispersion in the fiber and may lead to substantial pulse compression, as mentioned previously (Bergano *et al.* 1996; Park *et al.* 1996).

9.3.2.2 Chirp-Free Depolarizer

Pure polarization modulation without the additional phase modulation can be generated by electrooptic polarization scramblers on z-propagation LiNbO$_3$, where the waveguide runs parallel to the optical axis (Thaniyavarn 1985). This waveguide orientation also has the advantage that the light does not experience the large crystal birefringence. However, polarization modulators on z-propagation LiNbO$_3$ require substantially higher drive voltages than conventional z-cut modulators because of the smaller electrooptic coefficients in this orientation ($r_{12} = -r_{22} \approx 3.4 \times 10^{-12}$ m/V), which renders high-speed modulation more difficult. Figure 9.25 shows an example of such a polarization scrambler on x-cut LiNbO$_3$. The desired polarization modulation is achieved by applying a voltage $V(t)$ across two drive electrodes that are located symmetrically on both sides of the wave-guide and that induce an electric field, E_y^a, parallel to the crystal surface. This external field generates an anisotropic index change in the waveguide in such a fashion that the optical phases of the TE- and TM-polarized modes are shifted symmetrically in opposite directions; this induces pure polarization modulation without any common phase modulation. The total TE–TM phase retardation generated in an electrode section of length L is given by

$$\Delta\Phi = \Gamma_y n_o^3 \, (r_{12} - r_{22}) \, E_y^a \frac{\pi}{\lambda_0} \, L, \qquad (9.29)$$

where Γ_y is the spatial overlap of the induced electric fields with the op-tical fields. The voltage–length product of this scrambler is typically $V_\pi L \simeq 50$ V · cm at a 1.5-μm wavelength, which is about twice that of z-cut scramblers. Figure 9.26 displays the output DOP of a 5-cm-long scrambler as a function of drive voltage amplitude for sinusoidal modulation at 20 kHz. The first DOP minimum occurs at a voltage amplitude of $V_0 = 7.3$ V, corresponding to 535-mW drive power from a 50-Ω source. The insertion

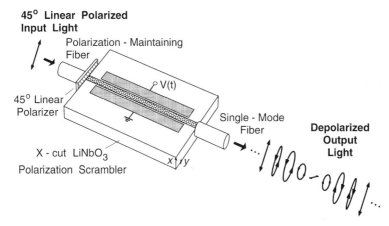

Fig. 9.24 Schematic diagram of a chirp-free polarization scrambler on x-cut, z-propagation lithium niobate, showing a sequence of modulated polarization states in the output light.

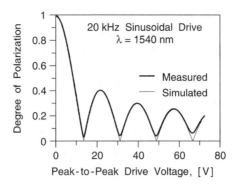

Fig. 9.26 Measured degree of polarization in the output light of a 5-cm-long polarization scrambler on x-cut lithium niobate as a function of drive voltage amplitude for sinusoidal modulation at 20 kHz.

loss and optical bandwidth of the scrambler are very similar to those of z-cut scramblers (Heismann *et al.* 1994).

Instead of modulating the relative TE–TM phase, one can also depolarize light by inducing variable TE $\leftrightarrow$ TM coupling (linear phase retardation at 45°). On conventional z-cut or x-cut LiNbO$_3$, however, TE $\longleftrightarrow$ TM coupling is considerably more difficult than TE–TM phase shifting because of the large difference in the phase indices of the two modes. Efficient TE $\leftrightarrow$ TM conversion can be attained by coupling the two modes through the electrooptic or acoustooptic effect at a periodic index grating that matches their beat period (about 21 μm at a 1.5-μm wavelength). Such periodic TE $\leftrightarrow$ TM conversion is typically limited to a narrow optical bandwidth of less than 1 nm and thus is not very attractive for polarization scramblers. However, narrow band TE $\leftrightarrow$ TM conversion has been utilized for tunable optical band-pass filters, as we discuss in Section 9.4. TE $\leftrightarrow$ TM mode conversion with considerably broader optical bandwidths can be obtained on z-propagation LiNbO$_3$ by means of the r_{61} electrooptic coefficient ($r_{61} = r_{12}$). However, these mode converters usually require an additional DC bias voltage to compensate for residual birefringence in the waveguide and hence do not offer any advantage over the simple TE–TM phase retarders described previously (Thaniyavarn 1985).

9.3.2.3   Polarization-Independent Depolarizer

The single-stage polarization scramblers described in the previous sections depolarize light in only certain, well-defined input polarization states. However, it is also possible to depolarize light in arbitrary polarization states using sophisticated multistage scramblers. Such a polarization-independent depolarizer can be constructed, for example, by cascading a TE–TM phase retarder with a TE ↔ TM mode converter (Heismann and Tokuda 1995a). In general, each scrambler stage should be modulated at a different frequency to decorrelate the various polarization transformations (Billings 1951). Although it is sufficient, in principle, to employ only two scrambling stages, in practice it is advantageous to add one or more extra stages to the depolarizer to improve its performance (Noé *et al.* 1994; Heismann and Tokuda 1995a). The modulation frequencies of the various stages and their relative phases have to be carefully chosen for optimal performance.

Arbitrarily polarized light can also be depolarized by combination cascades of rotating quarter-wave plates (QWPs) alternated with rotating half-wave plates (HWPs). Figure 9.27 shows the electrooptic implementation of such a rotating wave-plate scrambler on z-propagation LiNbO$_3$ (Heismann and Tokuda 1995b). This scrambler employs a modified three-electrode structure to induce simultaneously two independently adjustable electric fields, E_x^a and E_y^a, in the waveguide, which generate a variable combination of TE–TM phase retardation and TE ↔ TM mode coupling (Walker *et al.* 1988). Each electrode section employs a narrow common ground electrode on top of the waveguide and two drive electrodes on both sides of the waveguide. TE–TM phase retardation (linear phase retardation at 0°) is generated by applying equal voltages of opposite polarity to the two drive electrodes and thus inducing an electric field E_y^a as described

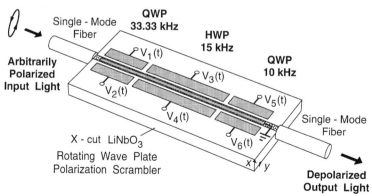

Fig. 9.27 Schematic diagram of a three-stage polarization scrambler on x-cut lithium niobate allowing depolarization of light independent of its state of polarization. The three stages act like three rotating wave plates: a first rotating quarter-wave plate (QWP) followed by a half-wave plate (HWP) and a second QWP.

previously, whereas TE $\leftrightarrow$ TM mode coupling (linear phase retardation at 45°) is generated by applying a common voltage to both electrodes, which induces an electric field, E_x^a, perpendicular to the crystal surface. The amount of induced phase retardation at 45° is given by

$$\Delta\Phi_{45°} = \Gamma_x n_o^3 \, r_{61} \, E_x^a \, \frac{\pi}{\lambda_0} \, L. \tag{9.30}$$

If the two electric fields are varied in such a fashion that the induced linear phase retardations at 0 and 45°, $\Delta\Phi_{0°}$ and $\Delta\Phi_{45°}$, are modulated in quadrature phase with $\Delta\Phi_{0°} = \Delta\Phi_{tot} \sin(\Omega t)$ and $\Delta\Phi_{45°} = \Delta\Phi_{tot} \cos(\Omega t)$, then the total amount of induced phase retardation, $\Delta\Phi_{tot}$, is always constant, but its angular orientation rotates continuously at angular velocity Ω (Noé and Smith 1988). The transfer function of a rotating wave plate can be represented by the matrix

$$T = \begin{bmatrix} \cos(\Delta\Phi_{tot}/2) - j\,\sin(\Omega t)\sin(\Delta\Phi_{tot}/2) & -j\,\cos(\Omega t)\sin(\Delta\Phi_{tot}/2) \\ -j\,\cos(\Omega t)\sin(\Delta\Phi_{tot}/2) & \cos(\Delta\Phi_{tot}/2) + j\,\sin(\Omega t)\sin(\Omega\Phi_{tot}/2) \end{bmatrix}, \tag{9.31}$$

which describes a rotating QWP when $\Delta\Phi_{tot} = \pi/2$ and a rotating HWP when $\Delta\Phi_{tot} = \pi$.

In practical devices, the shallow indiffused waveguide usually introduces a small amount of birefringence that gives rise to a slight difference in the phase velocities of the TE- and TM-polarized modes and thus inhibits complete TE $\leftrightarrow$ TM conversion. This undesired waveguide birefringence, Δn, can usually be compensated for by inducing an exactly opposite amount of birefringence by means of the r_{12} and r_{22} coefficients with $\Delta n = \Gamma_y n_o^3 r_{22} E_y^a$, which is accomplished by adding a constant DC offset, V_T, to the drive voltages inducing TE–TM phase retardation (Thaniyavarn 1985). Thus, the HWP section in Fig. 9.27 is driven by two voltages of the form

$$V_3 = V_\pi' \sin(\Omega t) + (V_\pi/2)\cos(\Omega t) + V_T/2$$
$$V_4 = V_\pi' \sin(\Omega t) - (V_\pi/2)\cos(\Omega t) - V_T/2, \tag{9.32}$$

where V_π' denotes the voltage for complete TE $\leftrightarrow$ TM conversion and V_π the voltage for inducing a TE–TM phase shift of π. A QWP section of the same length requires only half the voltage amplitudes in the sin (Ωt) and cos(Ωt) terms. In Fig. 9.27 the QWP sections are designed half as long as the HWP section (1 cm instead of 2 cm) to obtain similar drive voltage amplitudes in all three sections, as shown in Eq. (9.33):

$$V_{\frac{1}{2}} = 18.8 \text{ V} \cdot \sin(\Omega_1 t) \pm 9.9 \text{ V} \cdot \cos(\Omega_1 t) \pm 19.6 \text{ V}$$
$$V_{\frac{3}{4}} = 19.3 \text{ V} \cdot \sin(\Omega_2 t) \pm 10 \text{ V} \cdot \cos(\Omega_2 t) \pm 13.3 \text{ V} \qquad (9.33)$$
$$V_{\frac{5}{6}} = 18.9 \text{ V} \cdot \sin(\Omega_3 t) \pm 10 \text{ V} \cdot \cos(\Omega_3 t) \pm 20 \text{ V}$$

The input and output ports of the scrambler are directly connected to standard single-mode fibers with typical fiber-to-fiber insertion losses around 3 dB. Figure 9.28 shows an example of the performance of such a scrambler for various general input SOPs, which are generated by passing linearly polarized light from a 1.55-μm laser source through a bulk-optic HWP followed by a QWP. Both wave plates are rotated independently through 180° in steps of 5°; this yields a total of 1369 DOP measurements. The three scrambler stages are operated with the drive voltages shown in Eq. (9.33) at modulation frequencies of $\Omega_1/2\pi = 33.3333$ kHz, $\Omega_2/2\pi = 15$ kHz, and $\Omega_3/2\pi = 10$ kHz. For all tested input SOPs the output light is nearly completely depolarized with a residual DOP of less than 0.03. Substantially larger variations in the DOP of up to 0.25 are observed when only one QWP section and the HWP section are operated (Heismann and Tokuda 1995b). These large variations arise most likely from undesired

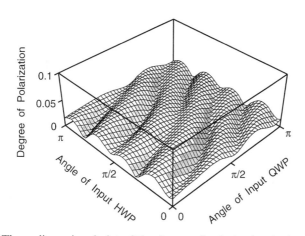

Fig. 9.28 Three-dimensional plot of the degree of polarization in the output light of a QWP–HWP–QWP polarization scrambler measured for various input polarization states, which are generated by varying the angular orientations of a cascaded bulk-optic HWP and QWP. The three stages are modulated at 33, 10, and 15 kHz.

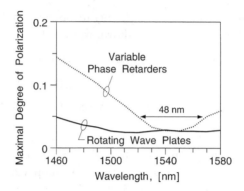

Fig. 9.29   Largest degree of polarization versus optical wavelength measured in the output of two three-stage polarization scramblers that are driven with fixed drive voltage amplitudes. The *solid line* shows the wavelength dependence of the QWP–HWP–QWP polarization scrambler of Fig. 9.29 and the *dotted line* that of a polarization scrambler using alternated TE — TM phase shifters and TE ↔ TM mode converters.

cross modulation between the induced TE–TM phase shifting and TE ↔ TM mode conversion. The additional QWP thus improves the performance of the scrambler substantially.

The optical bandwidth of this QWP–HWP–QWP scrambler is considerably wider than that of a scrambler employing cascaded TE–TM phase retarders and TE ↔ TM mode converters (Heismann and Tokuda 1995a). Figure 9.29 displays the largest residual output DOP of all tested input SOPs as a function of optical wavelength at fixed drive voltage amplitudes. For the QWP–HWP–QWP scrambler, the output DOP is less than 0.05 over the entire measured wavelength range of 120 mm.

9.3.3   POLARIZATION CONTROLLERS

The QWP–HWP–QWP polarization scrambler described in the previous section may also be employed as an automatic polarization controller to stabilize the random polarization fluctuations at the end of a transmission fiber. In fact, it was originally developed to serve as a polarization controller in coherent lightwave systems. For automatic polarization control, the modulator is driven by an electric feedback circuit that probes the SOP at some location downstream from the controller and adjusts the drive voltages for maximal (or minimal) optical power in the desired (undesired) polarization state. In many applications it is sufficient to probe the SOP with just a simple polarization discriminator, such as a polarizer or a polarization splitter, instead of analyzing the SOP with a complicated polarimeter. In this case, the controller simply varies the output SOP until the optical power after the discriminator is at a maximum (or minimum). Moreover, the polarization-sensitive element following the controller can often serve as the polarization discriminator itself. In coherent receivers, for example, the controller matches the output SOP of the local oscillator laser to that of the incoming optical signal by maximizing the heterodyne beat signal in the receiver (Okoshi 1985). As another example, Fig. 9.30 shows the simple arrangement used for automatic polarization demultiplexing of two orthogonally polarized soliton pulse trains in a 10-Gb/s transmission experiment (Mollenauer, Gordon, and Heismann 1995). In this example, the electrooptic polarization controller transforms the two arbitrarily polarized, time-interleaved soliton pulse trains into two orthogonal linear polarization states, which are then separated by a conventional polarization splitter and coupled into two 5-Gb/s receivers. The feedback signal for the polarization controller is obtained from one of the receivers by monitoring the intensity of a 5-GHZ frequency component in the detector signal. The time-interleaved soliton pulses are properly demultiplexed if the 5-GHz signal is at a maximum.

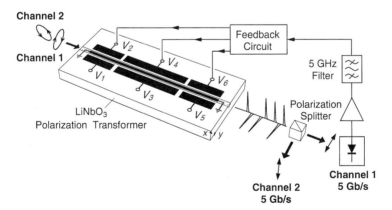

Fig. 9.30 Arrangement for automatic polarization demultiplexing in a 10-Gb/s polarization-multiplexed soliton transmission system, using the QWP–HWP–QWP polarization modulator of Fig. 9.27 as a fast polarization controller. Also shown are the two linearly polarized, time-interleaved soliton trains before they are separated by the polarization splitter.

As mentioned before, the polarization controller must not interrupt at any time, the transmission of the optical signals through the system, and should therefore be fast enough to allow instantaneous compensation of the random and potentially large polarization fluctuations in the fiber. The QWP–HWP–QWP polarization controller in Fig. 9.30 exhibits an inherently unlimited transformation range, provided that the orientations of the induced linear birefringence in three stages are endlessly rotatable. Moreover, it requires only a simple and hence fast control algorithm to transform any arbitrarily varying general input SOP continuously into any (varying) general output SOP. In fact, this electrooptic controller is readily capable of control speeds of up to 4900 rad/s, limited by the speed of the feedback circuit, which is about 10 times faster than natural polarization fluctuations in single-mode fibers, including those resulting from manual handling of the fiber (Heismann and Whalen 1992).

It is well known that a combination cascade consisting of a rotatable QWP followed by a HWP and a second QWP allows endless polarization transformations between two arbitrary polarization states (Walker and Walker 1990). The first QWP, for example, may transform a general elliptical input SOP into a linear SOP, which is then converted into another linear SOP by the HWP. The second QWP finally transforms the linearly polarized light into any desired elliptical output SOP. It can be shown mathematically that the transformation range of such an arrangment is inherently unlimited, i.e., that the orientations of the three wave plates do not require sudden resets, even if the input and output polarization states both fluctuate randomly. The transformation matrix T of the QWP–HWP–QWP controller is given by

$$T = \begin{bmatrix} A - jB & -C - jD \\ C - jD & A + jB \end{bmatrix}, \tag{9.34}$$

with

$$
\begin{aligned}
A &= -\cos \tilde{\gamma}(t)\cos\{[\delta(t) - \alpha(t)]/2\} \\
B &= -\sin \tilde{\gamma}(t)\sin\{[\delta(t) + \alpha(t)]/2\} \\
C &= -\cos \tilde{\gamma}(t)\sin\{[\delta(t) - \alpha(t)]/2\} \\
D &= +\sin \tilde{\gamma}(t)\cos\{[\delta(t) + \alpha(t)]/2\} \\
\tilde{\gamma}(t) &= \gamma(t) - [\delta(t) - \alpha(t)]/2,
\end{aligned}
\tag{9.35}
$$

where $\alpha/2$ characterizes the angular orientation of the first QWP, $\gamma/2$ that of the HWP, and $\delta/2$ the orientation of the second QWP (Heismann 1994).

In the electrooptic implementation of the controller, the three angles correspond to three adjustable electrical phases, α, γ, and δ, in the drive voltages, as follows:

$$V_{\frac{1}{2}} = 17.0 \text{ V} \cdot \sin \alpha(t) \pm 9.9 \text{ V} \cdot \cos \alpha(t) \pm 18.0 \text{ V}$$

$$V_{\frac{3}{4}} = 16.7 \text{ V} \cdot \sin \gamma(t) \pm 9.8 \text{ V} \cdot \cos \gamma(t) \pm 15.3 \text{ V} \qquad (9.36)$$

$$V_{\frac{5}{6}} = 17.8 \text{ V} \cdot \sin \delta(t) \pm 9.9 \text{ V} \cdot \cos \delta(t) \pm 14.1 \text{ V},$$

where the voltage amplitudes are given for the example of a controller operating at 1.3-μm wavelength. The three electrical phases, α, γ, and δ, are the only control variables in the drive voltages. They can be varied over an essentially unlimited range because they are not proportional to any physical quantity. Moreover, because the voltage amplitudes of the sine and cosine terms are fixed, the drive voltages never exceed their range limits — in stark contrast to polarization controllers that employ alternated TE–TM phase retarders and TE ↔ TM mode converters, where the drive voltages require periodic resets when they exceed certain limits (Heidrich et al. 1989).

It has further been shown that, in principle, it is not necessary to vary all three phases independently of one another (Heismann 1994). The number of independent control parameters can be reduced to only two, without limiting the transformation range of the controller, by rotating the two QWPs synchronously instead of independently — i.e., by keeping them at a fixed relative orientation, in particular at $\delta(t) = \alpha(t)$ or $\delta(t) = \alpha(t) \pm \pi$. In this case, the operation of the controller can easily be visualized from its three-dimensional transfer function, as shown in the example of Fig. 9.31. This graph displays the measured optical power in the selected output SOP as a function of the two independent control phases, α and γ for a fixed input SOP to the controller and $\delta = \alpha + \pi$. Within the parameter ranges plotted in Fig. 9.31 we find four absolute maxima, where all light is in the desired output SOP, as well as four absolute minima, where all light is in the orthogonal cross SOP. These four maxima and four minima move across the parameter space as the input and/or output SOP changes, but they always remain absolute maxima and minima. Moreover, there are no secondary maxima or minima in the transfer function. Thus, a particular polarization transformation can be obtained by simply maximizing the optical power in the selected output SOP, or, alternatively, by minimizing the power in the orthogonal cross SOP.

Such a simple maximum search algorithm can be implemented with an analog electronic circuit that searches for the nearest maximum in the

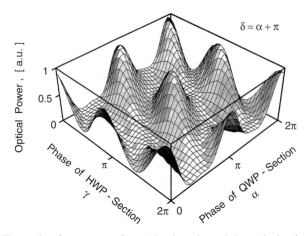

Fig. 9.31 Example of a measured transfer function of the polarization controller of Fig. 9.30 in the case of $\delta = \alpha + \pi$. This three-dimensional graph shows the optical power in the selected output polarization state as a function of the drive voltage phases α and γ for an arbitrary input polarization state.

transfer function and then tracks it continuously as the input and/or output SOP changes (Heismann and Whalen 1991). The six drive voltages may be generated by either analog or digital synthesizers that produce the sine and cosine functions of the three phase parameters, α, γ, and δ. However, it is essential that the synthesizers allow unlimited adjustment of the three phases to higher values as well as to lower values. A maximum (or minimum) in the transfer function can be identified by dithering the phases independently around their current values, typically by ± 0.05 rad or less, and by measuring the resulting intensity modulation after the polarization discriminator with a simple phase-sensitive detector. If the power in the desired SOP is not at a maximum, the circuit detects a local gradient in the transfer function and adjusts α, γ, and δ accordingly. For certain transformations, the four isolated maxima in Fig. 9.31 degenerate into two infinitely long ridges of maximal height. In these cases the control circuit can choose between an infinite number of combinations of α and γ that yield the desired output polarization state (Heismann 1994).

This simple algorithm allows automatic polarization control at speeds that are at least 10 times faster than the rapid polarization fluctuations resulting from accidental manual handling of an optical fiber. Figure 9.32 shows an example of such rapid polarization fluctuations, which are generated by manual bending and twisting of a short length of standard single-

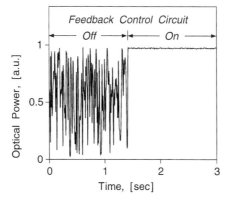

Fig. 9.32 Automatic stabilization of rapid polarization fluctuations using the electrooptic QWP–HWP–QWP polarization controller of Fig. 9.30 in combination with a fast electronic feedback control circuit. The fast polarization variations of up to 350 rad/s are generated by manually twisting and bending of a short length of single-mode fiber.

mode fiber. In this example, the optical phase retardation changes at speeds of up to 350 rad/s. If the polarization controller is turned off, as in the left part of Fig. 9.32, the fast polarization changes are converted into large intensity fluctuations by the polarization discriminator following the controller. However, when the automatic polarization controller is activated, as in the right part of Fig. 9.32, the intensity fluctuations disappear nearly completely, and most of the light is in the selected polarization state with typically less than -15 dB of optical power in the undesired orthogonal cross SOP.

In practice, however, it is advantageous to adjust the three phase parameters, α, γ, and δ, independently of one another. The additional third control variable does not affect the stability of the control loop if the three variables are adjusted sequentially in separate time intervals. However, the additional degree of freedom in the drive voltages can substantially improve the performance of the polarization controller when it is driven with improperly adjusted voltage amplitudes, or, equivalently, when it is operated at a different optical wavelength. This is demonstrated in Fig. 9.33, where the controller was driven with only 80% of the required voltage amplitudes. In Fig. 9.33a, the controller is turned off to show the periodic variations in the input SOP of the controller, which are generated by a polarization scrambler that modulates the phase retardation at speeds of up to

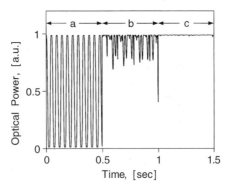

Fig. 9.33 Automatic compensation of periodic polarization variations using a controller driven with intentionally detuned voltage amplitudes: (a) feedback control circuit turned off, (b) with automatic feedback control of two independently adjustable variables (γ and $\alpha = \delta + \pi$), (c) with all three variables controlled independently.

225 rad/s. In Fig. 9.33b, the controller is operated with only two independent control variables — i.e., γ and $\alpha = \delta + \pi$. In this case, the controller is not able to completely stabilize the induced polarization fluctuations because of the improperly adjusted drive voltages. The performance of the system improves substantially, when the third phase, δ, is adjusted independently of α and γ, as shown in Fig. 9.33.

Another advantage of introducing an independent third control variable is that the controller is able to avoid situations where small changes in the input or output SOP require large changes in the control variables (Heismann 1994). The third control variable allows the controller to select a different path for the polarization transformation to the selected output SOP by inducing a different combination of linear and/or circular phase retardation. With $\delta = \alpha$, for example, the transformer produces only linear birefringence of a variable amount and orientation, whereas with $\delta = \alpha + \pi$, it generates a variable combination of linear and circular birefringence. Thus, by adjusting α and δ independently of each other, the control system can select the optimal combination of linear and circular birefringence.

9.4 Electrooptic and Acoustooptic Wavelength Filters

Tunable narrow band optical filters are key components in multiwavelength optical networks, where they can serve as tunable channel selectors or reconfigurable wavelength routers. The first generations of wavelength-division multiplexed lightwave systems will use only a moderate number of wavelength channels (typically about eight), which are spaced in frequency by 100–200 GHz (0.8–1.6 nm in wavelength). Aside from conventional microoptic interference filters, such as etalons, Fabry–Perot resonators, or diffraction gratings, there is considerable interest in low-loss guided-wave optical filters and multiplexers–demultiplexers. Although multiplexing and demultiplexing of several fixed-wavelength channels can be readily achieved with arrayed waveguide gratings or distributed Bragg reflectors (cf. Chapters 7 and 8 of Volume IIIB), there remains a need for rapidly tunable optical band-pass filters to serve as fast channel selectors in local distribution networks (Hood *et al.* 1993; Misono *et al.* 1996). Electrooptic and acoustooptic TE $\leftrightarrow$ TM mode converters on z-cut or x-cut LiNbO$_3$ are promising candidates for these applications. They typically exhibit narrow bandwidths of the order of 1 nm and can be rapidly tuned at speeds of several microseconds (Nuttall, Croston, and Parsons 1994; Misono *et al.* 1996).

Although the electrooptic and acoustooptic mode converters are both inherently polarization dependent, they can be made to operate independently of the input SOP by employing two identical TE $\leftrightarrow$ TM converters in a polarization-diversity arrangement, as shown schematically in Fig. 9.34 for an electrooptic mode converter filter (Warzansky *et al.* 1988; Smith *et al.* 1990). This polarization-independent filter employs two waveguide TE–TM polarization splitters in the input and output of the filter, which demultiplex the TE- and TM-polarized components of the input light and then route them separately through two parallel and identical TE $\leftrightarrow$ TM converters before they are recombined at the output. Because the wavelength dependence of TE $\rightarrow$ TM conversion is identical to that of TM $\rightarrow$

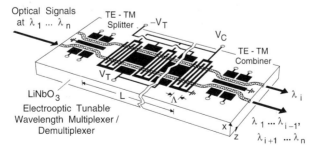

Fig. 9.34 Schematic of a polarization-independent electrooptic tunable wavelength filter on x-cut lithium niobate, employing two identical narrow band TE $\leftrightarrow$ TM mode converters in a polarization-diversity arrangement.

TE conversion, the TE- and TM-polarized components experience the same wavelength-selective polarization conversion. The polarization-converted components are spatially separated from the unconverted components by the output polarization splitter, The filter has, therefore, two complementary output ports, one serving as a band-pass filter and the other as a notch filter. It can thus be used as a reconfigurable add–drop multiplexer, where the second input port to the first polarization splitter allows simultaneous adding of a new wavelength channel (d'Alessandro, Smith, and Baran 1994).

Acoustooptic and electrooptic mode converter filters both can be tuned by means of an external electrical signal. However, acoustooptic filters can be tuned over extremely wide wavelength ranges of up to 200 nm (Heffner *et al.* 1988), in contrast to electrooptic filters, which allow tuning over only 10–30 nm (Nuttall *et al.* 1994), but which do not consume electric drive power during operation. Even though both filter types have been employed in prototype transmission systems, their performance needs further improvement, in particular with respect to interchannel cross talk.

9.4.1 ELECTROOPTIC TUNABLE FILTERS

The electrooptic tunable filter (EOTF) of Fig. 9.34 is fabricated on x-cut, y-propagation $LiNbO_3$ and employs a periodic electrode structure with interdigital finger electrodes of period Λ, which induce a spatially periodic electric field, E_x^a, in the waveguide that couples the TE- and TM-polarized modes by means of the r_{51} electrooptic coefficient ($r_{51} \approx 28 \times 10^{-12}$ m/V). The TE $\leftrightarrow$ TM mode conversion is most efficient when the period of the electrode fingers, Λ, matches the spatial beat period of the two modes, which propagate at substantially different phase velocities in this crystal orientation. Phase matched coupling is obtinted for input light at a free-space wavelength, λ_0, given by

$$\lambda_0 = \Lambda \, |\Delta n_p|, \tag{9.37}$$

where $\Delta n_p(\lambda_0)$ is the effective phase index difference of the two modes with $|\Delta n_p| \approx 0.074$ at a 1.55-μm wavelength. The 3-dB optical bandwidth, $\Delta\lambda_{3\,dB}$, of this periodic mode coupling is inversely proportional to the overall interaction length L,

$$\Delta\lambda_{3\,dB} = 0.8 \, \frac{\lambda_0^2}{|\Delta n_g| \, L}, \tag{9.38}$$

where

$$\Delta n_g = \Delta n_p(\lambda_0) - \lambda_0 \left[\frac{\partial \Delta n_p(\lambda)}{\partial \lambda}\right]_{\lambda=\lambda_0} \tag{9.39}$$

is the effective group index difference at λ_0 with $\Delta n_g \approx 1.1 \, \Delta n_p$ (Heismann, Buhl, and Alferness 1987). The center wavelength λ_0, of the EOTF can be tuned electrooptically by varying the birefringence, Δn_p, in the mode converters by means of the r_{13} and r_{33} electrooptic coefficients. In the EOTF of Fig. 9.34 this is achieved by interleaving periodically short sections of TE–TM phase shifter electrodes between short sections of mode converter electrodes. The phase shifter electrodes are all driven by a common voltage V_T and induce a nearly uniform electric field, E_z^a, in the waveguide. Likewise, the mode converter electrodes are all driven by a common voltage V_C. If a large number of alternating mode converter and phase shifter sections is used, the effect of such an arrangement is similar to one in which the fields for polarization conversion, E_x^a, and birefringence tuning, E_z^a, are induced simultaneously in each waveguide section (Heismann and Alferness 1988). The electrooptically induced shift in the center wavelength, $\Delta \lambda_T$, is given by

$$\Delta \lambda_T \approx \Gamma_z \frac{n_e^3 r_{33} - n_o^3 r_{13}}{2} E_z^a \zeta \frac{\lambda_0}{|\Delta n_g(\lambda_0)|}, \tag{9.40}$$

where ζ is the ratio of the total length of birefringence tuning electrodes to the entire interaction length. Typical tuning rates of EOTFs are of the order of 0.05 nm/V (Heismann, Buhl, and Alferness 1987). In addition, the center wavelength of the EOTF may be tuned by temperature with a fairly large tuning coefficient of about -1 nm/°C (Booth et $al.$ 1984). The voltage–length product for electrooptic TE $\leftrightarrow$ TM conversion depends strongly on the widths and gaps of the interdigital finger electrodes and can easily vary between 15 and more than 50 V · cm (Heismann, Buhl, and Alferness 1987; Warzanskyj, Heismann, and Alferness 1988; Heismann, Divino, and Buhl 1991).

The input and output polarization splitters in the polarization-independent EOTF on Fig. 9.34 are specially designed waveguide directional couplers, which are fine-tuned using $\Delta\beta$-reversal electrodes (Alferness and Buhl 1984; Habara 1987). The splitters couple TM-polarized light with less than 20 dB of cross talk into the crossover waveguide while leaving TE-polarized light in the straight-through waveguide. Completely passive polarization splitters with equal or better performance have also been demonstrated using either specially designed zero-gap directional couplers or waveguide branches with proton-exchanged waveguides (Tian *et al.* 1994; Baran and Smith 1992).

Figure 9.35 shows an example of the bandwidth and tuning range of such an EOTF. The period of the finger electrodes in this particular device is $\Lambda = 21$ μm, and the total coupling length of wavelength-selective mode conversion is 4.2 cm, which yields a narrow transmission band with a 3-dB bandwidth of only 0.6 nm (Heismann, Divino, and Buhl 1991). The center wavelength of the EOTF can be tuned continuously over a total range of 10 nm by applying tuning voltages between -100 and $+100$ V to the phase shifter electrodes. The extinction of the optical signal in the complementary notch filter output is typically better than 15 dB, because the amount of mode conversion can be adjusted externally by means of V_C (Warzanskyj, Heismann, and Alferness 1988). The bandwidth and tuning range shown in Fig. 9.35 allow, in principle, selective demultiplexing of at least nine multiplexed wavelength channels spaced about 1.2 nm apart. However, the cross talk from adjacent wavelength channels would be unacceptably high with this EOTF because of the large side lobes in the filter response on both sides of the main transmission peak. It has been demonstrated that these side-lobe levels can be reduced substantially, from about -10 to less than -25 dB, by spatially weighting the strength of the TE $\leftrightarrow$ TM conversion

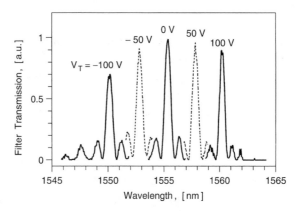

Fig. 9.35 Measured optical band-pass response of a polarization-independent electrooptic wavelength filter for various tuning voltages, V_T, between -100 and $+100$ V.

along the device — e.g., by varying the number of finger electrodes in the various sections (Croston *et al.* 1993). This important technique for side-lobe reduction results in only a small increase in the 3-dB filter bandwidth of about 30%.

Furthermore, the limited wavelength tunability of the EOTF can be extended by employing two or more independent TE $\leftrightarrow$ TM converter electrode systems of slightly different period (Nuttall, Croston, and Parsons 1994), which also allows simultaneous adding and dropping of two or more wavelength channels with a single EOTF. Typical fiber-to-fiber insertion losses of polarization-independent EOTFs are between 5 and 8 dB.

9.4.2 ACOUSTOOPTIC TUNABLE FILTERS

Tunable filters based on acoustooptic TE $\leftrightarrow$ TM conversion exhibit optical frequency characteristics similar to those of EOTFs, except the TE- and TM-polarized modes are coupled by a surface acoustic wave through the elastooptic effect. Such acoustooptic tunable filters (AOTFs) offer extremely wide tuning ranges and exhibit the unique ability to add or drop several wavelength channels simultaneously. Figure 9.36 shows an example of such a filter on x-cut, y-propagation $LiNbO_3$, where an acoustic Rayleigh wave is generated by an interdigital piezoelectric transducer on the crystal surface. Such transducers usually employ about 10 pairs of finger electrodes

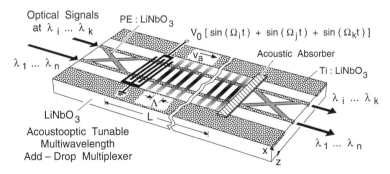

Fig. 9.36 Schematic of a polarization-independent acoustooptic tunable wavelength filter on x-cut lithium niobate, featuring an acoustic waveguide (PE : $LiNbO_3$ denotes proton-exchanged waveguides, and Ti : $LiNbO_3$ titanium-diffused regions). This arrangement allows multiwavelength add–drop multiplexing (d'Alessandro, Smith, and Baran 1994).

of period Λ and are driven by a sinusoidal voltage of frequency $f = v_a/\Lambda$, where $v_a \approx 3720$ m/s is the velocity of the acoustic wave. With $\Lambda \approx 21$ μm at a 1.5-μm wavelength, the drive frequency is usually around 180 MHz. Most efficient TE $\leftrightarrow$ TM conversion occurs when the period of the acoustic wave matches the beat period of the two coupled modes — i.e., at a wavelength $\lambda_0 = |n_p| \Lambda$. Complete conversion is obtained when the intensity of the acoustic wave, I_a, in the optic waveguide is equal to

$$I_a = \frac{\lambda_0^2 v_a^3 \rho}{2\Gamma_a n_e^3 n_o^3 p_{41}^2 L^2},\qquad(9.41)$$

where $p_{41} = 0.15$ is the relevant elastooptic coefficient, $\rho \approx 4700$ kg/m^3 is the material density, and Γ_a is an overlap parameter of the optical fields with the acoustic wave. The x-cut, y-propagation orientation is usually preferred over other crystal orientations because of a stronger electromechanical coupling coefficient for the excitation of the acoustic wave and an enhanced elastooptic effect through the strong vertical shear S_{12} of the acoustic wave (Heffner et al. 1988; Boyd and Heismann 1989).

The optical frequency response of the AOTF is very similar to that of EOTFs, except the output light of AOTFs is usually shifted in frequency because of an optical Doppler effect that occurs when two waves of different propagtion velocities are coupled at a traveling index grating. In the arrangement of Fig. 9.36, the TE $\rightarrow$ TM converted light is shifted to higher frequencies by the amount of the acoustic frequency, and TM $\rightarrow$ TE converted light is shifted down in frequency by the same amount. As with EOTFs, the optical bandwidth of AOTFs is inversely proportional to the overall interaction length of mode conversion, which extends from the interdigital transducer at the input to an acoustic absorber at the output. But unlike EOTFs, which can be tuned over only narrow optical wavelength ranges of typically $\Delta\lambda_T/\lambda_0 \sim 0.01$, AOTFs can be readily tuned over broad wavelength ranges of the order of $\Delta\lambda_T/\lambda_0 \sim 0.1$ by simply tuning the electric drive frequency f, and hence the period of the coupling acoustic wave, $\Lambda = v_a/f$. A 5% change in acoustic wavelength, for example, will cause approximately the same relative change in optical wavelength — e.g., about 75 nm around 1.5 μm. The electrical bandwidth of the interdigital transducers is usually broad enough to allow tuning across the entire 1.5-μm wavelength window (Heffner et al. 1988). Like their electrooptic counterparts, acoustooptic TE $\leftrightarrow$ TM converters allow polarization-independent filtering by arranging two parallel optical waveguides in a polarization-diversity scheme, as displayed in Fig. 9.36 (Smith *et al.* 1990).

Moreover, AOTFs can readily demultiplex several wavelength channels at the same time by applying drive signals of different frequencies simultaneously to the transducer, such that each frequency, f_i, selects a different optical wavelength, $\Lambda_i = v_a |n_p|/f_i$ (Cheung *et al.* 1989). A polarization-independent AOTF, as shown in Fig. 9.36, can thus serve as a rapidly reconfigurable multiwavelength add–drop multiplexer (d'Alessandro, Smith, and Baran 1994), or it may be used as an adjustable gain equalizer in optical amplifiers (Su *et al.* 1992). In practice, the number of simultaneously selected wavelength channels is limited by the acoustic drive power required for complete TE $\leftrightarrow$ TM conversion. AOTFs without lateral confinement of the acoustic wave need up to 100 mW of acoustic power per wavelength channel and about 500 mW of electric power (Heffner *et al.* 1988). These relatively high drive powers have recently been reduced by almost a factor of 10 through the introduction of low-loss acoustic waveguides, which confine the acoustic wave to a narrow (typically 100-μm-wide) region around the optical waveguides. As shown in Fig. 9.36, acoustic waveguides are usually fabricated by diffusion of titanium into the crystal surface (Frangen *et al.* 1989).

Acoustic waveguides are further useful for reducing the side lobes in the optical wavelength response by tapering the coupling strength along the acoustooptic interaction length, similar to the weighted coupling in EOTFs. Specially designed acoustic waveguide directional couplers, for example, have been employed to gradually transfer the acoustic wave from a parallel acoustic waveguide to the desired acoustooptic interaction region (Smith and Johnson 1992; Herrmann, Schäfer, and Sohler 11994). Moreover, additional acoustic attenuators placed along the interaction length can produce a substantially flat optical passband (Jackel, Baran, d'Alessandro, *et al.* 1996). Such optical passband shaping also helps to avoid undesired channel-to-channel interaction between neighboring transmission bands (Jackel, Baran, Chang, *et al.* 1995). Thus far, however, these techniques have not succeeded in reducing the side-lobe levels to less than -15 dB. These relatively high residual side lobes are attributed to undesired variations in the optical birefringence along the acoustooptic interaction length. Side-lobe suppressions of up to 20 dB have been demonstrated by using a cascade of two nearly identical AOTFs with an intermediate polarization filter (Smith *et al.* 1989). Such a double-stage filter also has the advantage of not shifting the frequency of the output light (Boyd and Heismann 1989).

9.5   Summary and Conclusions

In this chapter we reviewed recent advances in the state of the art of the lithium niobate integrated optic technology that have taken place in the areas of optical modulation, polarization control, and wavelength filtering. Worldwide research and devleopment in this field have produced dramatic improvement in device figures of merit, demonstrated new and significant functionality and applications, and established the technology as a benchmark of performance. Beyond these important achievements, another has been the first deployment of $LiNbO_3$ integrated optic components in commercial long-distance optical fiber telecommunications systems, which is a singular milestone in the evolution of any technology. Although the future of no technology is guaranteed for long, the extraordinary versatility of lithium niobate integrated optics provides it a unique advantage. We expect its story to continue to be written and its home in the marketplace to continue to grow.

Acknowledgments

The progress we summarized in this chapter implicitly represents the work of many researchers and developers around the world, and in this short space we were unable to do proper justice to even a fraction of their contributions. We are especially grateful to our colleagues working in this field because each has made a contribution to advancing the technology and has had an influence on our own work. It is our sincere pleasure to acknowledge, in particular, our closest collaborators of many years at Bell Laboratories who have made direct and immeasurable contributions to the advances in the device technologies that were discussed. Among them are the following: Thomas F. Adda, Rod C. Alferness, Lawrence L. Buhl, Robert Commozoli, M. D. Divino, Charles H. Joyner, Carl T. Kemmerer, Charles A. Mattoe, William J. Minford, David T. Moser, Timothy O. Murphy, Henry O'Brian, Joseph Schmulovich, Robert W. Smith, Ofer Sneh, Kathleen L. Tukuda, James E. Watson, and Matt S. Whalen. They have made the endeavor not only rewarding and exciting, but also truly enjoyable. We also acknowledge the members of the Lithium Niobate Team at the Lucent Technologies Optoelectronic Center for their commitment to excellence. Were it not for all these dedicated individuals, we would not have been able to write this chapter. Finally, we are most grateful for the un-wavering support and encouragement of our families and friends.

References

Alferness, R. C., and L. L. Buhl. 1984. Low-crosstalk waveguide polarization multi-plexer/demultiplexer for λ = 1.32 μm. *Opt. Lett.* 10:140–142.

Alferness, R. C., C. H. Joyner, L. L. Buhl, and S. K. Korotky. 1983. High-speed traveling-wave directional coupler switch/modulator for λ = 1.32 μm. *IEEE J. Quantum Electron.* QE-19:1339.

Alferness, R. C., V. R. Ramaswamy, S. K. Korotky, M. D. Divino, and L. L. Buhl. 1982. Efficient single-mode fiber to titanium diffused lithium niobate waveguide coupling for λ = 1.32 μm. *IEEE J. Quantum Electron.* QE-18:1807–1813.

Alferness, R. C., and J. N. Walpole, eds. 1986. (Special issue on integrated optics.) *IEEE J. Quantum Electron.* QE-22:803–988.

Baran, J. E., and D. A. Smith. 1992. Adiabatic 2×2 polarization splitter on LiNbO$_3$. *IEEE Photon. Tech. Lett.* 4:39–40.

Barnett, W. C., H. Takahira, J. C. Baroni, and Y. Ogi. 1996. The TPC-5 cable network. *IEEE Commun. Mag.* 34:36–40.

Becker, R. A. 1984. Traveling-wave electro-optic modulator with maximum bandwidth-length product. *Appl. Phys. Lett.* 45:1168.

Becker, R. A. 1985. Circuit effect in LiNbO$_3$ channel-waveguide modulators. *Opt. Lett.* 10:417–419.

Bellcore. 1993. Network equipment–building system (NEBS) generic equipment requirements. Document No. TR-EOP-000063. *Bellcore Tech. Ref.* (No. 5, Sep-tember).

Bergano, N. S. 1994. The time dynamics of polarization hole burning in an erbium-doped fiber amplifier. In *Technical Digest Conference on Optical Fiber Communi-cation, San Jose*, vol. 4, 305–306. Paper FF4. Washington, DC: Optical Society of America.

Bergano, N. S., C. R. Davidson, and F. Heismann. 1996. Bit-synchronous polarisa-tion and phase modulation scheme for improving the performance of optical amplifier transmission systems. *Electron. Lett.* 32:52–54.

Bergano, N. S., C. R. Davidson, and T. Li. 1993. A two-wavelength depolarized transmitter for improved transmission performance in long-haul EDFA systems. In *Technical Digest annual meeting of the IEEE Lasers and Electrooptics Society, San Jose*. Postdeadline paper PD2.2.

Billings, B. H. 1951. A monochromatic depolarizer. *J. Opt. Soc. Am.* 41:966–975.

Booth, R. C., B. E. Daymond-John, P. E. Sturges, and M. G. F. Wilson. 1984. Temperature tuning of LiNbO$_3$ electro-optic waveguide TE/TM mode convertors. *Electron. Lett.* 20:1045–1047.

Born, M., and E. Wolf. 1975. *Principles of optics*. New York: Pergamon, 554.

Boyd, G. D., and F. Heismann. 1989. Tunable acoustooptic reflection filters in LiNbO₃ without a Doppler shift. *J. Lightwave Tech.* 7:625–631.

Boyd, J. T., ed. 1988. (Special issue on integrated optics.) *J. Lightwave Tech.* 6:741–1159.

Bulmer, C. H., and W. K. Burns. 1984. Linear interferometric modulators in Ti: LiNbO₃. *J. Lightwave Tech.* LT-2:512–521.

Burns, W. K., A. B. Lee, and A. F. Milton. 1976. Active branching waveguide modulator. *Appl. Phys. Lett.* 29:790. *See also* Burns, W. K., M. M. Howerton, and R. P. Moeller. 1992. Performance and modeling of proton exchanged LiTaO₃ branching modulators. *J. Lightwave Tech.* 10:1403–1408.

Burns, W. K., R. P. Moeller, C. H. Bulmer, and A. S. Greenblatt. 1991. Depolarized light source for fiber-optic applications. *Opt. Lett.* 16:381–383.

Cheung, K-W., D. A. Smith, J. E. Baran, and B. L. Heffner. 1989. Multiple channel operation of integrated acousto-optic tunable filter. *Electron. Lett.* 25:375–376.

Childs, R. B., and V. A. O'Byrne. 1990. Multichannel AM video transmission using a high-power Nd:YAG laser and linearized external modulator. *IEEE J. Select. Areas Commun.* 8:1369.

Chipman, R. A. 1968. *Theory and problems of transmission lines*. Schaum's Outline Series. New York: McGraw-Hill.

Chung, H., W. S. C. Chang, and E. L. Adler. 1991. Modeling and optimization of traveling-wave LiNbO₃ interferometric modulators. *IEEE J. Quantum Electron.* 27:608–617.

Cox, C. H., III, G. E. Betts, and L. M. Johnson. 1990. An analytic and experimental comparison of direct and external modulation in analog fiber-optic links. *IEEE Trans. Microwave Theory Tech.* MTT-38:501–509.

Croston, I. R., A. D. Carr, N. J. Parsons, S. N. Radcliffe, and L. J. St. Ville. 1993. Lithium niobate electro-optic tunable filter with high sidelobe suppression. *Electron. Lett.* 29:157–159.

d'Alessandro, A., D. A. Smith, and J. E. Baran. 1994. Multichannel operation of an integrated acousto-optic wavelength routing switch for WDM systems. *IEEE Photon. Tech. Lett.* 6:390–393.

Doi, M., S. Taniguchi, M. Seino, G. Ishikawa, H. Ooi, and H. Nishimoto. 1996. 40 Gb/s integrated OTDM Ti:LiNbO₃ modulator. In *Technical Digest International Topical Meeting on Photonics in Switching, Sendai, Japan.* Paper PThB1.

Dolfi, D. W., and T. R. Ranganth. 1992. 50 GHz velocity-matched broad wavelength LiNbO₃ modulator with multimode active section. *Electron. Lett.* 28:1197–1198.

Eisenstein, G., S. K. Korotky, L. W. Stulz, J. J. Veselka, R. M. Jopson, and K. L. Hall. 1985. Antireflection coatings on lithium niobate waveguide devices using electron beam evaporated yttrium oxide. *Electron. Lett.* 21:363.

Evangelides, S. G., L. F. Mollenauer, J. P. Gordon, and N. S. Bergano. 1992. Polarization multiplexing with solutions. *J. Lightwave Tech.* 10:28–35.

Evangelides, S. G., Jr., B. M. Nyman, G. T. Harvey, L. F. Mollenauer, P. V. Mamyshev, M. L. Saylors, S. K. Korotky, U. Koren, T. A. Strasser, J. J. Veselka, J. D. Evankow, A. Lucero, J. Nagel, J. Sulhoff, J. Zyskind, P. C. Corbett, M. A. Mills, and G. Fergusson. 1996. Soliton WDM transmission with and without guiding filters. *IEEE Photon. Technol. Lett.* 8:1409–1411.

Ezekiel, S., and E. Udd, eds. 1991. Fiber optic gyros: 15th anniversary conference. *SPIE Proc.* 1585.

Fishman, D. A. 1993. Design and performance of externally modulated 1.5 μm transmitter in the presence of chromatic dispersion. *J. Lightwave Tech.* LT-11:624.

Fishman, D. A., J. A. Nagel, and S. M. Bahsoun. 1991. 1991 Roaring Creek field trial: Transmission results. Unpublished manuscript. *See* Fishman (1993).

Frangen, J., H. Herrmann, R. Ricken, H. Seibert, W. Sohler, and E. Strake. 1989. Integrated optical, acoustically tunable wavelength filter. *Electron. Lett.* 25:1583–1584.

Gee, C. M., G. D. Thurmond, and W. H. Yen. 1983. 17 GHz bandwidth electro-optic modulator. *Appl. Phys. Lett.* 43:998.

Giles, C. R., E. Desurvire, J. L. Zyskind, and J. R. Simpson. 1989. Noise performance of erbium-doped fiber amplifier pumped at 1.49 μm and application to signal preamplification at 1.8 Gbits/s. *IEEE Photon. Tech. Lett.* 1:367–369.

Giles, C. R., and S. K. Korotky. 1988. Stability of Ti:LiNbO$_3$ waveguide modulators in an optical transmission system. In *Technical Digest Topical Meeting on Integrated and Guided-Wave Optics, Sante Fe,* 115–118, Paper ME5. Washington, DC: Optical Society of America.

Glance, B. 1987. Polarization independent coherent optical receiver. *J. Lightwave Tech.* LT-5:274–276.

Gnauck, A. H., A. R. Chraplyvy, R. W. Tkach, J. L. Zyskind, J. W. Sulhoff, A. J. Lucero, Y. Sun, R. M. Jopson, F. Foghieri, R. M. Derosier, C. Wolf, and A. R. McCormick. 1996. One terabit/s transmission experiment. In *Technical Digest Conference on Optical Fiber Communication, San Jose.* Paper PD20.

Gnauck, A. H., S. K. Korotky, B. L. Kasper, J. C. Campbell, J. R. Talman, J. J. Veselka, and A. R. McCormick. 1986. Information-bandwidth-limited transmission at 8 Gb/s over 68.3 km of single-mode optical fiber. In *Technical Digest Conference on Optical Fiber Communication, Atlanta.* Postdeadline paper PD9.

Gnauck, A. H., S. K. Korotky, J. J. Veselka, J. Nagel, C. T. Kemmerer, W. J. Minford, and D. T. Moser. 1991. Dispersion penalty reduction using an optical modulator with adjustable chirp. *IEEE Photon. Tech. Lett.* 3:916–918.

Gopalakrishnan, G. K., C. H. Bulmer, W. K. Burns, R. W. McElhanon, and A. S. Greenblatt. 1992. 40 GHz, low half-wave voltage Ti:LiNbO$_3$ intensity modulator. *Electron. Lett.* 28:826–827.

Gopalakrishnan, G. K., W. K. Burns, R. W. McElhanon, C. H. Bulmer, and A. S. Greenblatt. 1994. Performance and modeling of broadband LiNbO$_3$ traveling wave optical intensity modulators. *J. Lightwave Tech.* 12:1807–1819.

Gunderson, D. R., A. Lecroat, and K. Tatekura. 1996. The Asia Pacific cable network. *IEEE Commun. Mag.* 34:42–48.

Gupta, K. C., R. Garg, and I. J. Bahl. 1979. *Microstrip lines and slotlines.* Dedham, MA: Artech House.

Habara, K. 1987. LiNbO₃ directional-coupler polarisation splitter. *Electron. Lett.* 23:614–616.

Haga, H., M. Izutsu, and T. Sueta. 1986. LiNbO₃ traveling-wave light modulator/switch with an etched groove. *IEEE J. Quantum Electron.* QE-22:902–906.

Hansen, P. B., L. Eskildsen, S. G. Grubb, A. M. Vengsarkar, S. K. Korotky, T. A. Strasser, J. Alphonsus, J. J. Veselka, D. J. DiGiovanni, D. W. Peckman, E. C. Beck, D. A. Truxal, W. Y. Cheung, S. Kosinski, D. S. Gasper, P. F. Wysocki, V. L. Da Silva, and J. R. Simpson. 1995. 2.4488 Gb/s unrepeatered transmission over 529 km using remotely pumped post- and preamplifiers, forward error correction, and dispersion compensation. In *Technical Digest Conference on Optical Fiber Communication, San Diego.* Postdeadline paper PD25. Washington, DC: Optical Society of America.

Heffner, B. L., D. A. Smith, J. E. Baran, A. Yi-Yan, and K-W. Cheung. 1988. Integrated-optic acoustically tunable infra-red optical filter. *Electron. Lett.* 24:1562–1563.

Heidrich, H., D. Hoffmann, C. H. V. Helmolt, and H. Ahlers. 1989. Lithium niobate guided-wave network for a coherent receiver. *Opt. Lett.* 14:99–101.

Heismann, F. 1994. Analysis of a reset-free polarization controller for fast automatic polarization stabilization in fiber-optic transmission systems. *J. Lightwave Tech.* 12:690–699.

Heismann, F. 1995. Polarization scrambling and multiplexing in optically amplified transoceanic communication systems. In *Proceedings of the 10th International Conference on Integrated Optics and Optical Fibre Communication, Hong Kong,* vol. 4, 92–93. Paper FD1, Hong Kong: The Chinese University Press.

Heismann, F., and R. C. Alferness. 1988. Wavelength-tunable electrooptic polarization conversion in birefringent waveguides. *IEEE J. Quantum Electron.* 24:83–93.

Heismann, F., A. F. Ambrose, T. O. Murphy, and M. S. Whalen. 1993. Polarization independent photonic switching system using fast automatic polarization controllers. *IEEE Photon. Tech. Lett.* 5:1341–1343.

Heismann, F., L. L. Buhl, and R. C. Alferness. 1987. Electro-optically tunable, narrowband Ti:LiNbO₃ wavelength filter. *Electron. Lett.* 23:572–574.

Heismann, F., M. D. Divino, and L. L. Buhl. 1991. Mirror-folded polarization-independent wavelength filter. *IEEE Photon. Tech. Lett.* 3:219–221.

Heismann, F., D. A. Gray, B. H. Lee, and R. W. Smith. 1994. Electrooptic polarization scramblers for optically amplified long-haul transmission systems. *IEEE Photon. Tech. Lett.* 6:1156–1158.

Heismann, F., P. B. Hansen, S. K. Korotky, G. Raybon, J. J. Veselka, and M. S. Whalen. 1993. Automatic polarization demultiplexer for polarization-multiplexed transmission systems. *Electron. Lett.* 29:1965–1966.

Heismann, F., and K. L. Tokuda. 1995a. Polarization-independent electrooptic depolarizer. *Opt. Lett.* 20:1008–1010.

Heismann, F., and K. L. Tokuda. 1995b. Polarization-independent electrooptic depolarizer. In *Technical Digest Topical Meeting on Integrated Photonics Research, Dana Point, CA,* vol. 7, 293–295. Paper ISaB2. Washington, DC: Optical Society of America.

Heismann, F., and M. S. Whalen. 1991. Broadband reset-free automatic polarization controller. *Electron. Lett.* 27:377–379.

Heismann, F., and M. S. Whalen. 1992. Fast automatic polarization control system. *IEEE Photon. Tech. Lett.* 4:503–505.

Henry, P. S. 1986. *Lightwave primer. IEEE J. Quantum Electron.* QE-21:1862–1879.

Herrmann, H., K. Schäfer, and W. Sohler. 1994 Polarization independent, integrated optical, acoustically tunable wavelength filters/switches with tapered acoustical directional coupler. *IEEE Photon. Tech. Lett.* 6:1335–1337.

Hill, P. M., R. Olshansky, and W. K. Burns. 1992. Optical polarization division multiplexing at 4 Gb/s. *IEEE Photon. Tech. Lett.* 4:500–502.

Hodgkinson, T. G., R. A. Harmon, and D. W. Smith. 1987. Polarisation-insensitive heterodyne detection using polarisation scrambling. *Electron. Lett.* 23:513–514.

Hood, K. J., P. W. Walland, C. L. Nuttall, L. J. St. Ville, T. P. Young, A. Oliphant, R. P. Marsden, J. T. Zubrzycki, G. Cannell, C. Bunney, J. P. Laude, and M. J. Anson. 1993. Optical distribution systems for television studio applications. *J. Lightwave Tech.* 11:680–687.

Hornak, L. A., ed. 1992. *Polymers for lightwave and integrated optics: Technology and applications.* New York: Marcel Dekker.

Howerton, M. M., and W. K. Burns. 1994. Depolarized source for high power remote operation of an integrated optical modulator. *IEEE Photon. Tech. Lett.* 6:115–117.

Hunsperger, R. G. 1982. *Integrated optics: Theory and technology.* New York: Springer-Verlag.

Hutcheson, L. D., ed. 1987. *Integrated optical circuits and components.* New York: Marcel Dekker.

Imai, T., K. Nosu, and H. Yamaguchi. 1985. Optical polarisation control utilising an optical heterodyne detection scheme. *Electron. Lett.* 21:52–53.

Ishikawa, G., H. Ooi, Y. Akiyama, S. Taniguchi, and H. Nishimoto. 1996. 80 Gb/s (2×40 Gb/s) transmission experiments over 667 km dispersion-shifted fiber using Ti : $LiNbO_3$ OTDM modulator and demultiplexer. In *Proceedings of the 22nd European Conference on Optical Communication, Oslo,* vol. 5. Paper ThC.3.3. Kjeller, Norway: Telenor R&D.

Izutsu, M., Y. Yamane, and T. Sueta. 1977. Broad-band traveling-wave modulator using a $LiNbO_3$ optical waveguide. *IEEE J. Quantum Electron.* QE-13:287.

Jackel, J. L., J. E. Baran, G-K. Chang, M. Z. Iqbal, G. H. Song, W. J. Tomlinson, D. Fritz, and R. Ade. 1995. Multichannel operation of AOTF switches: Reducing channel-to-channel interaction. *IEEE Photon. Tech. Lett.* 7:370–372.

Jackel, J. L., J. E. Baran, A. d'Alessandro, and D. A. Smith. 1995. A passband-flattened acousto-optic filter. *IEEE Photon. Tech. Lett.* 7:318–320.

Jordan, E. C., and K. G. Balmain. 1968. *Electromagnetic waves and radiating systems.* Englewood Cliffs, NJ: Prentice-Hall.

Jungerman, R. L., and D. W. Dolfi. 1991. Frequency domain optical network analysis using integrated optics. *IEEE J. Quantum Electron.* 27:580.

Jungerman, R. L., C. Johnsen, D. J. McQuate, K. Salomaa, M. P. Zurakowski, R. C. Bray, G. Conrad, D. Cropper, and P. Hernday. 1990. High-speed optical modulator for application in instrumentation. *J. Lightwave Tech.* 8:1363–1370.

Kaminow, I. P. 1963. Splitting of Fabry–Perot rings by microwave modulation of light. *Appl. Phys. Lett.* 2:41–42.

Kaminow, I. P. 1987. Linear electrooptic materials. In *CRC handbook of laser science and technology,* vol. IV, ed. M. J. Weber, 253–278. Boca Raton, FL: CRC Press.

Kaminow, I. P., ed. 1974. *An introduction to electro-optic devices.* New York: Academic Press.

Kaminow, I. P., T. J. Bridges, and M. A. Pollack. 1970. A 964 GHz traveling-wave electrooptic light modulator. *Appl. Phys. Lett.* 16:416–418.

Kaminow, I. P., and E. H. Turner. 1966. Electrooptic light modulators. *Appl. Opt.* 5:1612.

Kawano, K., T. Kitoh, H. Jumonji, T. Nozawa, and M. Yanagibashi. 1989. New traveling-wave electrode Mach–Zehnder optical modulator with 20 GHz bandwidth and 4.7 V driving voltage at 1.52 μm wavelength. *Electron. Lett.* 25:1382–1383.

Kersey, A. D., and A. Dandridge. 1987. Monomode fibre polarisation scrambler. *Electron. Lett.* 23:634–636.

Kincaid, B. E. 1988. Coupling of polarization-maintaining fibers to Ti:LiNbO$_3$ waveguides with angled interfaces. *Opt. Lett.* 13:425–427.

Kobayashi, K., and I. Mito. 1989. High speed and tunable semiconductor lasers. In *Technical Digest Conference on Optical Fiber Communication, Houston,* 397–456. Tutorial ThH1. Washington, DC: Optical Society of America.

Kobota, K. J. Noda, and O. Mikami. 1980. Traveling-wave optical modulator using a directional coupler LiNbO$_3$ waveguide. *IEEE J. Quantum Electron.* QE-16:754.

Koch, T. L., and J. E. Bowers. 1984. Nature of wavelength chirping in directly modulated semiconductor lasers. *Electron. Lett.* 20:1038.

Korotky, S. K. 1986. Three-space representation of phase-mismatch switching in coupled two-state optical systems. *IEEE J. Quantum Electron.* QE-22:952–958.

Korotky, S. K. 1989. Optimization of traveling-wave integrated-optic modulators. In *Technical Digest Topical Meeting on Numerical Simulation and Analysis in Guided-Wave Optics and Optoelectronics, Houston,* vol. 3, 102–105. Paper SF2. Washington, DC: Optical Society of America.

Korotky, S. K., and R. C. Alferness. 1983. Time- and frequency-domain response of directional-coupler traveling-wave optical modulators. *J. Lightwave Tech.* LT-1:244–251.

Korotky, S. K., and R. C. Alferness. 1987. Ti:LiNbO₃ integrated optic technology: Fundamentals, design considerations, and capabilities. In *Integrated optical circuits and components*, ed. L. D. Hutcheson, 169–227. New York: Marcel Dekker.

Korotky, S. K., and R. C. Alferness. 1988. Waveguide electrooptic devices for optical fiber communication. In *Optical fiber telecommunications II*, ed. S. E. Miller and I. P. Kaminow, 421–465. Boston: Academic Press.

Korotky, S. K., J. C. Campbell, and H. Nakajima, eds. 1991. (Special issue on photonic devices and circuits.) *IEEE J. Quantum Electron.* 27:516–849.

Korotky, S. K., G. Eisenstein, R. C. Alferness, J. J. Veselka, L. L. Buhl, G. T. Harvey, and P. H. Read. 1985. Fully connectorized high-speed Ti:LiNbO₃ switch/modulator for time-division multiplexing and data encoding. *J. Lightwave Tech.* LT-3:1–6.

Korotky, S. K., G. Eisenstein, A. H. Gnauck, B. L. Kasper, J. J. Veselka, R. C. Alferness, L. L. Buhl, C. A. Burrus, T. C. D. Huo, L. W. Stulz, N. K. Ciemiecki, L. G. Cohen, R. W. Dawson, and J. C. Campbell. 1985. 4 Gb/s transmission experiment over 117 km of optical fiber using a Ti : LiNbO₃ external modulator. *J. Lightwave Tech.* LT-3:1027–1031.

Korotky, S. K., G. Eisenstein, R. S. Tucker, J. J. Veselka, and G. Raybon. 1987. Optical intensity modulation to 40 GHz using a waveguide electro-optic switch. *Appl. Phys. Lett.* 50:1631–1633.

Korotky, S. K., P. B. Hansen, L. Eskildsen, and J. J. Veselka. 1995. Efficient phase modulation scheme for suppressing stimulated Brillouin scattering. In *Proceedings of the International Conference on Integrated Optics and Optical Communications, Hong Kong,* 110–111. Paper WD-1. Hong Kong: The Chinese University Press.

Korotky, S. K., and J. J. Veselka. 1990. Efficient switching in a 72 Gb/s Ti : LiNbO₃ binary multiplexer/demultiplexer. In *Technical Digest Conference on Optical Fiber Communication, San Francisco,* 32. Paper TuH2. Washington, DC: Optical Society of America.

Korotky, S. K., and J. J. Veselka. 1993. Versatile pulse generator for optical soliton transmission. In *Technical Digest Conference on Lasers and Electro-Optics, Baltimore.* Paper CPD25. Washington, DC: Optical Society of America.

Korotky, S. K., and J. J. Veselka. 1994. RC circuit model of long term Ti:LiNbO₃ bias stability. In *Techical Digest Topical Meeting on Integrated Photonics Research, San Francisco,* 187–189. Paper FB3. Washington, DC: Optical Society of America.

Korotky, S. K., and J. J. Veselka. 1996. RC network analysis of long term Ti : LiNbO₃ bias stability. *J. Lightwave Technol.* 14.

Korotky, S. K., J. J. Veselka, A. S. Hou, G. Raybon, and R. S. Tucker. 1990. Optical time-division demultiplexing of pulses with spacing corresponding to 72 Gb/s. Unpublished manuscript.

Korotky, S. K., J. J. Veselka, C. T. Kemmerer, W. J. Minford, D. T. Moser, J. E. Watson, C. A. Mattoe, and P. L. Stoddard. 1991. High-speed, low power optical modulator with adjustable chirp parameter. In *Technical Digest Topical Meeting on Integrated Photonics Research, Monterey, CA,* 53–54. Paper TuG2. Washington, DC: Optical Society of America.

Koyama, K., and K. Iga. 1988. Frequency chirping in external modulators. *J. Lightwave Tech.* 6:87.

Kuwata, N., H. Nishimoto, T. Harimatsu, and T. Touge. 1990. Automatic bias control circuit for Mach–Zehnder modulator. (in Japanese). In *IEICE, Spring National Convention Record.* Paper B-976.

Leonberger, F. J., T. K. Findakly, and P. G. Suchoski. 1989. LiNbO$_3$ and LiTaO$_3$ integrated optic components for fiber optic sensors. In *Optical fiber sensors,* ed. H. J. Arditty, J. P. Dakin, and R. T. Kersten, 5–9. New York: Springer-Verlag.

Marcuse, D. 1989. Electrostatic field of electrooptic modulators computed with the point matching method. In *Technical Digest topical meeting on numerical simulation and analysis in guided-wave optics and optoelectronics,* 98–101. Houston, Paper SF1. Washington, DC: Optical Society of America.

Marz, R. 1995. *Integrated optics design and modeling.* Boston: Artech House.

Mazurczyk, V. J., and J. L. Zyskind. 1994. Polarization dependent gain in erbium doped fiber amplifiers. *IEEE Photon. Tech. Lett.* 6:616–618.

McCaughan, L., and S. K. Korotky. 1986. Three-electrode Ti:LiNbO$_3$ optical switch. *J. Lightwave Tech.* LT-4;1324.

McConaghy, C., M. Lowry, R. A. Becker, and B. E. Kincaid. 1994. Characterization of lithium niobate electrooptic modulators at cryogenic temperatures. *Proc. SPIE, Design Simulation, and Fabrication of Optoelectronic Devices and Circuits.* 2150:283–291.

Miller, S. E. 1954. Coupled-wave theory and waveguide applications. *Bell Syst. Tech. J.* 33:661–719.

Miller, S. E., and A. G. Chynoweth, eds. 1979. *Optical fiber telecommunications I.* New York: Academic Press.

Miller, S. E., and I. P. Kaminow, eds. 1988. *Optical fiber telecommunications II.* Boston: Academic Press.

Misono, M., N. Henmi, T. Hosoi, and M. Fujiwara. 1996. High-speed wavelength switching of an acoustooptic tunable filter for WDM network in broadcasting stations. *IEEE Photon. Tech. Lett.* 8:572–574.

Mollenauer, L. F., J. P. Gordon, and F. Heismann. 1995. Polarization scattering by soliton–soliton collisions. *Opt. Lett.* 20:2060–2062.

Mollenauer, L. F., P. V. Mamyshev, and M. J. Neubelt, 1996. Demonstration of soliton WDM transmission at up to 8 × 10 Gb/s error-free over transoceanic distances. In *Technical Digest Conference on Optical Fiber Communication, San Jose.* Postdeadline paper PD22. Washington, DC: Optical Society of America.

Morioka, T., H. Takara, S. Kawanishi, O. Kamatani, K. Takiguchi, K. Uchiyama, M. Saruwatari, H. Takahashi, M. Yamada, T. Kanamori, and H. Ono. 1996. 100 Gbit/s × 10 channel OTDM/WDM transmission using a single supercontinuum WDM source. In *Technical Digest Conference on Optical Fiber Communication, San Jose.* Postdeadline paper PD21.

Noé, R., H. Heidrich, and D. Hoffmann. 1988. Endless polarization control systems for coherent optics. *J. Lightwave Tech.* LT-6:1199–1208.

Noé, R., M. Rehage, C. Harizi, and R. Ricken. 1994. Depolariser based on acousto-optical TE–TM converters for suppression of polarisation holeburning in long-haul EDFA links. *Electron. Lett.* 30:1500–1501.

Noé R., and D. A. Smith. 1988. Integrated-optic rotating waveplate frequency shifter. *Electron. Lett.* 24:1348–1349.

Noguchi, K., O. Mitomi, and H. Miyazawa. 1996. Low-voltage and broadband Ti:LiNbO$_3$ modulators operating in the millimeter wavelength region. In *Technical Digest Conference on Optical Fiber Communication, San Jose,* 205–206. Paper ThB2. Washington, DC: Optical Society of America.

Noguchi, K., O. Mitomi, K. Kawano, and M. Yanagibashi. 1993. Highly-efficient 40 GHz bandwidth Ti:LiNbO$_3$ optical modulator employing ridge structure. *IEEE Photon. Technol. Lett.* 5:52–54.

Nuttall, C. L., I. R. Croston, and N. J. Parsons. 1994. Electro-optic tunable filters for multi-wavelength networks. In *Technical Digest 20th European Conference on Optical Communication, Firenze,* vol. 2, 767–770. Genova, Italy: Istituto Internazionale delle Comunicazioni, Paper We.C.4.5.

Nyman, B. M., S. G. Evangelides, G. T. Harvey, L. F. Mollenauer, P. V. Mamyshev, M. Saylors, S. K. Korotky, U. Koren, V. Mizrahi, T. A. Strasser, J. J. Veselka, J. D. Evankow, A. J. Lucero, J. A. Nagel, J. W. Sulhoff, J. L. Zyskind, P. C. Corbett, M. A. Mills, and G. A. Ferguson. 1995. Soliton WDM transmission of 8 × 2.5 Gb/s, error free over 10 Mm. In *Technical Digest Conference on Optical Fiber Communication, San Diego.* Postdeadline paper PD21. Washington, DC: Optical Society of America.

O'Donnell, A. 1995. Packaging and reliability of active integrated optical components. In *Proceedings of the 7th European Conference on Integrated Optics. Delft,* 585–590. Paper ThC4. Delft: Delft University Press.

Okiyama, T., H. Nishimoto, T. Touge, M. Seino, and H. Nakajima. 1987. Optical fiber transmission over 132 km at 4 Gb/s using a Ti:LiNbO$_3$ Mach–Zehnder modulator. In *Proceedings of the European Conference on Optical Communication, Helsinki,* 55. Postdeadline paper.

Okoshi, T. 1985. Polarization-state control schemes for heterodyne or homodyne optical fiber communications. *J. Lightwave Tech.* LT-3:1232–1237.

Onaka, H., H. Miyata, G. Ishikawa, K. Otsuka, H. Ooi, Y. Kai, S. Kinoshita, M. Seino, H. Nishimoto, and T. Chikama. 1996. 1.1 Tb/s WDM transmission over a 150 km 1.3 μm zero-dispersion single-mode fiber. In *Technical Digest Conference on Optical Fiber Communication, San Jose.* Postdeadline paper PD19. Washington, DC: Optical Society of America.

Park, Y. K., T. V. Nguyen, O. Mizuhara, C. D. Chen, L. D. Tzeng, P. D. Yeates, F. Heismann, Y. C. Chen, D. G. Ehrenberg, and J. C. Feggeler. 1966. Field demonstration of 10-Gb/s line-rate transmission on an installed transoceanic submarine lightwave cable. *IEEE Photon. Tech. Lett.* 8:425–427.

Parsons, N. J., A. C. O'Donnell, and K. K. Wong. 1986. Design of efficient and wideband traveling-wave modulators. *Proc. SPIE, Integrated Optical Circuit Engineering III* Paper 24. 651.

Ramaswamy, R. V., M. D. Divino, and R. D. Standley. 1978. Balanced bridge modulator switch using Ti-diffused LiNbO$_3$ strip waveguides. *Appl. Phys. Lett.* 32:644–646.

Ramaswamy, R. V., and R. D. Standley. 1976. A phased, optical, coupler-pair switch. *Bell Syst. Tech. J.* 55:767–775.

Rumbaugh, S. H., M. D. Jones, and L. W. Casperson. 1990. Polarization control using nematic liquid crystals. *J. Lightwave Tech.* 8:459–465.

Sawaki, I., H. Nakajima, M. Seino, and K. Asama. 1986. Thermally stabilized z-cut Ti:LiNbO$_3$ waveguide switch. In *Technical Digest Conference on Lasers and Electro-Optics, Anaheim.* Paper MF2. Washington, DC: Optical Society of America.

Seino, M., N. Mekada, T. Namiki, and H. Nakajima. 1989. 33GHz-cm broadband Ti:LiNbO$_3$ Mach–Zehnder modulator. In *Proceedings of the European Conference on Optical Communication, Gothenburg.* Paper ThB22-5.

Seino, M., T. Nakazawa, Y. Kubota, M. Doi, T. Yamane, and H. Hakogi. 1992. A low DC-drift Ti:LiNbO$_3$ modulator assured over 15 years. In *Technical Digest Conference on Optical Fiber Communication, San Jose.* Postdeadline paper PD3.

Seino, M., T. Shiina, N. Mekada, and H. Nakajima. 1987. Low-loss Mach–Zehnder modulator using mode coupling Y-branch waveguide. In *Proceedings of the European Conference on Optical Communication, Helsinki,* 113.

Sheehy, F. T., W. B. Bridges, and J. H. Schaffner. 1993. 60 GHz and 94 GHz antenna-coupled LiNbO$_3$ electrooptic modulators. *IEEE Photon. Tech. Lett.* 5:307–310.

Skeath, P., C. H. Bulmer, S. C. Hiser, and W. K. Burns. 1986. Novel electrostatic mechanism in the thermal instability of z-cut LiNbO$_3$ interferometers. *Appl. Phys. Lett.* 49:1221.

Smith, D. A., J. E. Baran, K-W. Cheung, and J. J. Johnson. 1990. Polarization-independent acoustically tunable optical filter. *Appl. Phys. Lett.* 56:209–211.

Smith, D. A., and J. J. Johnson. 1992. Sidelobe suppression in an acousto-optic filter with a raised-cosine interaction strength. *Appl. Phys. Lett.* 61:1025–1027.

Smith, D. A., J. J. Johnson, B. L. Heffner, K-W. Cheung, and J. E. Baran. 1989. Two-stage integrated-optic acoustically tunable optical filter with enhanced sidelobe suppression. *Electron. Lett.* 25:398–399.

Smith, D. A., H. Rashid, R. S. Chakravarthy, A. M. Agboatwalla, A. A. Patil, Z. Bao, N. Imam, S. W. Smith, J. E. Baran, J. L. Jackel, and J. Kallman. 1996. Acousto-optic switch with near rectangular passband for WDM systems. *Electron. Lett.* 32:542–543.

Stephens, W. E., and T. R. Joseph. 1987. System characteristics of directly modulated and externally modulated RF fiber-optic links. *J. Lightwave Tech.* LT-5:380–387.

Su, S. F., R. Olshansky, G. Joyce, D. A. Smith, and J. E. Baran. 1992. Gain equalization in multiwavelength lightwave systems using acoustooptic tunable filters. *IEEE Photon. Tech. Lett.* 4:269–271.

Suche, H., I. Baumann, D. Hiller, and W. Sohler. 1993. Model-locked Er:Ti:LiNbO$_3$-waveguide laser. *Electron. Lett.* 29:1111–1112.

Suchoski, P. G., and G. R. Boivin. 1992. Reliability and accelerated aging of LiNbO$_3$ integrated optic fiber gyro circuits. *SPIE, Fiber Optic and Laser Sensors X* 1795:38–47.

Sueta, T., and M. Izutsu. 1982. High speed guided-wave optical modulators. *J. Opt. Commun.* 3:52–58.

Takada, T., and M. Saruwatari. 1988. 100 Gbit/s optical signal generation by time-division multiplication of modulated and compressed pulses from gain-switched distributed feedback (DFB) laser diode. *Electron. Lett.* 24:1406–1408.

Tamir, T., ed. 1979. *Integrated optics.* 2d ed. New York: Springer-Verlag.

Tamir, T., ed. 1988. *Guided-wave optoelectronics.* New York: Springer-Verlag.

Taylor, M. G. 1993. Observation of new polarization dependence effects in long haul optically amplified systems. *IEEE Photon. Tech. Lett.* 5:1244–1246.

Taylor, M. G., and S. J. Penticost. 1994. Improvement in performance of long haul EDFA link using high frequency polarisation modulation. *Electron. Lett.* 30:805–806.

Thaniyavarn, S. 1985. Wavelength independent, optical damage immune z-propagation LiNbO$_3$ waveguide polarization converter. *Appl. Phys. Lett.* 47:674–677.

Tian, F., C. Harizi, H. Herrmann, V. Reimann, R. Ricken, U. Rust, W. Sohler, F. Wehrmann, and S. Westenhöfer. 1994. Polarization-independent integrated optical, acoustically tunable double-stage wavelength filter in LiNbO$_3$. *J. Lightwave Tech.* 12:1192–1197.

Trischitta, P., M. Colas, M. Green, G. Wuzniak, and J. Arena. 1996. The TAT-12/13 cable network. *IEEE Commun. Mag.* 34:24–28.

Tsang, T., and V. Radeka. 1995. Electro-optical modulators in particle detectors. *Rev. Sci. Instrum.* 66:3844.

Tucker, R. S., G. Eisenstein, and S. K. Korotky. 1988. Optical time-division multiplexing for very high bit-rate transmission. *J. Lightwave Tech.* 11:1737–1749.

Uehara, S. 1978. Calibration of optical modulator frequency response with application to signal level control. *Appl. Opt.* 17:68–71.

Vasilev, P. P., and A. B. Sergeev. 1989. Generation of bandwidth-limited 2 ps pulses with 100 GHz repetition rate from multi-segmented injection laser. *Electron. Lett.* 25:1049–1050.

Veselka, J. J., and S. K. Korotky. 1986. Optimization of Ti:LiNbO$_3$ optical waveguides and directional coupler switches for 1.56 μm wavelength. *IEEE J. Quantum Electron.* QE-22:933–938.

Veselka, J. J., and S. K. Korotky. 1966. Pulse generation for soliton systems using lithium niobate modulators, *IEEE J. Sel. Top. Quantum Electron.* 2, June, 1996.

Veselka, J. J., S. K. Korotky, P. V. Mamyshev, A. H. Gnauck, G. Raybon, and N. M. Froberg. 1966. A soliton transmitter using a CW laser and an NRZ driven Mach–Zehnder modulator. *IEEE Photon. Technol. Lett.* 8:950–952.

Veselka, J. J., S. K. Korotky, C. T. Kemmerer, W. J. Minford, D. T. Moser, and R. W. Smith. 1992. Sensitivity to RF drive power and the temperature stability of Mach–Zehnder modulators. In *Technical Digest topical meeting on integrated photonics research, New Orleans.* Paper TuG4. Washington, DC: Optical Society of America.

Wadell, B. C. 1991. *Transmission line design handbook.* Boston: Artech House.

Walker, G. R., and N. G. Walker, 1990. Polarization control for coherent communications. *J. Lightwave Tech.* 8:438–458.

Walker, G. R., N. G. Walker, J. Davidson, D. C. Cunningham, A. R. Beaumont, and R. C. Booth. 1988. Lithium niobate waveguide polarisation convertor. *Electron. Lett.* 24:103–105.

Warzanskyj, W., F. Heismann, and R. C. Alferness. 1988. Polarization-independent electro-optically tunable narrow-band wavelength filter. *Appl. Phys. Lett.* 53:13–15.

Welsh, T., R. Smith, H. Azami, and R. Chrisner. 1996. The FLAG cable system. *IEEE Commun. Mag.* 34:30–35.

Yamada, S., and M. Minakata. 1981. DC drift phenomena in LiNbO$_3$ optical waveguide devices. *Jpn. J. Appl. Phys.* 20:733–737.

Chapter 10 | Photonic Switching

Edmond J. Murphy

Lucent Technologies, Bell Laboratories, Breinigsville, Pennsylvania

10.1 Introduction

The use of lightwave transmission systems for interoffice transport and access will continue to grow both in speed and bandwidth utilization. The transparency offered by optical fiber will lead to the merger of transport and switching functions, which will result in faster and more flexible network capabilities. Just as time and space switches are needed in today's electronic cross-connects to access and reconfigure capacity, optical space switches, wavelength multiplexers, and time-delay switches will be required in future optical systems. The capabilities of such devices will greatly enhance system performance and functionality over the current point-to-point systems. The optical transparency of a guided-wave network allows flexible transport and routing of signals between different network endpoints. Optical cross-connect systems simplify provisioning and operations by allowing network-ing at the optical layer. In this chapter, we describe the technologies and architectures needed for optical switching and routing systems. Many new concepts have been realized since the last volume of this book was published in 1988 (Miller and Kaminow 1988). System and device attributes have been studied in an effort to simultaneously optimize device and system design. Significant demonstrations of guided-wave switching technology and guided-wave switching systems have been made. We begin with an in-depth discussion of the state of the art in optical switching. Our main focus is on guided-wave switches in lithium niobate and semiconductor materials. We first describe progress in basic switch elements and then review the state of the art in integration of these elements. By discussing examples of switching system demonstrations, we describe the broad range of potential

463

OPTICAL FIBER TELECOMMUNICATIONS,
VOLUME IIIB

Copyright © 1997 by Lucent Technologies.
All rights of reproduction in any form reserved.
ISBN: 0-12-395171-2

applications and the system advantages associated with switching in the optical domain. Our discussion is not intended to be an exhaustive review but rather to be illustrative of the state of the art.

Even today, optical switching systems approach new frontiers. The rapid deployment of wavelength multiplexing will lead to new opportunities for application of optical switches. Indeed, a network infrastructure in which protection, restoration, and provisioning are provided at the transmission data rate on a per-wavelength basis is imagined. The confluence of wavelength and space switching technologies offers the exciting possibility of line-rate cross-connects and wavelength provisioning. Furthermore, they offer the potential for rapid restoration and automatic protection, again, at the line rate. An optical space switching capability will be required in nodes to switch signals between fibers and between wavelengths. This is a particularly appropriate application for optical switching that will likely accelerate the deployment of systems. Reconfiguration at the line rate requires much smaller switch fabrics than needed for today's electronic crossconnect and central office switches — fabric sizes that can readily be built with today's switch technology.

10.2   Optical Switching Overview

In this section, we describe some of the basic attributes of optical switching systems. We define important terminology, describe high-level system architectures, and discuss some of the fundamental system constraints.

The necessary terminology can be introduced in a three-step hierarchical approach. Figure 10.1 depicts both a 1×2 and a 2×2 *switch element*. The switch element is the basic building block for switch fabrics. The most common elements are the 1×1 amplifier gate, the 1×2 Y switch, and the 2×2 directional coupler. The switch element is characterized by its excess insertion loss, cross talk or extinction ratio, control voltage or current, switching speed, polarization sensitivity, and physical length. The actual values of these parameters depend upon the fabrication technology and are detailed in the following discussions. Figure 10.2 shows a representative *switch module*, in which a number of switch elements have been integrated. The columns of switch elements are connected with a passive waveguide interconnect web. The switch module is the packaged device from which fabrics are built. It is characterized by several physical size and performance parameters. Important size parameters (and typical values) include the

a

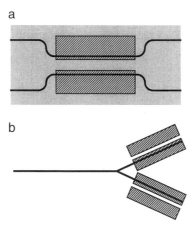

b

Fig. 10.1   (a) Two-by-two directional coupler and (b) 1 × 2 Y-switch elements.

number of input and output (I/O) waveguides (≤20), the number of electrical control leads (≤250), the number of columns of switches (≤8), and the number of integrated switch elements (≤150). The performance parameters include insertion loss (or gain), loss uniformity, control voltage or current uniformity, polarization dependence, and end-face reflectivity. Figure 10.3 shows a representative *switching fabric* composed of stages or columns of switch modules interconnected by a fiber interconnect mesh. The switching fabric is characterized by its connectivity (i.e., the number of I/O ports and the probability of blocking a desired connection), overall insertion loss (or gain), loss uniformity, system-level cross-talk at each output port, and the complexity of the fiber interconnect.

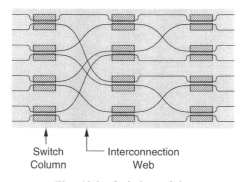

Switch   Interconnection
Column   Web

Fig. 10.2   Switch module.

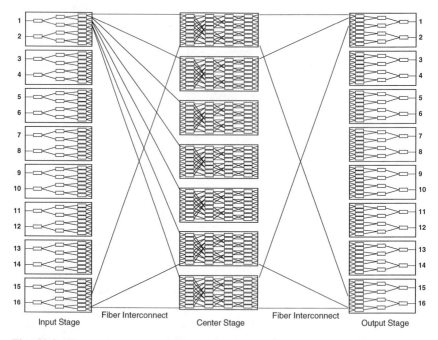

Fig. 10.3 Example of a 16 × 16 switch fabric showing three stages of switch modules and a portion of the fiber interconnect. The actual system is pictured in Fig. 10.14.

Figure 10.3 also illustrates a basic switching system functionality — that of an optical space switch. This fabric allows the physical routing of the signals on each of the inputs to each of the outputs, which effectively repositions the signals in space. If the switch elements inside the fabric are simple space switches (directional couplers or Ys), the space switch fabric has the important property that it is independent of the data rate or format through the switch. That is, once a path has been set through the switch, any data rate or signal format may be passed through the switch just as it is passed through an optical fiber. In fact, each input fiber could contain a number of wavelength multiplexed signals. This attribute is in contrast to electronic switches, in which the switch elements themselves must toggle at the data rate. An important consequence of this property is the low power consumption of optical switching systems relative to electronic systems — especially when setting the switch state requires only charging an electric field across a dielectric material.

Figure 10.4 depicts architectures in which higher speed switch modules are used for switching bits or packets of data. The time slot interchange system in Fig. 10.4a illustrates the functions of multiplexing — time slot interchange (i.e., variable optical delay) and demultiplexing — that can be accomplished in optical switches. The net effect of the time slot interchange system is to physically reroute the input signals on a bit-by-bit basis. Figure 10.4b illustrates a switching function in a time multiplexed network in which a rapid, reconfigurably nonblocking switch fabric is used to reroute synchronous signals on a frame-by-frame basis. Unlike the basic space switching system just discussed, time division switching is inherently linked with the data or packet rate through the switch.

Figure 10.5 shows an architecture that takes advantage of the combination of wavelength-division multiplexing transport and optical switching reconfiguration. The system is capable of cross-connecting, on a per-wavelength basis, $N \times M$ optical signals carried on M optical fibers and N wavelengths. The system is subject to the constraint that no wavelength can be represented more than once on any output fiber. The switching function is implemented by first demultiplexing the optical signals and routing them to λ *layers* within the switching fabric. Space switches are used to reroute the signals on each λ layer before they are multiplexed onto the output fibers. This network is inherently blocking because signals at the same wavelength on different input fibers cannot be switched to the same output fiber. An alternative technology, the acoustooptic tunable filter, which eliminates the need for separate multiplexing and switching

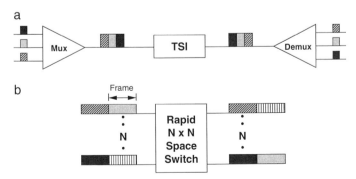

Fig. 10.4 Time multiplexed switching fabrics. (a) Time slot interchange (TSI) system. (b) Time multiplexed network.

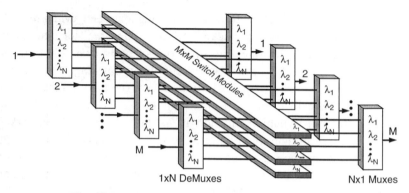

Fig. 10.5 Wavelength-layered space cross-connect.

devices, is discussed later in this chapter. Figure 10.6 shows a more general wavelength–space–wavelength cross-connect, which is made strictly non-blocking with the addition of wavelength converters on each of the $N \times M$ switch channels. The increase in connectivity comes at the expense of an increase in switch fabric size. The switch fabric grows from N layers of $M \times M$ fabric required in the first example to a single layer of dimension $(N \times M) \times (N \times M)$.

Optical switch modules differ in characteristics and constraints from their electronic counterparts so new system architectures are needed to leverage their particular attributes. Deployment of the systems also requires knowledge of these attributes to ensure use in appropriate applications. Several attributes dominate the differences between optical and electronic devices.

First, the current level of integration of guided-wave optical devices is limited to approximately 150 switch elements in a single packaged module. This limit results from the magnitude of the underlying physical effects, the available substrate size, and, ironically, the number of electrical leads. Large systems are built by interconnecting smaller functional modules, but

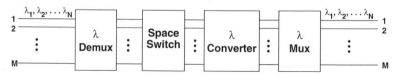

Fig. 10.6 Full connectivity wavelength–space cross-connect.

care must be taken to minimize the cost of the optical interconnection between stages — a cost that can dominate the system cost. The mapping of fiber interconnects between stages must also be done in a way that allows simple fiber routing and thus a sound system physical design.

Second, the size of many switch architectures is limited by insertion loss suffered by the signals passing through the fabric (Spanke 1987). Sources of loss within a fabric are fiber–waveguide coupling, bends, intersections, and scattering, all of which increase as the length of the fabric increases. Thus, a proper design will minimize the length of the fabric even at the expense of more depth (i.e., fan out) because this minimizes the optical loss and maximizes the system size.

Finally, the buildup of optical cross talk through the system must be understood in the design of switch fabrics (Spanke 1987). Cross-talk avoiding architectures can be designed that guarantee that all cross-talk paths pass through at least two OFF-state switches and thus achieve system cross-talk levels that are better than the crosstalk at individual switching elements.

10.3   Technology Advances

10.3.1   *REQUIREMENTS FOR OPTICAL SWITCH ELEMENTS*

Improvements in optical switch performance have generated increased interest in system demonstrations, which in turn has generated new and more difficult device performance requirements. Polarization dependence, loss, cross talk, level of integration, control signal magnitude, tolerance to control signal variations, and switching speed are among the important parameters. These parameters are dependent variables, and any given device design necessarily requires trade-offs among them. In this section, we first describe the impact of several of these parameters, and afterward we describe a sampling of recent advances.

Material and waveguide birefringence, electrooptic coefficients, and other parameters influence the polarization characteristics of integrated optical switches. In turn, the polarization characteristics of optical switches have an important impact on switching systems. Because the state of polarization is not maintained in standard single-mode fiber, optical signals coupled into a switching system will have an unknown and time-varying state of polarization. The systems must be designed to either operate independent of polarization, operate with a polarization controller on the input signal (Heismann 1993), or accept the expense and complication of polarization-maintaining fiber. Designs for polarization-independent devices are described later in this chapter.

Guided-wave switches have great utility because they serve to route signals through switching fabrics without modifying the signal. Thus they are transparent to the format and bit rate of the transmitted data. A consequence of this transparency is that optical power loss accrues in the fabric. Despite the development and prevalence of optical amplifiers, loss can limit system size and capabilities. Loss minimization is still an area of fruitful study. Waveguide, process, and material engineering have generated significant decreases in fiber coupling, propagation, bend and intersection losses.

For systems that are not loss limited, cross talk is the most likely parameter to limit system size and performance. Cross talk occurs when signals mix within a switch matrix as a result of imperfect switching and is defined as the ratio of the power of unwanted signals in a particular output to the power of the desired signal at that output. Cross talk can be defined at the final output of a switch fabric and at the output of individual switch elements within a matrix. For applications in which cross talk is a particularly demanding parameter, system cross talk can be improved beyond the cross-talk level of the switch elements by choosing dilated architectures (Padmanhabman and Netravali 1987) or by constraining switches to have only one active input (Richards and Hwang 1990; E. J. Murphy, Murphy, et al. 1996). Recent systems experiments using many narrowly spaced wavelength channels and narrow-linewidth lasers place greater requirements on switch cross talk due to coherent effects (Goldstein, Eskildsen, and Elrefaie 1994; Legg et al. 1994; Blumenthal, Granestrand, and Thylen 1995).

10.3.2 POLARIZATION-INDEPENDENT SWITCH ELEMENTS IN LITHIUM NIOBATE

The desire for polarization-independent switch elements and the difficulty in fabricating them are chronicled in the literature. There have been many attempts to obtain polarization independence involving variations in crystal, waveguide, and electrode geometry. In this section we briefly review some early devices but then quickly focus on the Y-shaped 1×2 "digital" optical switch. It is important to note that, in most cases, switch elements are polarization independent only at their endpoints. A switch can be set to route an input to one of two output ports. Polarization-independent switches have some operating voltage at which light of either polarization will be routed to the same output port. Any light appearing at the second port will be less than a specified extinction ratio for both polarizations. Similarly, at some other operating voltage, the light will be routed to the second port. However, as shown in Fig. 10.7, the optical transfer curves for each polarization generally differ. Thus the switches are not independent of polarization at intermediate voltages. For this discussion, polarization-independent switches are considered to be those that have loss and extinction ratio characteristics that are reasonably stable with respect to polarization changes at a given applied switching voltage.

Various techniques to obtain polarization independent directional couplers have been reported. These techniques include tapered waveguide coupling (Alferness 1979; Ramer, Mohr, and Pikulski 1982; Watson, Milbrodt, and Rice 1986), multiple electrode sections (Ramer, Mohr, and Pikulski 1982; Tsukada and Nakayama 1981; Kuzuta and Takakura 1991; Granestrand 1992), weakly confined waveguide couplers (Kondo *et al.* 1987; Nishimoto, Suzuki, and Kondo 1988; Nishimoto *et al.* 1990), polarization diversity (Duthie and Wale 1991), two-mode interference (Ctyroky, Janta, and Proks 1991), three guide couplers (Okayama, Ushikubo, and Kawahara 1991b), and $\Delta\beta$, $\Delta\kappa$ switch elements (Granestrand, Thylen, and Stoltz 1988). Although these techniques successfully produce polarization-independent devices, they generally involve tight fabrication tolerances or complex electrode geometries, both of which make them impractical for use in high-yield, highly integrated switch modules.

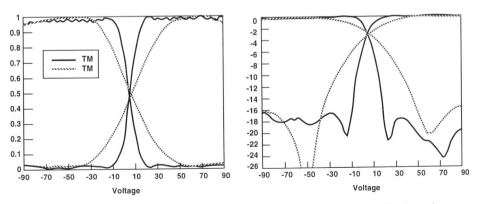

Fig. 10.7   Switching curve of a 7.5-mm Y switch on linear and logarithmic scales.

Figure 10.8a shows a digital Y-shaped switch element. Typical transfer characteristics on linear and logarithmic axes are shown for a 7.5-mm-long device in Fig. 10.7. The term *digital* is derived from the shape of the transfer curve on the linear axis. In practice, the transfer curve is not truly digital, but the design does provide sufficient suppression of secondary maxima in the transfer curve. This switch element structure was originally proposed in 1976 (Burns, Lee, and Milton 1976; Sasaki and de la Rue 1976) but not pursued seriously until a demonstration in 1987 (Silberberg, Perlmutter, and Baran 1987). The saturating behavior of the device with respect to voltage produces performance that shows little dependence on polarization, wavelength, or fabrication variations. The broad optical bandwidth of these devices is evidenced in Fig. 10.9, which shows typical cross-talk performance of a switch versus wavelength. Several theoretical and experimental studies (Okayama, Ushikubo, and Kawahara 1991a; Okayama and Kawahara 1993a; Burns 1992; Thylen *et al.* 1989) have concluded that the most voltage-efficient switch structure is a shaped Y, as shown in Fig. 10.8b, rather than the constant-angle structure of Fig. 10.8a. As shown in Fig. 10.10 (T. O. Murphy 1994), shaping the Y lowers the operating voltage at the expense of less "digital" character. The curves show experimental data for Ys shaped after Burns (1992). The shaping parameter, α, ranges from unity for a constant-angle Y, in which mode coupling varies in the branch, to infinity

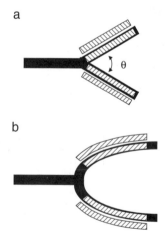

Fig. 10.8   (a) Constant-angle Y switch. (b) Shaped Y switch.

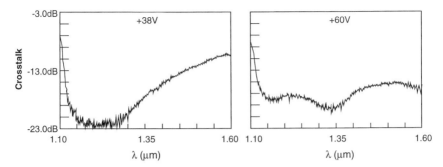

Fig. 10.9 Cross-talk of a Y switch versus wavelength for unpolarized input light.

for a fully shaped Y, in which the angle changes to maintain constant mode coupling. For shaping parameters as high as 1.5, low values of cross talk are maintained over a broad enough range of voltage to allow the devices to operate with all the advantages of the Y switch. The use of these switches in matrices is described later in this chapter. However, it is worthwhile

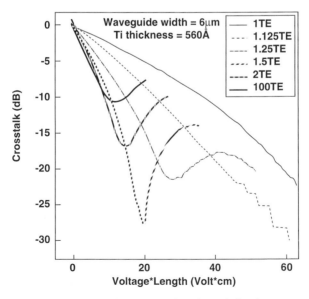

Fig. 10.10 Data on Y switches as a function of shaping parameter.

noting here that use of a 1×2 Y rather than a 2×2 directional coupler as a basic switch element does not severely limit switch fabrics. In fact, 1 $\times$ 2 switch elements are a natural choice for the cross-talk avoidance architectures mentioned previously.

10.3.3 POLARIZATION-INDEPENDENT SWITCH ELEMENTS IN SEMICONDUCTORS

Polarization-independent switch structures have also been developed in semiconductor materials, and, again, the most recent advances have taken advantage of the attributes of Y-switch designs. PIN heterojunction devices that can operate either in forward or reverse bias configurations have been reported (Cavailles *et al.* 1991; Stoll *et al.* 1992; Vinchant, Renaud, *et al.* 1993). Under forward bias the devices show well-behaved switching curves and cross talk of less than -15 dB for currents of 10 mA. Reverse bias devices show electroabsorption effects; as the bias voltage increases, so do the absorption losses. The optimum operating voltage is a compromise between excess loss and improved cross talk; typical reported values for little excess loss are 11 V and -13 dB of cross talk. The advantage of the reverse bias switch is switching speed. For this work, the reversed bias devices had an RC limited rise time of 300 ps, whereas the injection mode devices were limited by recombination times to 130-ns fall times.

The utility of any switch element is determined in part by its extinction ratio and overall size. Size is of particular importance in semiconductor-based devices because of the high propagation loss. Thus there is a premium on devices that minimize device length by maximizing the separation angle in the Y while also maintaining low cross-talk levels. Studies motivated by the shaping of Ys in lithium niobate referenced above have shown that improved device performance is also obtained by shaping the semiconductor Y. Khan *et al.* (1994) showed that using a high initial separation angle (0.5°) followed by a lower coupling region angle (0.2°) results in an improved voltage-length product (4.4 V · mm versus 14 V · mm) and improved cross-talk levels (-10 dB versus -7 dB) over a constant 0.2°-angle device. In recent work on forward biased PIN structures, a polarization independent extinction ratio of -20 dB has been obtained on a constant-angle (0.5°) device (Nelson *et al.* 1994). The same device was characterized at 1.3 and 1.5 μm. Extinction ratios of -20 dB were obtained at 1.3 and 1.5 μm for 50 and 100 mA of current, respectively, although the device showed some loss penalty at these operating currents.

10.3.4 POLARIZATION-INDEPENDENT ACOUSTOOPTIC WAVELENGTH–SPACE SWITCHES

In addition to the electrically controlled space switches discussed previously, acoustically driven space–wavelength switches have been demonstrated (Cheung et al. 1989; Smith et al. 1990; d'Alessandro, Smith, and Baran 1994). These devices are particularly attractive for multiwavelength optical networks because they operate on each wavelength independently, allowing reconfiguration on a per-wavelength basis without the need for multiplexers and demultiplexers. A two-input–two-output device can serve as an add–drop filter in which any wavelength or combination of wavelengths can be switched. A system application of these devices is discussed later.

Device operation is based on acoustooptic-driven polarization conversion. In the simplest case, an acoustic wave is launched along a straight waveguide that is located between two polarization beam splitters. Wavelengths matched to the acoustic wavelength are converted to the orthogonal polarization and separated from the main signal by the beam splitter. Polarization-independent devices have been fabricated by integrating the acoustooptic switch between polarization splitters and combiners. More advanced designs allow for broader passbands and lower cross talk (Tian et al. 1994; Hermann, Schafer, and Sohler 1994; Jackel, Baran, Chang, et al. 1995; Jackel, Baran, d'Alessand, et al. 1995). Typically, the devices require 10–100 mW of drive power per wavelength channel and achieve cross talk values in the −10- to −20-dB range.

10.3.5 LOW-LOSS WAVEGUIDE INTERCONNECTS

Integrated optical switch modules require waveguide interconnection webs between the switch stages (see Fig. 10.2). Although often considered second in importance to switch element characteristics, these interconnect columns typically occupy half the available substrate length and introduce more loss than the switches themselves. More complex interconnection columns allow for more complex switch fabrics and higher levels of integration in a given module. We consider passive waveguide components such as straight waveguides, width tapers, bends, and intersections as elements of these interconnect columns that must be optimized with respect to loss, loss uniformity, cross talk, and physical length. For polarization-independent devices, the loss uniformity requirement must account for polarization differences.

In general, different waveguide widths are needed for optimum fiber-to-waveguide coupling, switch design, and bend loss. These differences require the switch module designer to engineer the details of the waveguide design as a function of position within a module. Waveguide width tapers must be designed with as short a transition region as possible consistent with minimum excess loss. Waveguide bends have been studied in great depth (Minford, Korotky, and Alferness 1982; Johnson and Leonberger 1981; Smit, Pennings, and Blok 1993; T. O. Murphy, Murphy, and Irvin 1994b; Al-hemyari *et al.* 1993; Aizawa *et al.* 1994), with the general result that radii of curvature are limited to 30–40 mm for lithium niobate devices and 2–10 mm for indium phosphide devices. There have also been several studies on loss and cross talk from intersecting waveguides (Bogert 1987; T. O. Murphy *et al.* 1989). Cross talk is negligible for angles greater than 5°, and excess loss reaches acceptably low levels for angles greater than 7°. T. O. Murphy, Murphy, and Irvin (1994b) showed that modifying the detailed waveguide structure in the vicinity of the intersection generates less loss for a given intersection angle.

The design of waveguide interconnects involves compromises among several parameters. For instance, low-loss bends require high radii of curvature and thus relatively gentle angles, whereas the excess loss from intersections decreases with increasing angle. Also, in lithium niobate, bend loss is smaller for the transverse magnetic (TM) polarization, whereas intersection loss is smaller for the transverse electric (TE) polarization. The inherent nature of the interconnect column necessitates that some paths through the switch will pass through more intersections and more bends than other paths will. Because loss uniformity is often as important a system requirement as absolute loss, additional bends and mock intersections are often added to paths with low numbers of intersections and bends. All these factors must be considered when one is designing a particular interconnect column. In addition, switch module architectures that distribute the waveguide intersections over several columns tend to be more efficient than those that require an interconnect column with a very large number of intersections. For larger architectures, consideration should be given to locating intersection-intensive interconnect columns in the fiber interconnect between switch modules.

The ideal interconnect column will have no excess loss from bends and intersections and will use very little of the valuable substrate real estate. The highest packing density is obtained when the interconnect guides run parallel to the switch guides, but it is difficult to obtain the required 90° turns. Such right-angle turns are difficult to fabricate in lithium niobate because of its resistance to etching but can be fabricated in semiconductor materials and have been used in a demonstration of a 4×4 switch matrix (Vinchant, Gouteller, *et al.* 1993). Another possible approach would involve hybrid integration of switches on an active substrate material with an interconnection board composed of glass waveguides and turning mirrors.

10.3.6 OTHER SEMICONDUCTOR ADVANCES

Optical switching in semiconductor materials can be effected by several physical phenomena that make this a fertile ground for research advances. Besides the electrooptic effect used in lithium niobate devices, switching occurs through carrier injection, carrier depletion, the Franz–Keldysh effect, the quantum-confined Stark effect, and quantum well electron transfer. A full review is beyond the scope of this chapter, but the field has recently been thoroughly reviewed (Shimomura and Shigehisa 1994). The use of quantum wells has led to devices with very low voltage–length figures of merit and high-speed performance. Traveling-wave velocity-matched 2×2 elements have been demonstrated with a 6-V switching voltage, -12 dB of cross talk, and a 35-GHz 3-dB bandwidth (Kappe, Bornholdt, and Hoffmann 1994). These devices, which used bulk InGaAsP, had a 3-mm interaction length to minimize voltage and are projected to have a 100-GHz bandwidth at a 1-mm length. However, the voltage requirements for efficient switching at this length would be prohibitive. In related work, a 2×2 Mach–Zehnder multiple quantum well high-speed switch has been demonstrated (Agrawal *et al.* 1995). This 0.5-mm-long, lumped electrode device operated at 10 GHz, and 6.8 V. However, this latter device exhibited a large insertion loss, in part because of absorption at the band edge. Finally, a digital optical switch has been reported with a 10-GHz bandwidth operating at 4 V (Khan *et al.* 1995).

Insertion loss is of particular importance in evaluating the performance of semiconductor switch elements. Although one can conceive of added gain through active regions in the semiconductor material, any addition of gain also increases noise that can, in the end, limit system performance. Besides propagation loss, which, as noted previously, varies with the choice of the specific material system and the switch physics, fiber–waveguide coupling loss and waveguide interconnect loss need to be studied. Because of the high effective index change in semiconductor waveguides, single-mode waveguides are required to have small physical dimensions. The resultant small mode sizes provide optimum device performance but poor coupling to standard single-mode fiber. Although it is possible to use small mode size fiber or lensed fiber, these approaches do not solve the problem of interfacing to actual fiber networks or of attaching arrays of fiber to switch modules. Recent work has focused on waveguide tapers to serve as transition regions between the optical fibers and the switch element guides. Ideal tapers will provide circularly symmetrical optical modes that match the fiber mode size. As such, they must alter the mode size in both axes without adding inordinate complications to the processing sequence. Wenger *et al.* (1994) described a waveguide taper process in which the waveguide mode is tapered sequentially in the horizontal and vertical dimensions. Shadow masking and directional ion beam etching introduce a vertical taper, whereas photolithography is used for the horizontal taper. These researchers achieved a mode size increase of $2\times$ and $10\times$ in the horizontal and vertical directions, respectively. This results in a 3.5-dB decrease in fiber–waveguide insertion loss relative to a fiber–lens– waveguide measurement. Larger mode sizes also result in a significant reduction of fiber alignment tolerances. Vinchant *et al.* (1994) proposed an alternative approach, in which a compromise is made between device performance and fiber coupling in the interest of maintaining fabrication simplicity. A triple-core waveguide was demonstrated that reduces the modulation efficiency by $3\times$ but yields improvements in fiber–waveguide coupling and alignment tolerances. However, this approach still requires the use of lensed fiber, and it remains to be seen if it can be modified to be effective with standard single-mode fiber.

To date, the amount of effort directed toward understanding and reducing bend and interconnection loss in semiconductor waveguides is small compared with the effort spent on device demonstrations. Most of the device work has not included attention to the details associated with packaging the device for use in practical systems. For practical devices, the input and output waveguides must be aligned to fibers. This requires waveguide spacings of at least 125 μm (the fiber diameter) as opposed to the 50-μm or less separation of typical device demonstrations. Aizawa *et al.* (1994) have investigated the loss associated with bends in InGaAsP/InP multiple quantum well waveguides. A low bend loss (<0.1 dB) has been measured for bend radii of approximately 5 mm, and a loss of less than 0.5 dB has been measured for radii as small as 2 mm. Bends with 2-mm radii were integrated with a directional coupler switch. The 2-mm bends were chosen because the sum of the bend and propagation loss for these bends was less than that for larger bends. For optimum performance, the waveguide mode in the directional coupler required less confinement than in the bend, so the upper cladding layer was removed in the bend region. Unfortunately, this results in a mode mismatch loss at the waveguide transition, which in this case was 1 dB.

10.3.7 PACKAGING

A detailed review of integrated optics packaging is beyond the scope of this chapter (see E. J. Murphy [1988] for a discussion of fiber attachment technology). However, its importance to the success of commercial photonic switching systems and even to medium- and large-scale system experiments cannot be underestimated. Systems demonstrations have reached a level of sophistication wherein tabletop experiments are no longer of great interest. Integration of optical switches into medium-size switch fabrics, inclusion of switch modules on circuit packs with associated drive electronics, and computer control of the entire fabric are required for meaningful demonstrations. Such demonstrations would not be possible without robustly packaged modules. There have been few reports on packaging technology (Stone and Watson 1989), but the commercial success of lithium niobate components (suppliers include AT&T, Integrated Optical Components, LTD (IOC), and United Technology Photonics) indicates the level of sophistication that has been achieved for this technology, at least for small levels of integration. Also, the system demonstrations mentioned in Section 10.5 indicate the packaging levels that have been achieved for switch modules with higher levels of integration.

Demonstrations of packaging technology for semiconductor waveguide devices lag behind those for lithium niobate because of the later device development and increased complexity. Tight alignment tolerances and fragile material render the task of optical interconnections to semiconductor waveguides more difficult. In addition, to minimize propagation loss, shorter die are preferred, which implies tighter fiber spacings to eliminate the bend length associated with wide fiber spacing. Moreover, the physical size, the fragility, and the need to maintain high-performance antireflection coatings mean that it is generally not possible to glue fiber arrays directly to the waveguide end faces. The difficulty of the task can be put in perspective by considering the need to align large numbers of fibers to both ends of a semiconductor substrate against the effort that has been devoted since the mid-1970s to aligning a single fiber to semiconductor lasers. Despite the degree of the challenge, progress has been made. There have been two recent reports on using flip-chip bonding for InP waveguides. The techniques are analogous to those suggested by Bulmer *et al.* (1980) for dielectric waveguides but are more appropriate in this case because of the reduced length of the substrate. In addition, advances in silicon etching precision and the improved understanding of submicron alignment technologies are used advantageously in the recent work. Acklin *et al.* (1995) demonstrated the self-alignment of fibers to an array of four 2 × 2 InP directional coupler switches. They used tapered waveguides to reduce alignment tolerances and alignment ribs of the InP for self-alignment. The average excess loss was 2.7 dB. Leclerc *et al.* (1995) used a similar flip-chip–alignment rib technique to align fibers to an array of four semiconductor amplifiers. These researchers estimated that the packaging insertion loss penalty is negligible.

10.4   Device Demonstrations

10.4.1   *LITHIUM NIOBATE SWITCH MODULES*

Many significant advances in lithium niobate switch modules have been reported in recent years; perhaps the most significant were in the areas of device integration, polarization-independent operation, and packaging. The principles of polarization-independent switch elements were described previously. In this section we focus on a review of recently reported switch modules and discuss several important characteristics of these devices.

For single polarization devices, switch matrices ranging in size from 8 × 8 to 32 × 32 have been reported. In some designs, large switch matrices have been fabricated on a single substrate. These include a 16 × 16 employing 56 directional couplers in a Benes architecture (Duthie and Wale 1991) and a 32 × 32 employing 80 directional couplers in a banyan architecture (Okayama and Kawahara 1994). The high density of these switches and the limits on available substrate size required some compromises in switch performance. The 16 × 16 had high switching voltages (35–60 V), an undilated architecture, and nonstandard fiber spacing. The 32 × 32 had an undilated blocking architecture, nonstandard fiber spacing, and a high insertion loss.

These problems can be solved by coupling two substrates end to end within a single package (Fig. 10.11a). Eight-by-eight (Watson *et al.* 1990; E. J. Murphy *et al.* 1995) and 16 × 16 (T. O. Murphy, Kemmerer, and Moser 1991) rearrangeably nonblocking, dilated Benes matrices using two substrates have been reported. The 8 × 8, which consisted of 48 directional

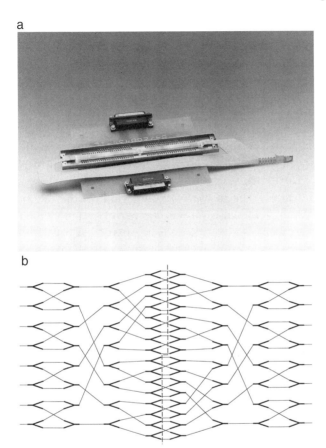

Fig. 10.11 (a) Photograph of photonic integrated 8 × 8 switch in which two substrates are butt coupled to increase the effective substrate length. (Reprinted with permission from Alferness, R. C. 1995, February. Advanced technologies pave the way for photonic switches. *Laser Focus World.* p. 109.) (b) Schematic of photonic integrated 8 × 8 switch. The *dark lines* indicate 1 × 2 switching elements, whereas the *lighter lines* indicate waveguide interconnect paths. The *dashed line* shows the division of switch elements and waveguides on each of the two identical substrates.

couplers had an average switching voltage of 9.4 V, an average loss of 9 dB, and an extinction ratio per switch element that was typically less than −30 dB. The 16 × 16 matrix was composed of 128 directional couplers — the largest number integrated into a single package. The average switching voltage was 12.4 V, and the worst case extinction ratio was −15 dB. In both cases, the dilated architecture yielded cross talk from unwanted signals on the output channels that was significantly lower than the switch element extinction ratios.

A natural extension of the integration of large matrices of optical switches into single packages is the addition of hybrid integration of control electronics or fiber interconnects into those packages. A 1 × 16 switch with transistor-transistor logic (TTL) control has been demonstrated (O'Donnell and Parsons 1991). The switch is packaged with an electronic drive circuit that sets bias and control voltages for each of the switch elements. The only required inputs are two power rails, a four-bit control word, and a clock pulse. The switch operates with a 4.5-ns switching speed. Techniques such as this will become increasingly important as the number of switch elements increases. Increases in the number of switch elements mean increases in the number of control leads in a particular package. Already we are approaching integration levels in which device integration is limited by pin-out density. Inclusion of hybrid electronics will minimize the number of package pins by requiring passing only logical control signals through the package wall. Algorithms implemented with integrated circuits within the package will then allow for rapid reconfiguration of the switch module.

Hybrid packaging of an optical fiber interconnect has been demonstrated in optical time-delay networks (Ackerman et al. 1992; E. J. Murphy, Adda et al. 1996). In this case, two separate adjacent substrates were interconnected with fiber loops. The fiber loops were manufactured to precise length tolerances. Selection of appropriate switch states allowed the choice of 1 of the 64 possible time delays. Successful demonstration of an optical interconnect in this time-delay network suggests that this approach could also be applied to complex interconnect columns in large switch matrices. Typically, a large matrix will have at least one interconnect column in which connectivity between distant rows within the matrix is required. If the entire matrix is to be built from several substrates, the large-offset interconnect columns could be placed at the periphery of each substrate so that they could be implemented in fiber rather than in waveguides. This removes the need to use long lengths of substrate for passive waveguides and eliminates the loss associated with the bends and intersections in those columns. These fiber meshes could be produced using the optical equivalent of flexible circuit boards (Shahid, Roll, and Shevchuck 1994; T. O. Murphy, Murphy, and Irvin 1994b).

Progress in the integration of polarization-independent switch elements into large switch arrays has been equally exciting. Switch modules have been reported with 1 × 16, 1 × 32, 4 × 4, and 8 × 8 arrays. The earlier work on polarization-independent switch modules focused on 2 × 2 switch elements (see the discussion in Section 10.3), but the more recent work has primarily utilized 1 × 2 Y-shaped structures. In this section we first review the 2 × 2 based devices, then devices based primarily on 1 × 2 switch elements.

The first reported integrated polarization-independent switch matrix was a 1 × 16 (Watson, Milbrodt, and Rice 1986). The device was based on directional couplers with weighted coupling (Alferness 1979), and it achieved less than −10-dB extinction ratios with approximately 70-V switching voltages. A 4 × 4 matrix based on 24 $\Delta\beta$, $\Delta\kappa$ switch elements (Granestrand, Thylen, and Stoltz 1988) and using a modified tree architecture has been reported (Granestrand *et al.* 1988). Bias voltages as large as 105 V were required, but the maximum dynamic switching voltage was 50 V and the inherent dilated nature of the tree architecture led to cross-talk levels of less than −35 dB. Nishimoto, Suzuki, and Kondo (1988) and Nishimoto *et al.* (1990) have reported 4 × 4 and 8 × 8 polarization-independent switch matrices based on directional couplers with weakly confined waveguides. These use 12 and 64 switch elements, respectively. The weakly confined guides allow the directional couplers to be designed for one coupling length for both polarizations. This allows efficient voltage operation at the expense of requiring very tight fabrication tolerances. For the 8 × 8, cross-talk levels of less than −18 dB were obtained with 85-V switching voltages.

Y-shaped "digital" optical switches are attractive for large switch matrices because of their polarization independence, less stringent fabrication tolerances, less stringent voltage control tolerances, and wavelength insensitivity. The first reported polarization-independent switch matrix based on Y-shaped switches was a 4 × 4 (Granestrand *et al.* 1990). This was quickly followed by 1 × 16 and 1 × 32 matrices (O'Donnell 1991). Studies on the optimum design of the Y shape led to lower voltage switch elements used in simplified tree structure 4 × 4 and 8 × 8 modules with 25- and 40-V switching voltages, respectively (Okayama and Kawahara 1993b). The largest Y-based switch matrices reported to date are a single substrate, strictly nonblocking tree structure 8 × 8 module with 112 switch elements (Granestrand *et al.* 1994) and a two-substrate, rearrangeably nonblocking 8 × 8 module with 80 switch elements (T. O. Murphy, Murphy, and Irvin 1994a). For the former device, the insertion loss was in the 8- to 14-dB range, the switching voltage was approximately 100 V, and the switch element extinction ratio ranged from −10 to −30 dB. For the latter device, Fig.

10.11b shows a schematic of the switch matrix. In this figure, each Y represents a 1 × 2 switching element, and the lines represent the waveguide interconnect between the columns of switch elements. This single-package device consists of 80 integrated switches. It operates with a uniform switching voltage of 38 V and system cross talk of less than −24 dB. The mean insertion loss is 11.5 dB. Figure 10.9 shows cross talk as a function of wavelength for a typical switch element used in the 8 × 8. Low cross-talk levels are achieved over a broad wavelength range, which makes these devices suitable for wavelength-division multiplexing systems. This device has been packaged with a new fiber routing technology (Shahid, Roll, and Shevchuck 1994) that greatly simplifies fiber handling (see Fig. 10.11a). The 16 input and output fibers are all routed to a single fiber array connector that can be mated to another connector through the electrical back plane of an equipment cabinet.

10.4.2 SEMICONDUCTOR WAVEGUIDE SWITCH MODULES

Several impressive switch modules have also been demonstrated in semiconductor materials. One of the earliest (Inoue *et al.* 1988) used a novel Y-branch switch structure and current injection to achieve polarization independence in InGaAsP/InP. Sixteen of these switch elements were integrated to form a nonblocking 4 × 4 crossbar matrix. Fiber-to-fiber insertion losses in the 20- to 30-dB range and cross-talk values in the −7- to −19-dB range were measured with 100 mA of drive current. Siemens has developed a rearrangeably nonblocking 4 × 4 switch composed of five 2 × 2 directional coupler switches in InP. These polarization-insensitive devices also use carrier injection for switching. Typical characteristics of the 2 × 2 switch elements include 15 dB of cross talk at 9 mA of current. Switching times of 2 ns were successfully demonstrated with these switches (Cada *et al.* 1992). A 4 × 4 strictly nonblocking tree structure matrix has been fabricated in InP using the Y-shaped "digital" optical switches described previously (Vinchant *et al.* 1992). The device exhibited approximately 13 dB of internal loss and switching currents of 7 mA. In later work (Vinchant, Goutelle, *et al.* 1993), a compact 4 × 4 using 24 digital optical switches was demonstrated. For this device, 32 integrated turning mirrors were used within the waveguide interconnect web to eliminate waveguide bends and allow fabrication of a compact structure. In this case, a fiber-to-fiber loss of 15 dB was achieved. Cross-talk values averaged −12 dB with 30 mA applied to the digital switches.

One of the difficulties of demonstrating semiconductor switch modules can be avoided by using GaAs materials, which have a low absorption loss in the 1.3- and 1.55-μm wavelength range. (However, this precludes the integration of sources, detectors, and gain sections in these devices.) Hamamoto *et al.* (1992) have demonstrated an 8 × 8 switch matrix in GaAs/AlGaAs using 64 switch elements in a strictly nonblocking tree topology. The electrooptically switched directional couplers had low cross talk (< -21 dB) and excellent switching voltage uniformity ($< \pm 4\%$). The minimum internal loss for the 26.5-mm-long chip was 8.7 dB. Later, Hamamoto *et al.* (1993) reported process, material, and waveguide design changes that lowered the minimum loss of a 17-mm-long 4 × 4 matrix to only 1.6 dB. In more recent work on GaAs, Jenkins *et al.* (1994) demonstrated 1 × 10 and 10 × 10 switches using self-imaging multimode waveguides separated by electrooptic phase shifters. The length of the multimode region of these devices is directly proportional to the effective index and inversely proportional to the wavelength. The latter dependence makes these devices narrow band — these researchers calculate that a variation of a few nanometers from the operating wavelength results in a 6-dB loss penalty. Meanwhile, the dependence on the effective index is likely to render the devices equally sensitive to polarization.

The potential for integrating active optical devices with switch elements is one of the true attractions of semiconductor integrated optics. The size of switch architectures is inevitably limited by loss and cross-talk accumulation. Integrating amplifiers within switch modules extends both of these limits. Obviously, loss can be compensated for by the addition of gain, but cross talk can also be improved by turning off gain sections in unused paths and absorbing cross-talk radiation before it couples back into a signal-carrying path. Achieving the integration of amplifiers requires the development of techniques for deposition and processing of materials that are efficient at both switching and amplifying functions. Materials that, when pumped, provide high gain at a given wavelength will have a high propagation (absorption) loss in regions where it is not pumped. Detailed waveguide engineering as a function of position on the wafer must be accomplished to achieve the desired integration. Other characteristics of semiconductor amplifiers, such as their polarization dependence and their sensitivity to end-face reflections, place additional constraints on the integration of gain with switch elements. Moreover, broadband antireflection coatings must be designed to eliminate ripple over wavelength ranges of interest to wavelength-division multiplexing system development. In recent work in this area, Glastre *et al.* (1993) have demonstrated the monolithic integration of a 2 × 2 directional coupler switch with an optical amplifier. The amplifier section was fabricated by adding an active waveguide stripe directly on top of the passive waveguide structure. The end faces of the device were angled cleaved to reduce gain ripple from reflections. The 2 × 2 switch was operated with no net loss for the TE polarization with a 140-mA drive current to the 1-mm-long amplifier. van Roijen *et al.* (1993) have demonstrated a 1 × 2 switch with an integrated amplifier. In this case, two epitaxial depositions and selective area etching were used to fabricate the passive and active regions, and antireflection coatings were used to reduce gain ripple. The amplifier region of this device was 0.5 mm long and was operated at 200 mA. For the TE polarization, the net fiber-to-fiber gain was measured over a 1510- to 1590-nm wavelength range with a maximum gain of approximately 10 dB at the shorter wavelengths. Finally, Kirihara *et al.* (1993) have demonstrated a 2 × 2 crossbar switch with integrated amplifiers. Lossless switching was demonstrated in some switch paths for amplifier currents of 250 mA.

These demonstrations mark important beginnings for semiconductor switch modules. Advances in processing reproducibility, loss, and packaging will be necessary to make them practical for system experiments.

10.4.3   GATE ARRAYS

As noted in the previous section, the integration of gain with waveguide switches also acts to improve cross talk by absorbing light in unpumped waveguides. This concept can be extended to the fabrication of switch modules formed by passively splitting an incoming signal and using not only a pumped amplifier gate to pass the signal and compensate for loss in the desired switch paths, but also an unpumped amplifier gate to attenuate the superfluous signals in the undesired paths. Figure 10.12 shows such a system for the simple case of a 4 × 4 spatial switch. Each of the four inputs is split four ways to an array of 16 optical gates. Light continues to propagate through the gates that are pumped and is extinguished in all other gates. The high level of absorption in the unpumped gates provides high extinction ratios in those channels. The remaining signals are then recombined by 4 × 1 passive combiners. The passive splitters and combiners could be discrete devices or could be formed by hybrid or monolithic integration. Effective amplifier gates should have low noise, low current, low ripple, low polarization sensitivity, a broad bandwidth, and enough gain to compen-

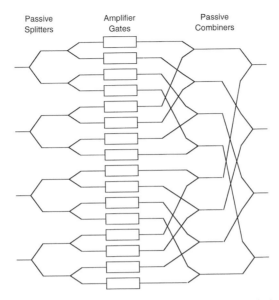

Passive Splitters   Amplifier Gates   Passive Combiners

Fig. 10.12   Four-by-four passive splitter–passive combiner switch with amplifier gates.

sate for the passive splitting losses. The dynamic range allowed for the input power, gain saturation, and maximum output power should also be considered. Finally, for practical device arrays, gain uniformity across the array must be attained.

There have been several demonstrations of gate arrays and of the semiconductor optical amplifier arrays needed to fabricate the switching devices. Recently, a very high-performance array has been reported (Leclerc *et al.* 1995). In this work, an array of four polarization-independent amplifiers were integrated on a single chip and packaged using self-aligning silicon optical bench techniques. With an 80-mA drive current, internal gains as high as 32 dB and fiber-to-fiber gains from 13 to 16 dB were achieved. This device was used to demonstrate the switching of asynchronous transfer mode (ATM)-like packets (Gavignet *et al.* 1995). A 424-bit cell of 2.5-Gb/s data was split, sent through two paths of the gate array, and alternately routed to a receiver. The array was successfully switched with 400-ps rise and fall times, thus requiring only a 2-bit guard band around the data packet. An earlier device (Davies *et al.* 1992) based on twin guide amplifiers showed lossless fiber-to-fiber operation for all four ports of a 1 × 4 switch for the TE polarization. However, on-chip variations resulted in up to 15-dB variations in the output power. A more recent 1 × 4 device (Ratovelomanana *et al.* 1995) used passive Ys for power splitting followed by four amplifier waveguide sections. This device showed a fiber-to-fiber gain ranging from 0 to +2.8 dB at 200 mA. The device had a 20-nm optical bandwidth, 40-dB extinction ratios, and less than 1 dB of polarization sensitivity. Koren *et al.* (1992) used a free radiation region to couple a single input to 16 outputs. By using an input amplifier and 16 output amplifiers, they measured an internal loss (excluding fiber coupling losses) of 2.5 dB for 11 of the 16 channels.

In addition to the device-oriented work just described, there have been several demonstrations of gate array switch fabrics. Burton *et al.* (1993) demonstrated a 2 × 2 switch with four integrated waveguide amplifiers. The device was polarization independent and exhibited 3–10 dB of net loss (fiber to fiber) for a 200-mA drive current. In another monolithic integration demonstration (Gustavsson *et al.* 1992), a 4 × 4 gate array has been reported. This device integrated four input booster amplifiers, passive waveguide splitters, 16 gate amplifiers, passive waveguide combiners, and 4 output booster amplifiers. At 60 mA, the device showed up to 6 dB of gain for the TE polarization, but there were significant path-to-path variations. Later, this device was used in a 2.5-Gb digital transmission experiment (Gustavsson, Janson, and Lundgren 1993). A 4 × 4 gate array built by the hybrid integration of glass waveguides on silicon and two arrays of 8 amplifiers has also been reported (Yamada *et al.* 1992).

10.4.4   NONLINEAR OPTICAL SWITCHING

The dielectric and semiconductor-based switches described previously operate on linear changes to the complex refractive index through Pockel's effect, charge injection, and other phenomena. Although modulation frequencies of many tens of gigahertz have been demonstrated, in the long term the operating speeds for practical devices will be limited by device size and switching energy. Researchers have been investigating the application of nonlinear optical phenomena for switching. Unlike the devices described previously, which rely on electrical control of the states of optical switches, these nonlinear devices rely on changes to the refractive index due to a change in the intensity of incident light. They are referred to as *nonlinear* because they depend upon the square of the electric field rather than having the linear dependence of the devices discussed previously.

Optically controlled optical switches are expected to attain higher switching speeds than will be practical to attain with electronics. Networks based on such switches will be able to route signals without requiring inefficient optical-to-electrical and electrical-to-optical conversions.

The major fundamental challenge in demonstrating such devices is the small size of the physical effects. The change in refractive index is proportional to the third-order dielectric susceptibility, the intensity of the radiation, and the length of the interaction. High intensities and long interaction lengths must be used to accommodate the weak material properties. These requirements make optical fibers a natural medium for nonlinear switches because the small mode size allows high-power density and because they are readily available in long lengths. However, improvements are needed to realize practical devices. Currently, very long fiber lengths (kilometers) and very high optical powers (hundreds of milliwatts) are required to achieve switching. Compact modules involving short lengths of nonlinear media and small, high-power lasers must be developed before this technology will have practical use in telecommunications systems.

A device that has received much attention is the nonlinear optical loop mirror (Fig. 10.13). In this device, the signal is split and counterpropagates in a fiber loop. Because the split signals travel the same path, they recombine at the directional coupler and the input is thus "reflected" from the loop "mirror." If a control pulse of a different polarization or frequency is coupled into the loop such that it propagates in only one direction, it affects the phase of only one arm of the propagating signals (through cross-phase modulation due to nonlinear changes in the refractive index). The resultant phase shift modulates the signal at the output port.

A full discussion of this technology is beyond the scope of this chapter. The reader is referred to Islam (1992, 1994) for more detail.

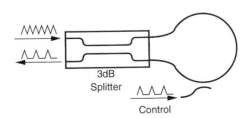

Fig. 10.13 Nonlinear optical loop mirror schematic.

10.5   System Demonstrations and Advances

As evidenced previously, guided-wave switch technology has attained a high level of performance and sophistication. The next step toward network utilization is to demonstrate system-level control and functionality. The switch module technology must move beyond laboratory prototypes to systems in which the optical devices are combined with printed circuit board level controllers and integrated into switch fabrics along with appropriate hardware and software controls. The demonstrations are essential to prove that the technology is mature enough to produce systems for practical applications. We have already seen impressive results from several such demonstrations. In this section, we first describe several laboratory-based space switching systems, then several field experiments, and finish with a description of wavelength–space switching fabrics currently under investigation.

Figure 10.14 is a photograph of a 16 × 16 optical space switching system (E. J. Murphy, Murphy, et al. 1996). Figure 10.3 shows a schematic of this strictly nonblocking, extended generalized shuffle (EGS) network (Richards and Hwang 1990). The fabric was designed as a three-stage EGS network, with each stage interconnected with optical fiber. The input and output stages consist of 16 dual 1 × 8 voltage-controlled switch modules. The center stage consists of seven 16-input–16-output modules with one path from each input to each output. This particular set of modules was chosen because the modules are basic building blocks for even larger fabrics. The directional couplers were fabricated as single polarization switch elements on lithium niobate. The fiber interconnection used polarization-maintaining fiber. The 448 switches had a mean switching voltage of 12.5 V, and the mean bias voltage was −9.4 V. The median cross talk was −32 dB; the median loss was −6 and −8 dB for the 1 × 8s and the center stage modules, respectively.

Each packaged module was mounted on a control board that supplied the necessary bias and switching voltages. The boards were mounted in three shelves of a vertical cabinet (Fig. 10.14). Fiber interconnect panels separated the device shelves. The entire fabric was controlled by a PC with a graphical user interface that allowed the user to initiate the path hunt and complete the connection setup by pointing and clicking on any I/O pair. This system clearly shows the low power consumption advantage of optical switching networks. The entire equipment cabinet consumed only 90 W — including 20 W for the indicator light-emitting diodes (LEDs).

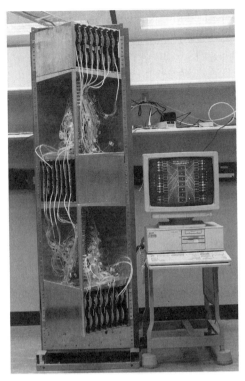

Fig. 10.14 Photograph of a 16 × 16 optical space switching system. [Reprinted with permission from Murphy, E. J., *et al.* 1996. 16 × 16 Strictly non-blocking guided-wave optical switching system. *J. Lightwave Tech.* 14(3):356. Copyright © 1996 IEEE.]

After the modules were connected, the system was tested for insertion loss and cross talk. The mean system loss was 22 dB (including 2 dB of loss attributable to the different mode sizes of the system and device fibers) and was consistent with the sum of the losses through the individual modules. Typically, the system cross talk from one input to any nonselected output was less than −50 dB and the worst case cross talk value was −26 dB. This performance is consistent with the device cross talk and shows the cross-talk-reducing advantage of the single-input active (dilated) architectures described in Section 10.3.1. As an indication of device stability, the system has operated continuously for more than 2 years.

In another laboratory demonstration, one-fourth of a 128×128 switching system was built (Sawano *et al.* 1995; Burke *et al.* 1992). The system was designed to provide strictly nonblocking connections as well as paths for less essential services with a low, but finite, probability of blocking. This five-stage fabric used 4×4, 4×8, and 8×8 polarization-independent switch modules (Nishimoto, Suzuki, and Kondo 1988; Nishimoto *et al.* 1990). As discussed previously, optical insertion loss can limit the ultimate size of switch fabrics. For this demonstration, traveling-wave optical amplifiers were used between the second and third switch stages and again between the third and fourth stages to compensate for the ($>$50-dB) loss. Another feature of this system is a novel physical design for the optical interconnect between circuit packs containing optical switches. Orthogonal mounting of neighboring switch shelves resulted in a direct mapping of outputs of one stage onto the inputs of the subsequent stage. The orthogonal physical layout easily implements the "shuffle" of interconnect paths without the need for complex fiber routing. The system was tested with simultaneous transmission of 150- and 600-Mb data. A bit error rate of 10^{-9} was achieved for the higher data rate with a 6-dB system margin.

The Research and Development (in) Advanced Communications Technologies in Europe (RACE) program has spawned several field experiments employing optical switching. In England, two commercial 4×4 switches (Granestrand *et al.* 1990) were used to route signals on nine routes over installed cables in the British Telecom network (Lynch *et al.* 1992). The network involved six nodes, two of which could be reconfigured from a centralized controller. The switching nodes included grating-based wavelength-division multiplexing components to separate the three network wavelengths (1.536, 1.548, and 1.560 μm). The latter wavelength was routed through the optical switches. Pseudo-random data at 622 Mb/s were sent over the network for distances up to 135 km with error rates less than 10^{-9}. This experiment demonstrated the potential of optical networks to provide wavelength routing, reconfiguration, and drop–add functions.

Optical switching and networking functions have also been demonstrated as part of the Stockholm Gigabit Network (Johansson, Almstrom, and Hubinette 1994). A wide range of devices has been deployed at two nodes as part of this network. The devices include 4×4 and 8×8 space switches, semiconductor amplifier gate arrays, acoustooptic tunable filters, and multigrating filters (Johansson *et al.* 1993; Hill *et al.* 1993). Experiments on the network have shown the basic capabilities of optical cross-connect systems and are being used to understand the limits of such networks and the requirements on devices used within them.

Finally, we describe two other photonic system experiments that focus equally on space and wavelength switching. Both programs are supported in part by the U.S. Advanced Research Projects Agency (ARPA). The size and complexity of these programs (and the aforementioned RACE programs) require the formation of consortia from a range of corporate and government groups to produce successful demonstrations. Clearly, the most meaningful results are obtained by integrating the architecture, network control, device, and manufacturing expertise from various organizations.

The Optical Network Technology Consortium (ONTC) was formed to study reconfigurable multiwavelength optical networks. The resulting network test bed used multiwavelength laser arrays, multichannel receiver arrays, and acoustooptic cross-connect switches. Simultaneous transmission of subcarrier multiplexed analog and synchronous optical network (SONET)–ATM digital signals on different wavelengths over the same fiber was demonstrated. Each of the four nodes in the network was capable of transmitting and receiving four wavelengths, spaced at 4 nm in the 1.55-μm band. Five wavelength-division multiplexing switches were used as cross-connects, four between the nodes and the fiber rings and one to interconnect the two rings. Network control was obtained by software accessed through a graphical user interface.

The Multiwavelength Optical Network (MONET) Consortium has begun a broad-ranging program. The program will study optimum architectures for transparent multiwavelength networks and the control of such networks. Technology studies include transmitter and receiver arrays, optical amplifiers, wavelength conversion, and wavelength and space switching devices. Systems integrators will build the network elements necessary for the field experiment. The system will operate with eight wavelength channels. Several add–drop filters and optical cross-connects will be built. The largest of these will be a strictly nonblocking eight-fiber, eight-wavelength cross-connect that provides connectivity between any input fiber–wavelength and any output fiber–wavelength. The program will culminate with a field experiment linking seven sites over a large geographic area.

10.6 Summary and Comments

We have described the state of the art in guided-wave optical switches. The applications, architectures, and devices described in this chapter represent the significant progress in this technology since the early 1990s. System concepts and device technology are both well enough understood to allow for moving the technology from the laboratory into demonstration applications. The system experiments clearly showcase the maturity of the technology and the high degree of functionality achievable in switched optical networks. The capabilities of the technology coupled with the ever-increasing bit rate and the use of wavelength-division multiplexing in transmission systems provide an opportunity to use this technology to provide significant system enhancements. The momentum for system demonstrations appears to be growing. The next few years should see significant advances in the demonstration of higher levels of device integration, in the size and extent of system demonstrations and in the understanding of the optimum application areas.

References

Ackerman, E., S. Wanuga, D. Kasemset, W. Minford, N. Thorsten, and J. Watson. 1992. Integrated 6-bit photonic true-time-delay unit for lightweight 3–6 GHz radar beamformer. *IEEE MTT-S Dig.* 681–684.

Acklin, B., J. Bellermann, M. Schienle, L. Stoll, M. Honsburg, and G. Muller. 1995. Self-aligned packaging of an optical switch array with integrated tapers. *IEEE Photon. Tech. Lett.* 7:406–408.

Agrawal, N., C. M. Weinert, H-J. Ehrke, G. G. Mekonnen, D. Franke, C. Bornholdt, and R. Langenhorst. 1995. Fast 2 × 2 Mach–Zehnder optical space switches using InGaAsP/InP multiquantum well structures. *IEEE Photon. Tech. Lett.* 7:644–645.

Aizawa, T., K. G. Ravikumar, Y. Nagasawa, T. Sekiguchi, and T. Watanabe. 1994. InGaAsP/InP MQW directional coupler switch with small and low-loss bends for fiber array coupling. *IEEE Photon. Tech. Lett.* 6:709–711.

Alferness, R. C. 1979. Polarization independent optical directional coupler switch using weighted coupling. *Appl. Phys. Lett.* 35:748–750.

Al-hemyari, K., G. F. Doughty, C. D. W. Wilkinson, A. H. Kean, and C. R. Stanley. 1993. Optical loss measurements on GaAs/GaAlAs single-mode waveguide Y-junctions and waveguide bends. *J. Lightwave Tech.* 11:272–276.

Blumenthal, D. J., P. Granestrand, and L. Thylen. 1995. A comparison of coherent crosstalk induced BER floors in four types of $N \times N$ space photonic switches. In *Technical Digest, Photonics in Switching Conference*, Washington, DC, 109–111. Paper PFA2.

Bogert, G. A. 1987. Ti : LiNbO$_3$ intersecting waveguides. *Electron. Lett.* 23:72–73.

Bulmer, C. H., S. K. Sheem, R. P. Moeller, and W. K. Burns. 1980. High efficiency flip-chip coupling between single mode fibers and LiNbO$_3$ channel waveguides. *Appl. Phys. Lett.* 37:351–353.

Burke, C., M. Fujiwara, M. Yamaguchi, H. Nishimoto, and H. Honmou. 1992. 128 line photonic switching system using LiNbO$_3$ switch matrices and semiconductor travelling wave amplifiers. *J. Lightwave Tech.* 10:610–615.

Burns, W. K. 1992. Shaping the digital switch. *IEEE Photon. Tech. Lett.* 4:861–863.

Burns, W. K., A. B. Lee, and A. F. Milton. 1976. Active branching waveguide modulator. *Appl. Phys. Lett.* 29:790–792.

Burton, J. D., P. J. Fiddyment, M. J. Robertson, and P. Sully. 1993. Monolithic InGaAsP-InP laser amplifier gate switch matrix. *IEEE J. Quantum Electron.* 29:2023–2027.

Cada, M., G. Muller, A. Greil, L. Stoll, and U. Wolff. 1992. Dynamic switching characteristics of a 4 × 4 InP/InGaAsP matrix switch. *Electron. Lett.* 28:2149–2150.

Cavailles, J. A., M. Renaud, J. F. Vinchant, M. Erman, P. Svensson, and L. Thylen. 1991. First digital optical switch based on InP/GaInAsP double heterostructure waveguides. *Electron. Lett.* 27:699–700.

Cheung, K. W., D. A. Smith, J. E. Baran, and B. L. Hefner. 1989. Multiple channel operation of integrated acousto-optic tunable filter. *Electron. Lett.* 25:375–376.

Ctyroky, J., J. Janta, and J. Proks. 1991. Two mode interference Ti : LiNbO$_3$ electro-optic polarization independent switch or polarization splitter. *Electron. Lett.* 27:965–966.

d'Alessandro, A., D. A. Smith, and J. E. Baran. 1994. Multichannel operation of an integrated acousto-optic wavelength routing switch for WDM systems. *IEEE Photon. Tech. Lett.* 6:390–393.

Davies, D. A. O., P. S. Mudhar, M. A. Fisher, D. A. H. Mace, and M. J. Adams. 1992. Integrated lossless InP/InGaAsP 1 to 4 optical switch. *Electron. Lett.* 28:1521–1522.

Duthie, P. J., and M. J. Wale. 1991. 16 × 16 Single chip optical switch array in lithium niobate. *Electron. Lett.* 27:1265–1266.

Gavignet, P., M. Sotom, J. C. Jacquinot, P. Brosson, D. Leclerc, W. Hunziker, and H. Duran. 1995. Penalty free 2.5 Gbit/s photonic switching using a semiconductor four-gate-array module. *Electron. Lett.* 31:487–488.

Glastre, G., D. Rondi, A. Enard, E. Lallier, R. Blondeau, and M. Papuchon. 1993. Monolithic integration of 2 × 2 switch and optical amplifier with 0 dB fiber to fiber insertion loss grown by LP-MOCVD. *Electron. Lett.* 29:124–126.

Goldstein, E. L., L. Eskildsen, and A. F. Elrefaie. 1994. Performance implications of component crosstalk in transparent lightwave networks. *IEEE Photon. Tech. Lett.* 6:657–659.

Granestrand, P. 1992. Four section polarization independent directional coupler with extremely relaxed fabrication tolerances. *IEEE Photon. Tech. Lett.* 4: 594–596.

Granestrand, P., B. Lagerstrom, P. Svensson, H. Olofsson, J-E. Falk, and B. Stoltz. 1994. Pigtailed, tree-structured 8 × 8 LiNbO$_3$ switch matrix with 112 digital optical switches. *IEEE Photon. Tech. Lett.* 6:71–73.

Granestrand, P., B. Lagerstrom, P. Svensson, L. Thylen, B. Stoltz, K. Bergvall, J-E. Falk, and H. Olofsson. 1990. Integrated optics 4 × 4 switch matrix with digital optical switches. *Electron. Lett.* 26:4–5.

Granestrand, P., B. Lagerstrom, P. Svensson, L. Thylen, B. Stoltz, K. Bergvall, and H. Olofsson. 1988. Tree structured polarization independent 4 × 4 switch matrix in LiNbO₃. *Electron. Lett.* 24:1198–1200.

Granestrand, P., L. Thylen, and B. Stoltz. 1988. Polarization independent switch and polarization splitter using $\Delta\beta$ and $\Delta\kappa$ modulation. *Electron. Lett.* 24:1142–1145.

Gustavsson, M., M. Janson, and L. Lundgren. 1993. Digital transmission experiment with monolithic 4 × 4 InGaAsP/InP laser amplifier gate switch array. *Electron. Lett.* 29:1083–1085.

Gustavsson, M., B. Lagerstrom, L. Thylen, M. Janson, L. Lundgren, A-C. Morner, M. Rask, and B. Stoltz. 1992. Monolithically integrated 4 × 4 InGaAsP/InP laser amplifier gate switch arrays. *Electron. Lett.* 28:2223–2225.

Hamamoto, H., T. Anan, K. Komatsu, M. Sugimoto, and I. Mito. 1992. First 8 × 8 semiconductor optical matrix switches using GaAS/AlGaAs electro-optic guided wave directional couplers. *Electron. Lett.* 28:441–443.

Hamamoto, H., S. Suguo, K. Komatsu, and M. Kitamura. 1993. Extremely low loss 4 × 4 GaAs/AlGaAs optical matrix switch. *Electron. Lett.* 29:1580–1582.

Heismann, F. 1993. Polarization independent photonic switching system using fast automatic polarization controllers. *IEEE Photon. Tech. Lett.* 5:1341–1343.

Herrmann, H., K. Schafer, and W. Sohler. 1994. Polarization independent, integrated optical, acoustically tunable wavelength filters/switches with tapered acoustical directional coupler. *IEEE Photon. Tech. Lett.* 6:1335–1337.

Hill, G. R., P. J. Chidgey, F. Kaufhold, T. Lynch, O. Sahlen, M. Gustavsson, M. Janson, B. Lagerstrom, G. Grasso, F. Meli, S. Johansson, J. Ingers, L. Fernandez, S. Rotolo, A. Antonielle, S. Tebaldini, E. Vezzoni, R. Caddedu, N. Caponio, F. Testa, A. Scavennec, M. J. O'Mahony, J. Zhou, A. Yu, W. Sohler, U. Rust, and H. Hermann. 1993. A transport network layer based on optical network elements. *J. Lightwave Tech.* 11:667–679.

Inoue, H., H. Nakamura, K. Morosawa, Y. Sasaki, T. Katsuyama, and N. Chinone. 1988. An 8mm length nonblocking 4 × 4 optical switch array. *IEEE J. Select. Areas Commun.* 6:1262–1265.

Islam, M. N. 1992. *Ultrafast fiber switching devices and systems.* Cambridge, England: Cambridge University Press.

Islam, M. N. 1994. Ultrafast switching with nonlinear optics. *Phys. Today* 5:34–40.

Jackel, J. L., J. E. Baran, G-K. Chang, M. Z. Iqbal, G. H. Song, W. J. Tomlinson, D. Fritz, and R. Ade. 1995. Multichannel operation of AOTF switches: Reducing channel-to-channel interaction. *IEEE Photon. Tech. Lett.* 7:370–372.

Jackel, J. L., J. E. Baran, A. d'Alessandro, and D. A. Smith. 1995. A passband flattened acousto-optic filter. *IEEE Photon. Tech. Lett.* 7:318–320.

Jenkins, R. M., J. M. Heaton, D. R. Wight, J. T. Parker, J. C. H. Birbeck, G. W. Smith, and K. P. Hilton. 1994. Novel 1 × N and N × N integrated optical switches using self-imaging multimode GaAs/AlGaAs waveguides. *Appl. Phys. Lett.* 64:684–686.

Johansson, S., A. Almstrom, and C. Hubinette. 1994. Demonstration of a multi-wavelength optical network layer in Stockholm gigabit network. In *Proceedings of the LEOS topical meeting on optical networks and their enabling technologies, Lake Tahoe*, 19–20. Paper M4.1. New York: IEEE.

Johansson, S., M. Lindblom, P. Granestrand, B. Lagerstrom, and L. Thylen. 1993. Optical cross-connect system in broadband networks: System concept and demonstrators' description. *J. Lightwave Tech.* 11:688–694.

Johnson, L. M., and F. J. Leonberger. 1981. Low-loss lithium niobate waveguide bends with coherent coupling. In *Proceedings of the 3rd International Conference on Integrated Optics and Optical Fiber Communication*. Paper TuM1. Washington, DC: Optical Society of America.

Kappe, F., C. Bornholdt, and D. Hoffmann. 1994. Ultra high speed photonic 2 × 2 space switch with traveling wave electrodes on InP. In *Proceedings of the Conference on Optical Fiber Communications*. Paper PD11-1.

Khan, M. N., J. E. Zucker, L. L. Buhl, B. I. Miller, and C. A. Burrus. 1995. Fabrication tolerant, low-loss, and high speed digital optical switches in InGaAsP/InP quantum wells. In *Proceedings of the 21st European Conference on Optical Communications*. Paper MoL3.2.

Khan, M. N., J. E. Zucker, T. Y. Chang, N. J. Sauer, and M. D. Divino. 1994. Weighted coupling Y-branch optical switch in InGaAs/InGaAlAs quantum well electron transfer waveguides. *IEEE Photon. Tech. Lett.* 3:394–397.

Kirihara, T., M. Ogawa, H. Inoue, and K. Ishida. 1993. Lossless and low crosstalk characteristics in an InP-based 2 × 2 optical switch. *IEEE Photon. Tech. Lett.* 5:1059–1061.

Kondo, M., Y. Ohta, Y. Tanisawa, T. Aoyama, and I. Ishikawa. 1987. Low drive voltage and low loss polarization independent lithium niobate optical waveguide switches. *Electron. Lett.* 23:1167–1169.

Koren, U., M. G. Young, B. I. Miller, M. A. Newkirk, M. Chien, M. Zirngibl, C. Dragone, B. Glance, T. L. Koch, B. Tell, K. Brown-Goebeler, and G. Rayborn. 1992. 1 × 16 Photonic switch operating at 1.55 micron wavelength based on optical amplifiers and a passive optical splitter. *Appl. Phys. Lett.* 61:1613–1615.

Kuzuta, K., and K. Takakura. 1991. Polarization independent lithium niobate optical devices with power splitting and switching functions. *Electron. Lett.* 27:157–158.

Leclerc, D., P. Brosson, F. Pommereau, R. Ngo, P. Doussiere, F. Mallecot, P. Gavignet, I. Wamsler, G. Laube, W. Hunziker, W. Vogt, and H. Melchior. 1995. High performance semiconductor optical amplifier array for self-aligned packaging using Si V-groove flip-chip technique. *IEEE Photon. Tech. Lett.* 7:476–478.

Legg, P. J., D. K. Hunter, I. Andonovic, and P. E. Barnsley. 1994. Inter-channel crosstalk phenomena in optical time division multiplexed switching networks. *IEEE Photon. Tech. Lett.* 6:661–663.

Lynch, T. G., P. J. Chidgey, E. G. Bryant, P. Brown, and M. Greatbanks. 1992. Experimental field demonstration of a managed multi-noded reconfigurable wavelength routed optical network. In *European Conference on Optical Communication, Berlin*, 609–612. Paper Th A12.4. Berlin.

Miller, S. E., and I. P. Kaminow. 1988. *Optical fiber telecommunications II*. Boston: Academic Press.

Minford, W. J., S. K. Korotky, and R. A. Alferness. 1982. Low-loss Ti:LiNbO$_3$ waveguide bends at $\lambda = 1.3$ μm. *IEEE J. Quantum Electron.* 18:1802–1806.

Murphy, E. J. 1988. Fiber attachment for guided wave devices. *J. Lightwave Tech.* 6:862–871.

Murphy, E. J., T. F. Adda, W. J. Minford, R. W. Irvin, E. I. Ackerman, and S. A. Adams. 1996. Guided wave optical time delay network. *IEEE Photon. Tech. Lett.* 8(4):545–547.

Murphy, E. J., C. T. Kemmerer, D. T. Moser, M. R. Serbin, J. E. Watson, and P. L. Stoddard. 1995. Uniform 8 × 8 lithium niobate switch arrays. *J. Lightwave Tech.* 13:967–970.

Murphy, E. J., T. O. Murphy, A. F. Ambrose, R. W. Irvin, B. H. Lee, P. Peng, G. W. Richards, and A. Yorinks. 1996. 16 × 16 Strictly non-blocking guided-wave optical switching system. *J. Lightwave Tech.* 14(3):352–358.

Murphy, T. O. 1994. Unpublished manuscript. AT&T Bell Laboratories.

Murphy, T. O., F. Hernandez-Gil, J. J. Veselka, and S. K. Korotky. 1989. Reduced waveguide intersection losses for large tree structured Ti:LiNbO$_3$ switch arrays. In *Topical meeting on photonic switching*. Paper PD13. Washington, DC: Optical Society of America.

Murphy, T. O., C. T. Kemmerer, and D. T. Moser. 1991. A 16 × 16 dilated Benes photonic switch module. In *Topical meeting on photonic switching*. Paper PD3. Washington, DC: Optical Society of America.

Murphy, T. O., E. J. Murphy, and R. W. Irvin. 1994a. An 8 × 8 Ti:LiNbO$_3$ polarization independent photonic switch. In *Photonics in Switching/European Conference on Optical Communications, Florence, Italy*.

Murphy, T. O., E. J. Murphy, and R. W. Irvin. 1994b. Uniform low-loss waveguide interconnects. In *Proceedings on Integrated Photonics Research*. Paper FB2.

Nelson, W. H., A. N. M. Masum-Choudhury, M. Abdalla, R. Bryant, E. Meland, and W. Niland. 1994. Wavelength and polarization independent large angle InP/InGaAsP digital optical switches with extinction ratios exceeding 20 dB. *IEEE Photon. Tech. Lett.* 6:1332–1334.

Nishimoto, H., M. Iwasaki, S. Suzuki, and M. Kondo. 1990. Polarization independent lithium niobate 8 × 8 matrix switch. *IEEE Photon. Tech. Lett.* 2:634–636.

Nishimoto, H., S. Suzuki, and M. Kondo. 1988. Polarization independent lithium niobate 4 × 4 matrix switch. *Electron. Lett.* 24:1122–1123.

O'Donnell, A. C. 1991. Polarization independent 1 × 16 and 1 × 32 lithium niobate optical switch matrices. *Electron. Lett.* 27:2349–2350.

O'Donnell, A. C., and N. J. Parsons. 1991. 1 × 16 Lithium niobate optical switch matrix with integral TTL compatible drive electronics. *Electron. Lett.* 27:2367–2368.

Okayama, H., and M. Kawahara. 1993a. Reduction of voltage–length product for Y-branch digital optical switch. *J. Lightwave Tech.* 11:379–387.

Okayama, H., and M. Kawahara. 1993b. Ti : LiNbO$_3$ digital optical switch matrices. *Electron. Lett.* 29:765–766.

Okayama, H., and M. Kawahara. 1994. Prototype 32 × 32 optical switch matrix. *Electron. Lett.* 30:1128–1129.

Okayama, H., T. Ushikubo, and M. Kawahara. 1991a. Low drive voltage Y-branch digital optical switch. *Electron. Lett.* 27:24–26.

Okayama, H., T. Ushikubo, and M. Kawahara. 1991b. Three guide directional coupler as polarization independent optical switch. *Electron. Lett.* 27:810–812.

Padmanhabman, K., and A. N. Netravali. 1987. Dilated networks for photonic switching. *IEEE Trans. Commun.* 35:1357–1365.

Ramer, O. G., C. Mohr, and J. Pikulski. 1982. Polarization independent optical switch with multiple sections of $\Delta\beta$ reversal and a Gaussian taper function. *IEEE Trans. Microwave Theory Techn.* 30:1760–1767.

Ratovelomanana, F., N. Vodjdani, A. Enard, G. Glastre, D. Rondi, and R. Blondeau. 1995. Active lossless monolithic one-by-four splitters/combiners using optical gates on InP. *IEEE Photon. Tech. Lett.* 7:511–513.

Richards G., and F. K. Hwang. 1990. Extended generalized shuffle networks: Sufficient conditions for strictly non-blocking operation. U.S. patent nos. 4,993,016 and 4,991,168.

Sasaki, H., and R. de la Rue. 1976. Electro-optic Y junction modulator/switch. *Electron. Lett.* 12:459–460

Sawano, T., S. Suzuki, M. Fujiwara, and H. Nishimoto. 1995. A high capacity photonic space division switching system for broadband networks. *J. Lightwave Tech.* 13:335–340.

Shahid, M. A., R. A. Roll, and G. J. Shevchuck. 1994. Connectorized optical fiber circuits. In *Proceedings of the 44th ECTC (Electronic Components and Technology Conference), Washington, DC.* New York: IEEE.

Shimomura, K., and A. Shigehisa. 1994. Semiconductor waveguide optical switches and modulators. *Fiber and integrated optics.* 13:65–100.

Silberberg, Y., P. Perlmutter, and J. E. Baran. 1987. Digital optical switch. *Appl. Phys. Lett.* 51:1230–1232.

Smit, M. K., E. C. M. Pennings, and H. Blok. 1993. A normalized approach to the design of low-loss optical waveguide bends. *J. Lightwave Tech.* 11:1737–1742.

Smith, D. A., J. E. Baran, K. W. Cheung, and J. J. Johnson. 1990. Polarization-independent acoustically tunable optical filter. *Appl. Phys. Lett.* 56:209–211.

Spanke, R. A. 1987. Architectures for guided wave optical space switching networks. *IEEE Commun.* 25:42–48.

Stoll, L., G. Muller, U. Wolff, B. Sauer, S. Eichinger, and S. Surgec. 1992. Compact and polarization independent optical switch on InP/InGaAsP. In *Proceedings of the European Conference on Optical Communications*, 337–340. Berlin.

Stone, F. T., and J. E. Watson. 1989. Performance and yield of pilot-line quantities of lithium niobate switches. *SPIE, OE/Fibers*. 988. Bellingham, WA: SPIE.

Thylen, L., P. Svensson, B. Lagerstrom, B. Stoltz, P. Granestrand, and W. K. Burns. 1989. Theoretical and experimental investigation of 1×2 switches. In *Proceedings of the European Conference on Optical Communications*, 240–243. Sweden: Novum Grafiska AB.

Tian, F., C. Harizi, H. Herrmann, V. Reimann, R. Ricken, U. Rust, W. Sohler, F. Wehrmann, and S. Westenhofer. 1994. Polarization-independent integrated optical, acoustically tunable double-stage wavelength filter in $LiNbO_3$. *J. Lightwave Tech.* 12:1192–1196.

Tsukada, N., and T. Nakayama. 1981. Polarization independent integrated optic switches: A new approach. *IEEE J. Quantum Electron.* 17:959–964.

van Roijen, R., J. M. M. van der Heijden, L. F. Tiemeijer, P. J. A. Thijs, T. van Dongen, J. J. M. Binsma, and B. H. Verbeek. 1993. Over 15 dB gain from a monolithically integrated optical switch with an amplifier. *IEEE Photon. Tech. Lett.* 5:529–531.

Vinchant, J-F., A. Goutelle, B. Martin, F. Gaborit, P. Pagnod-Rossiaux, J-L. Peyre, J. le Bris, and M. Renaud. 1993. New compact polarization insensitive 4×4 switch matrix on InP with digital optical switches and integrated mirrors. In *Proceedings of the European Conference on Optical Communications*. Paper ThC12.4

Vinchant, J-F., P. Pagnod-Rossiaux, J. le Bris, A. Goutelle, H. Bissessur, and M. Renaud. 1994. Low-loss fiber-chip coupling by InGaAsP/InP thick waveguides for guided wave photonic integrated circuits. *IEEE Photon. Tech. Lett.* 6:1347–1349.

Vinchant, J-F., M. Renaud, M. Erman, J. L. Peyre, P. Jarry, and P. Pagnod-Rossiaux. 1993. InP digital optical switch: Key element for guided wave photonic switching. *IEEE Proc.* 140:301–307.

Vinchant, J-F., M. Renaud, A. Goutelle, J-L. Peyre, P. Jarry, M. Erman, P. Svennson, and L. Thylen. 1992. First polarization insensitive 4×4 switch matrix on InP with digital optical switches. In *Proceedings of the European Conference on Optical Communications*. Paper TuB7.3. Berlin.

Watson, J. E., M. A. Milbrodt, K. Bahadori, M. Dautartas, C. T. Kemmerer, D. T. Moser, A. W. Schelling, J. J. Veselka, and D. A. Herr. 1990. A low voltage 8×8 $Ti:LiNbO_3$ switch with a dilated Benes architecture. *J. Lightwave Tech.* 8:794–801.

Watson, J. E., M. A. Milbrodt, and T. C. Rice. 1986. A polarization independent 1×16 guided wave optical switch integrated on lithium niobate. *J. Lightwave Tech.* 4:1717–1721.

Wenger, G., L. Stoll, B. Weiss, M. Schienle, R. Muller-Nawarth, S. Eichinger, J. Muller, B. Acklin, and G. Muller. 1994. Design and fabrication of monolithic optical spot size transformers (MOST's) for highly efficient fiber-chip coupling. *J. Lightwave Tech.* 12:1782–1790.

Yamada, Y., H. Terui, Y. Ohmori, M. Yamada, A. Himeno, and M. Kobayashi. 1992. Hybrid integrated 4 × 4 optical gate matrix switch using silica based optical waveguides and LD array. *J. Lightwave Tech.* 10:383–390.

Index

503